About Island Press

Since 1984, the nonprofit Island Press has been stimulating, shaping, and communicating the ideas that are essential for solving environmental problems worldwide. With more than 800 titles in print and some 40 new releases each year, we are the nation's leading publisher on environmental issues. We identify innovative thinkers and emerging trends in the environmental field. We work with world-renowned experts and authors to develop cross-disciplinary solutions to environmental challenges.

Island Press designs and implements coordinated book publication campaigns in order to communicate our critical messages in print, in person, and online using the latest technologies, programs, and the media. Our goal: to reach targeted audiences—scientists, policymakers, environmental advocates, the media, and concerned citizens—who can and will take action to protect the plants and animals that enrich our world, the ecosystems we need to survive, the water we drink, and the air we breathe.

Island Press gratefully acknowledges the support of its work by the Agua Fund, Inc., Annenberg Foundation, The Christensen Fund, The Nathan Cummings Foundation, The Geraldine R. Dodge Foundation, Doris Duke Charitable Foundation, The Educational Foundation of America, Betsy and Jesse Fink Foundation, The William and Flora Hewlett Foundation, The Kendeda Fund, The Forrest and Frances Lattner Foundation, The Andrew W. Mellon Foundation, The Curtis and Edith Munson Foundation, Oak Foundation, The Overbrook Foundation, the David and Lucile Packard Foundation, The Summit Fund of Washington, Trust for Architectural Easements, Wallace Global Fund, The Winslow Foundation, and other generous donors.

The opinions expressed in this book are those of the authors and do not necessarily reflect the views of our donors.

A Pivotal Moment

To Bristow, Ben, and Sam

A Pivotal Moment

POPULATION, JUSTICE, AND
THE ENVIRONMENTAL CHALLENGE

••••••••••••••••••••••••

Edited by
Laurie Mazur

WASHINGTON, D.C.
COVELO, CALIFORNIA

Library of Congress Cataloging-in-Publication Data

A pivotal moment : population, justice, and the
environmental challenge / edited by Laurie Mazur.
    p.   cm.
  Includes bibliographical references and index.
ISBN-13: 978-1-59726-661-1 (cloth : alk. paper)
ISBN-10: 1-59726-661-2 (cloth : alk. paper)
ISBN-13: 978-1-59726-662-8 (pbk. : alk. paper)
ISBN-10: 1-59726-662-0 (pbk. : alk. paper)
    1. Population—Environmental aspects.   2. Social
justice.   3. Sustainable development.   I. Mazur,
Laurie.

HB849.415.P58   2010

304.6—dc22   2009026573

Contents

PART III. LOOKING BACK, MOVING FORWARD

PART IV. THOUGHTS FOR THE JOURNEY

Foreword

TIMOTHY E. WIRTH

Fifteen years ago, Laurie Mazur asked me to write the foreword to another contributed volume, *Beyond the Numbers: A Reader on Population, Consumption and the Environment.* The year was 1994; I was serving as Undersecretary of State for Global Affairs under President Clinton.

It was a heady time. The polarized thinking of the Cold War was receding into history, and a broader concept of national security—one that encompasses environmental sustainability—was ascendant. The much anticipated "peace dividend" promised new resources for a broad range of measures to improve human well-being.

And it was the eve of the UN International Conference on Population and Development (ICPD), in Cairo, a time when the old divisions over population issues—divisions between north and south, between advocates for population stabilization and advocates for women's rights and health—gave way to consensus and collaboration. In Cairo, the world's nations agreed that rapid population growth and unsustainable resource consumption threaten our common future. They approved a groundbreaking plan to address those problems by ensuring access to family planning and reproductive health services, alleviating poverty, and ensuring the human rights of girls and women. For those of us who took part in the Cairo conference (I had the privilege of leading the

Timothy E. Wirth is the President of the United Nations Foundation and Better World Fund.

U.S. delegation to that meeting), there was a sense that a significant revolution had begun.

The revolution, it seems, was postponed.

The world's nations have not kept the promises made in Cairo. In the United States, the very success of the Cairo conference invigorated conservative groups who oppose family planning and wider opportunities for women. With the advent of a conservative Congress in 1994, followed by the Bush victory in 2000, those groups amplified their political power. As a result, the United States has decreased its support for international population programs by almost 40 percent since 1995.[1] At the same time, the sheer magnitude of the AIDS epidemic overshadowed the Cairo agenda, while the introduction of promising new antiretroviral therapies in the mid-1990s shifted resources from prevention to treatment, often at the expense of comprehensive prevention strategies that include reproductive health and family planning. While the reasons are complex, the impact is clear: By neglecting the Cairo agenda, the world's nations missed opportunities to slow population growth and improve human well-being. Opportunities were missed on the environmental side of the equation as well: Failure to grapple with the looming threat of climate change has left us with precious little time to head off unthinkable consequences.

NEW CHALLENGES AND OPPORTUNITIES

Those missed opportunities have undoubtedly intensified the challenges we face today. But, reading the essays in this new volume, I am struck by the extraordinary opportunities we have now to construct a future that is sustainable *and* just—opportunities we cannot afford to miss.

Demographically, the challenges are clear. Countless women and men lack the power to make real choices about childbearing because of crushing poverty or persistent gender discrimination. And neglect of the Cairo agenda means that over 200 million women in developing countries who wish to delay or prevent pregnancy still lack access to family planning services.[2] Less than a quarter of developing countries offer free reproductive health services in public health facilities and maternal health services in remote areas.[3] Earlier gains have been lost: As Carmen Barroso and Steven W. Sinding observe in Chapter 19, in many countries community-based family planning programs have closed their doors, leaving a void that has not been filled by any other form of health care. Indeed, because of funding declines, progress toward universal reproductive health care, including family planning, is moving at a slower pace than in the 1980s and 1990s.

As a result, huge disparities remain in reproductive health. Barroso and Sinding note that in parts of sub-Saharan Africa, women have a 1 in 6 chance of dying in childbirth, while in parts of North America and Europe, that risk is as low as 1 in 8,700. There are between 70 and 80 million unintended pregnancies in the developing world each year,[4] 46 million of which end in abortion.[5] Unsafe abortions kill some 74,000 women every year.[6]

> A smaller world population is not a panacea for the environmental and social problems before us—but it will lower the hurdles we must leap.

Progress toward smaller families is also uneven. In more than a third of the world's nations, fertility rates are now below the "replacement level" of 2.1 children per woman. But fertility remains above 3.0 in another third of countries; one-fifth still average more than 5 children per woman.[7] Fertility declines have stalled altogether in six sub-Saharan states.[8]

The world has chosen precisely the wrong moment to neglect family planning and reproductive health. As several authors in this volume observe, the largest generation in history is now coming of age. The need for health care, education, and jobs will increase dramatically as the young men and women of this generation move into adulthood.

By meeting the reproductive health needs of this largest generation, we have a chance to stabilize world population at the lower end of the projected spectrum—at 8 billion rather than 11 billion. A smaller world population is not a panacea for the environmental and social problems before us—but it will lower the hurdles we must leap.

Meeting the needs of these young men and women will not be easy. Nor will it be easy to tackle the deeply rooted injustices that constrain their choices. But, by so doing, we will make an essential investment in the future. The young men and women of the twenty-first century, who enter adulthood saddled with the legacies of the twentieth century—including debt and climate change—deserve no less.

Missed opportunities mean that the environmental crises we face today are much more serious than most of us imagined in 1994. Our profligate use of fossil fuels is changing the climate so quickly that many events that were expected to happen much later this century are happening right now. The Arctic Ocean—engine of the Northern Hemisphere's weather—could be ice-free in summer within five years.[9] The polar ice caps are melting so quickly that the world's oceans are rising more than twice as fast as they were in the 1970s.[10] In accepting the 2007 Nobel Peace Prize, Al Gore said, "We, the human species, are confronting a planetary emergency, a threat to the survival of our civilization,"[11] and he was right.

Civilization was built around the climate we have—along coastlines that may be washed away by storms and rising sea levels, around farmland and forests that may become less productive as water supplies diminish, and away from lowlands infested with insect-borne disease. Changing the climate puts the very organization of modern civilization at risk.

But here, too, daunting challenges contain the seeds of change. Climate change—perhaps more than any other environmental issue—illustrates the profound interdependence of people and the planet. It is a problem that transcends national borders and links the fates of the world's people. Recognition of that interdependence has galvanized a dynamic global movement to address climate change. The energy and ingenuity of that movement inspire hope that a sustainable future is possible.

There are other reasons for hope. Notably, the Obama administration has ushered in a new era of U.S. leadership in addressing these challenges. Early in his presidency, Obama signaled his determination to begin a "fresh conversation on family planning" and to find common ground in addressing the needs of women and families around the world.[12] And he is moving swiftly to address the looming threats of climate change and other environmental problems.

COMPLEX PROBLEMS CAN BE SOLVED

The demographic and environmental challenges we face are bewilderingly complex, linked through intricate chains of cause and effect. Nonetheless, as the authors of this volume demonstrate, there are clear steps we can take to address this knot of interrelated problems.

We must first ensure that the young women and men of the largest generation have the means and the power to make real choices about childbearing. That requires access to voluntary family planning as part of comprehensive reproductive health services. It requires education and employment opportunities, especially for women. And it requires tackling the inequities—both gender and economic—that are associated with rapid population growth. In short, it requires keeping the promises we made in Cairo.

In policy terms, this means the United States must more than double its contribution for family planning through its bilateral assistance program and through the multilateral UN Population Fund (UNFPA), consistent with the levels agreed to at ICPD. We must also mobilize the resources to meet our commitments to the Millennium Development Goals, which set measurable targets on poverty reduction, education, maternal and child health, gender equality, and environmental sustainability.

At the same time, those of us in the industrialized world must dramatically reduce our consumption of resources—we must do more with less. It can be done: McKinsey & Company, the consulting firm, estimates that the United States could shave nearly 30 percent off the amount of greenhouse gases we emit at fairly modest cost and with only small technological innovations.[13] Sustainable development means meeting the needs of current generations without stealing from the future. That is a standard we must meet.

ECONOMIC CRISIS AND FALSE CHOICES

Against the backdrop of demographic and environmental challenges, we now face a global financial crisis, the severity and duration of which we cannot fathom. It is unclear what impact that crisis will have on mobilizing the resources and political will to address the demographic, humanitarian, and environmental challenges outlined here. But the crisis will likely embolden those who argue that we cannot afford to address these problems—that development assistance and environmental protection are luxury items to be purchased later, when we can afford them.

In fact, we cannot afford *not* to make these investments. An investment in the next generation is a down payment on future peace, prosperity, and progress. All of the problems we seek to address will be more costly later if we fail to invest now. And these investments are our fundamental moral obligation to the future.

Pitting prosperity against environmental sustainability is likewise a false choice. Ecological systems are the very foundation of our economy. Five essential biological systems—croplands, forests, grasslands, oceans, and fresh waterways—supply all the raw materials for industry and provide all our food. Stated in the jargon of the business world, the economy is a *wholly owned subsidiary* of the environment. When we pollute, degrade, and irretrievably compromise that ecological capital, we destroy the very foundations of our economic well-being.

Other false choices may present themselves in troubled times. The severity of climate change and other environmental crises may revive calls for "population control." Some will argue that human rights are secondary to the urgent need to slow population growth. But history shows that human rights and slower population growth are not mutually exclusive; they are, in fact, mutually reinforcing. Where people have the rights and the power to make real choices about childbearing, they have fewer children, and they invest more in each child.

> The choice before us is not between respecting rights or protecting the environment—respecting rights *will* protect the environment.

The old thinking that pitted individual rights against the common good was wrong. The choice before us is not between respecting rights or protecting the environment—respecting rights *will* protect the environment. The choice is whether to continue on our current path toward global disaster or to empower all individuals to make informed decisions about whether and when to bear children—with full knowledge of the impact those decisions will have on their lives, their communities, and the world.

IDEAS FOR CHANGE

The economist Milton Friedman once said, "Only a crisis—actual or perceived—produces real change. When that crisis occurs, the actions that are taken depend on the ideas that are lying around."[14]

The future will offer no shortage of crises, real and perceived. The ideas in this book are the ones to heed. Its authors call for a nuanced understanding of the relationship between human numbers and environmental harm—and the inequitable patterns of consumption that mediate that relationship. They offer solutions that are grounded in human rights and social justice—solutions that can meet human needs today without diminishing the prospects of future generations. These ideas can spark the change we need.

REFERENCES

1. Adjusted for inflation; 39 percent decrease from FY 1995 ($542 million) to FY 2008 ($461 million).

2. UNFPA Web site, http://www.unfpa.org/rh/planning.htm.

3. UNFPA, 2004, *Investing in People: National Progress in Implementing the ICPD Program of Action 1994–2004,* UNFPA, New York.

4. UNFPA Web site, http://www.unfpa.org/swp/2007/english/chapter_6/social_development.html.

5. Daulaire, Nils, Global Health Council, 2007, written testimony to the Senate Committee on Appropriations, Subcommittee on State, Foreign Operations and Related Operations, April 18.

6. UNFPA Web site, http://www.unfpa.org/mothers/facts.html.

7. UN Population Division, World Population Prospects: The 2006 Revision.

8. Bongaarts, John, 2007, *Fertility Transition in the Developing World: Progress or Stagnation?* Population Association of America.

9. Amos, Jonathan, 2007, Arctic summers ice-free "by 2013," BBC NEWS, Dec. 12, http://news.bbc.co.uk/2/hi/science/nature/7139797.stm.

10. Leake, Jonathan, 2009, Polar icecaps melting faster, *The Sunday Times (London)*, February 8, http://www.timesonline.co.uk/tol/news/environment/article5683655.ece.

11. Gore, Albert, 2007, U.S., China must lead fight against "planetary emergency," CNN, Dec. 10, http://www.cnn.com/2007/WORLD/europe/12/10/gore.nobel/index.html.

12. Obama, Barack, 2009, Statement of President Barack Obama on Rescinding the Mexico City Policy, The White House, Washington, DC, January 23.

13. Wald, Matthew L., 2007, Study details how U.S. could cut 28% of greenhouse gases, *The New York Times*, November 30.

14. Friedman, Milton, 1962, *Capitalism and Freedom*, University of Chicago Press, Chicago, Introduction.

Introduction

LAURIE MAZUR

We are living in a pivotal moment.

Even a casual glance at the headlines reveals this to be a pivotal moment *environmentally*: In 2007, scientists warned that we have less than a decade left to head off catastrophic climate change.[1] Now it seems that even those dire forecasts were optimistic: Recent evidence shows that the climate is changing much more quickly than predicted just a few years ago.[2] And it's not just the climate. From acidifying oceans to depleted aquifers, the natural systems we depend upon are nearing "tipping points," beyond which they may not recover.

But it is less well known that this is a pivotal moment *demographically*. While the rate of population growth has slowed in most parts of the world, rapid growth is hardly a thing of the past. Our numbers still increase by 75 to 80 million every year, the numerical equivalent of adding another United States to the world every four years.[3] And while a certain amount of future growth is virtually inevitable—an echo of the great boom of the late twentieth century—the ultimate size of the human population will be decided in the next decade or so.

That's because right now the largest generation of young people in human history is coming of age. Nearly half the world's population—some 3 billion people—is under the age of twenty-five.[4] Those young people will, quite literally, shape the future. The decisions they make, especially about sexuality and childbearing, will have a great impact on their own lives and on the well-being of their families and communities. Collectively, their choices will determine whether human numbers—now at 6.8 billion—climb to anywhere between 8 and nearly 11 billion by mid-century.

These phenomena—environmental crisis and population growth—are connected, in ways that are profound and complex. Of course, the relationship between population growth and environmental harm is not simple or linear. Human numbers are not a primary cause of environmental degradation, but they do magnify the harmful effects of unsustainable production and consumption. Some people have much greater environmental impact than others; we in the industrialized countries use about thirty-two times the resources—and emit thirty-two times as much waste—as our counterparts in the developing world.[5]

> Nearly half the world's population—some 3 billion people—is under the age of twenty-five. Collectively, their choices will determine whether human numbers—now at 6.8 billion—climb to anywhere between 8 and nearly 11 billion by mid-century.

Nonetheless, the evidence presented here suggests that it would be better for human beings and the environment if world population peaks at 8 or 9 billion rather than almost 11 billion. Of course, slowing population growth is not *all* we must do. Our numbers are almost certain to grow by more than 30 percent by 2050. That means we must swiftly reduce our collective impact by a third just to maintain the disastrous status quo; a truly sustainable future will require sweeping social and technological changes. Slowing population growth won't feed the hungry or eradicate poverty, either; that will require a wholesale rethinking of development, trade, and other economic policies.

But slower population growth could help give us a fighting chance to meet these challenges. It could reduce pressure on natural systems that are reeling from stress. And it could help give families and nations a chance to make essential investments in education, health care, and sustainable economic development.

Yet while the goal of a smaller world population is an important one, it matters—a lot—how we get there. As the authors of this volume show, the best way to slow population growth is not with top-down "population control" but by ensuring that all people have the means and the power to make real choices about childbearing. That means, first, that all people must have access to family planning and other reproductive health services.* Second, it means addressing the economic and gender inequities that limit choices for so many of the world's people. Finally, and urgently, it means investing in the young men and women of the largest generation. All of these steps are worth taking for their own sakes, as a matter of human rights and social justice. Together, they will shape a sustainable, equitable future.

THE DAWN OF THE ANTHROPOCENE

The twentieth century saw the human enterprise—and its environmental impact—grow as never before. It took our species about 200,000 years to go from a few thousand individuals to a billion around 1800.[6] We rounded the corner of the twentieth century at 1.65 billion and then took off. Our numbers nearly doubled—to 3 billion—by 1960 and then doubled again—to 6 billion—by 1999.[7]

We hit our peak rate of growth—2.1 percent per year—in the second half of the twentieth century. That half century also marked a seismic shift in our relationship with the natural world: In just fifty years, we altered the planet's ecosystems more than in all of human history combined.[8] In previous millennia, the Earth was transformed by massive forces of nature—the advance of glaciers, volcanic eruptions, the clash of continents. But in the late twentieth century, the "Anthropocene" era dawned as human beings became the most powerful force of environmental change.[9] Humanity has developed the capacity to transform—and destroy—the fundamental processes of nature that sustain us.

Climate change offers a dramatic example. Our emissions of heat-trapping gases—from burning fossil fuels, agriculture, deforestation, and other activities—have already warmed the planet's land areas by about 2 degrees Fahrenheit.[10] This seemingly small change has, in fact, thrown a switch on the slow-moving machinery of the climate. As a result, the Intergovernmental Panel on Climate Change (IPCC) predicts the twenty-first century will bring widespread famine in Africa and elsewhere, more violent storms, and the extinction of nearly a third of the Earth's species.[11]

While no region will be completely spared, the worst effects of climate change will fall on the poorest people in the tropics and subtrop-

*Reproductive health services include family planning; prenatal, obstetric, and postnatal care; and prevention and treatment of reproductive tract infections, including sexually transmitted diseases like HIV/AIDS.

ics. In other words, says UN Secretary-General Ban Ki-moon, "Those who have done the least to cause the problem bear the gravest consequences."[12] For the world's most marginalized people, those consequences "could be apocalyptic," according to the United Nations Development Program (UNDP).[13]

Climate change threatens the natural systems that undergird human civilization. Consider this: One in six people on the planet gets their drinking water from glaciers and snowpack on the world's great mountain ranges—including the Hindu Kush, the Himalayas, and the Andes. Those glaciers are receding worldwide; when they are gone, so is the essential water they provide.[14]

And the effects of climate change are not consigned to the future—they are being felt today by the thousands killed and tens of millions displaced in the record-breaking 2007 south Asian monsoons,[15] by the 15,000 people in France who died in the heat wave of 2003,[16] by children in east Africa who are dying from malaria as disease-carrying mosquitoes move into new latitudes and altitudes.[17]

While climate change is beginning to get the attention—if not the action—it deserves, few are aware that human activities are threatening the planet's life-support systems in more direct ways. Nearly two-thirds of the planet's ecosystems—including freshwater and fisheries—are being used in ways that simply cannot be sustained.[18] As we have transformed natural systems to meet human needs, billions have been lifted from poverty and bare subsistence. But according to the Millennium Ecosystem Assessment, a five-year study that involved 1,360 scientists worldwide, those improvements in human well-being have come at great cost to the complex systems of plants, animals, and biological processes that make the planet habitable. As a result, "the ability of the planet's ecosystems to sustain future generations can no longer be taken for granted."[19]

Climate change, says environmental journalist Andrew Revkin, is not the story of our time; it is "a subset of the story of our time, which is that we are coming of age on a finite planet and only just now recognizing that it is finite."[20]

IS POPULATION GROWTH THE PROBLEM?

It is tempting to ascribe these environmental problems to the growth of human population. Surely it is no coincidence that our environmental impact accelerated just as our numbers took off. It makes sense that more people require more food, more electricity, more freshwater—

more of everything we take from the Earth. That is true, but the story is not that simple.

First, a focus on aggregate numbers masks vast disparities in the human condition—and in environmental impact. The gains of the last half century were distributed very unevenly. Some 2.5 billion people—40 percent of the world's population—live on less than $2 per day, accounting for just 5 percent of global income. One billion of those live on the knife edge of survival at less than $1 per day. Meanwhile, the richest 10 percent of the world's population, almost all of whom live in industrialized countries, claim more than half of all income. Huge disparities exist within, as well as between, countries. Even before the financial crisis that began in 2007, income inequality was *increasing* in countries that are home to more than 80 percent of the world's population.[21]

Not surprisingly, those at the top of the income curve have a disproportionate impact on the global environment. Again, climate change offers an example. The average American or Canadian produces more than 18 metric tons of CO_2 emissions every year, while each sub-Saharan African produces little more than a ton. Clearly, the less populous countries of the global north bear far more responsibility for creating the climate crisis than the populous south (though that may change in the future; read on). If we measure the environmental "footprint" of each nation, the United States is surely Bigfoot: Per capita, Americans use more resources and produce more pollution and waste than anyone in the world. In fact, if everyone in the world lived as Americans do, we would need five Earths to support us.[22]

Moreover, our footprints can be seen on every continent. In an ever more integrated global economy, consumption in the affluent countries drives environmental destruction worldwide. The food we eat, the clothes we wear, our children's toys—all leave a trail of harm that spans the globe. The shrimp served at my local Red Lobster began life in a man-made lagoon in Thailand; to meet the growing demand for their products, shrimp farmers are destroying mangroves that provide essential protection from storm surges and tidal waves—like the 2004 tsunami.[23] And, about a third of all Chinese carbon dioxide emissions are the result of producing goods for export, mostly to the United States and other affluent countries.[24]

LOCAL IMPACTS, GLOBAL CAUSES

Even where rapid population growth and environmental degradation coexist, the damage may be driven by consumers on another continent.

Take Ghana, for example. Ghana has very high rates of population growth—at 2.2 percent a year, double the average for the world as a whole.[25] It also has significant environmental problems: In less than fifty years, Ghana's primary rain forest has been reduced by 90 percent; a quarter of its forest cover was lost just between 1990 and 2005.[26] When the forests are lost, flooding and erosion degrade the land, often leaving it unsuitable for any productive use.[27] As a result, according to the United Nations Environment Programme, a third of Ghana's land is in danger of turning to desert. [28]

At first glance, you might think that growing human numbers are to blame for deforestation in Ghana. But it is not, for the most part, the citizens of Ghana who are cutting down the forests—or profiting from their sale. Most logging is done by the timber industry, whether legal (multinational corporations) or illegal (wildcat loggers)—for sale on the global market.[29]

Ghana's story is typical of developing countries. Beginning in the 1980s, the World Bank and International Monetary Fund (IMF) required Ghana to deregulate its economy and develop export-oriented industries, such as timber and mining, in exchange for loans. Regulations were relaxed; foreign investment was courted with generous incentives. As money poured in, Ghana's resources flowed out: Mahogany logs, gold, and cocoa beans filled ships bound for Europe and North America.[30] The success of export industries sparked robust economic growth that reached 6.3 percent in 2007.[31] However, that growth has come at a terrible environmental cost, and it has not delivered the promised benefits to Ghana's people.

Ask Awudu Mohammed, who was just finishing high school when a gold mine opened in his village of Sanso in central Ghana. "There was gold under [our] farm so they wanted to mine the place though my father disagreed with them," Mohammed told the BBC. "They went and brought police, all of them holding guns." Mohammed's father was given the equivalent of $50 to compensate for the loss of his livelihood. Now his family lives in a squatter's camp, eking out a living by illegally panning for gold. Mohammed's family is hardly alone: It is estimated that hundreds of thousands of people in Ghana have been evicted from farms to make way for multinational mining interests in the last twenty years.[32]

Some may view such dislocation as the "creative destruction" that inevitably accompanies the transition from subsistence agriculture to a more prosperous modern industrialized economy. But many of Ghana's people are not making that transition. While there have been reductions in poverty overall, the profits from mining and other export industries are not "trickling down" to poor, rural Ghanaians. The UNDP found that a large number of people are not benefiting from Ghana's robust growth and are thus socially excluded.[33] Even the World Bank's independent evaluation department has questioned whether mining has brought real benefit to Ghana's people; in 2003, Bank analysts concluded that "it is unclear what the true net benefits to Ghana from large-scale gold mining are."[34]

Those who were left out of Ghana's economic boom are, in fact, paying for the wholesale liquidation of resources that fueled it. "Resource degradation accounts, to a large extent, for the persistence or worsening of social exclusion among some social groups," says the UNDP. "Depletion of forests and mangroves, soil erosion, degeneration of soil fertility, drying rivers and streams, desertification have become common features of environments in which the poor eke out an existence."[35]

> The twenty-first century presents two equally urgent imperatives: to lessen human impact on the environment and to reduce the glaring inequalities that divide humanity. Slowing population growth is central to both.

So, does Ghana have a population problem? Certainly, high rates of population growth are a challenge for displaced subsistence farmers who are moving into forest reserves and cutting trees for firewood, or overcultivating marginal land and depleting the soil. The poorest Ghanaian women have more than six children, on average.[36] Nearly one in four would like to prevent or delay having another child but lacks access to contraceptives.[37] But rapid population growth is not to blame for Ghana's poverty or environmental problems. And while the people of Ghana need family planning and other reproductive health services, that is certainly not all they need. Most urgently, they need trade and development policies that encourage broad-based sustainable development.

In Ghana, and around the world, people need alternatives to an economic system that—in the words of Adriana Varillas—"promises prosperity but ignores the natural world on which prosperity depends; . . . which delivers wealth for a few, and grinding, inescapable poverty for many more." That system, a brand of capitalism forged in the United States and exported throughout the world, has spawned great wealth and even greater inequality, while laying waste to the natural resources and processes that make the planet habitable. It is, therefore, unsustainable—environmentally, economically, and morally. It is also, at this writing, a system in crisis.

The challenge, writes James Gustave Speth, is to change "the operating instructions for the modern world economy," to reorient economic and social systems toward sufficiency rather than excess; to reward the protection, rather than the destruction, of the natural resource base. Fundamentally, the twenty-first century presents two equally urgent imperatives: to lessen human impact on the environment and to reduce the glaring inequalities that divide humanity. Slowing population growth is central to both.

8 VERSUS 11 BILLION

In the words of the authors of the Millennium Ecosystem Assessment, "the living machinery of Earth has a tendency to move from gradual to catastrophic change with little warning. Such is the complexity of the relationships between plants, animals, and microorganisms that these 'tipping points' cannot be forecast by existing science." We don't know when such tipping points will be reached, but the pressure that human societies have placed on natural systems "makes it likely that more will occur in the future."[38]

In this context, it matters whether we add 1 billion, 2 billion, or nearly 4 billion people to the world's population by 2050. (Those are, respectively, the UN's low, medium, and high population projections.) While there are great disparities in environmental impact among the world's citizens, everyone has *some* impact. We all share an inalienable right to food, water, shelter, and the makings of a good life. If we take seriously the twin imperatives of sustainability and equity, it becomes clear that it would be easier to provide a good life—at less environmental cost—for 8 rather than almost 11 billion people.

The dual challenges of sustainability and equity are vividly illustrated by the climate crisis. In Chapter 5, Climate Change and Population Growth, Brian C. O'Neill concludes that slower population growth could make a significant contribution to solving the climate problem. To put that contribution in perspective, imagine a pie divided into slices—each representing an action begun today that would eliminate a billion tons of CO_2 per year by 2050—for example, energy efficiency and renewable energy. Seven slices are needed to avert catastrophic climate change. O'Neill estimates that stabilizing world population at 8 billion, rather than 9 or more, would provide one—or even two—"slices" of emissions reductions.

Now, wait a minute, you might say. Climate change is mostly caused by greenhouse gas emissions in the industrialized world, where population growth rates are low. And most population growth is taking place in the developing countries, where per capita emissions are a fraction of ours. So how can slower population growth help solve the climate crisis?

The answer lies in the future. The developing countries are where the lion's share of population growth will occur, and they are also where development *must* occur for half of humanity to escape from grinding poverty. The affluent countries can reduce emissions by reducing the vast amounts of waste in our systems of production and consumption. But the developing countries are not likely to raise their standards of living without more intensive use of resources and higher emissions.

In fact, the aggregate emissions of developing countries are already beginning to overtake those of what we call the industrialized countries. The biggest emitters of carbon dioxide, in absolute terms, are not the affluent countries of the north but the rapidly emerging economies of the global south. According to the Center for Global Development (CGD), which tracks CO_2 emissions from 50,000 power plants worldwide, rapidly rising emissions would put developing countries on track to produce their own climate crisis in just twenty years—even without emissions from the high-income countries.[39]

Of course, absolute numbers gloss over huge inequities in *per capita* emissions. China now emits more CO_2 overall than the United States, but the average Chinese citizen produces a tenth as much carbon dioxide as the average American.[40] Those inequities have stymied efforts to broker a global treaty to cap emissions, because developing countries do not want to be "capped" at a level that precludes growth. After all, since fossil fuel emissions are closely correlated with per capita wealth, "what sane government would agree to cap its citizens' per capita emissions in perpetuity below those of other countries?" asks Robert Engelman in Chapter 6, Fair Weather, Lasting World. Any global climate treaty that secures the support of the developing countries—and a treaty without that support would be worthless—must find a way to lower emissions without locking in those inequities (see Chapter 6 for how we might do this).

Slowing population growth is a piece of the "pie"—it is part of what we must do to avert catastrophic climate change. And, compared with the other things we must do to stabilize the climate, this is a relatively easy one. Everything we must do to slow population growth—ensuring access to reproductive health services, improving the status of girls and women, alleviating poverty—is something we should be doing anyway. And slowing population growth in this way is surprisingly cost-effective. For example, the developed countries' share of the cost to provide reproductive health services for every woman on Earth is $20 billion—about what the bankers on Wall Street gave themselves in bonuses in 2008.[41]

THE INEQUALITY CONNECTION

Like the population-environment connection, the relationship between population growth and poverty is equally complex. But it is fair to say that rapid population growth is both a cause and an effect of poverty and inequality, and slower growth can help close the gaps that divide men and women, rich and poor.

Population growth rates have fallen in most of the world but remain high where poverty and gender inequality are most intractable. Poverty can be an engine of population growth: For example, in poor communities where many children don't live to see their fifth birthday, couples may have many children in order to make sure that some survive. Conversely, high fertility can worsen poverty—for families and for entire nations—by diminishing the per capita resources available for education, health care, and productive investment. Many families and communities are caught in what Rachel Nugent calls a "high-fertility poverty trap," in which poverty exacerbates high fertility rates and vice versa. As we have seen, the trap is often set by economic policies that deepen inequality. But high fertility and rapid growth can make it harder to escape.

Gender discrimination also fuels population growth. Where women are denied education, secure livelihoods, property ownership, credit—in short, the full legal and social rights of citizenship—they are forced to rely on childbearing for survival, status, and security. And in many parts of the world, girls and women are forced into early marriage and/ or childbearing, either by the subtle pressure of social norms or by the overt coercion of threats and violence. Nearly half of young women in south Asia are married before the age of eighteen.[42] Girls who are married in their teens have more children, on average, and both they and their children fare worse than children of older mothers.

Here, too, the connection is a two-way street. High fertility can exacerbate gender inequality—and poverty—by limiting women's ability to pursue education and employment, which diminishes prospects for women and their children. The vicious cycle of high fertility and gender inequality is one reason that women constitute 70 percent of the world's poor.[43]

At the same time, slower population growth is part of a "virtuous circle" that can help promote equality. Where family planning is available, where couples are confident their children will survive, where girls go to school, where young women and men have economic opportunity, couples will have healthier and smaller families—and the gaps that divide men and women, rich and poor, will diminish.

POPULATION POLITICS AND THE UNFINISHED REVOLUTION

Rapid population growth has not ended. And population growth is connected—albeit indirectly—to the crucial issues of our time, such as climate change and poverty. Yet, until quite recently, population issues had all but fallen off the public agenda. When world leaders gathered in 2005 to affirm the Millennium Development Goals—an ambitious plan to eradicate poverty—population growth was not mentioned. Why?

The answer is a long one, but here it is in broad strokes. In the 1960s and 1970s, there was a broad political consensus in the United States and elsewhere about the need to address rapid population growth. That consensus launched the international family planning movement—bilateral and multilateral programs that brought contraceptive services to the developing world and sparked a revolution in reproductive behavior. Between the mid-1960s and the mid-1990s, contraceptive use increased from less than 10 percent to nearly 60 percent.[44] The movement also helped reshape the hockey stick–shaped curve of population growth: During that period, average fertility in the developing world fell from six children per woman to about three, and growth rates fell accordingly.[45]

But family planning programs soon became embroiled in controversy. Some—notably in India and China—flagrantly abused human rights with coercive practices such as forced sterilization and abortion (which continue to this day in China). And many first-generation programs focused more on demographic "targets" than on individual needs. With limited choice and poor quality of care, these programs

BOX i.1

Beware the Techno-Fix

Laurie Mazur and Shira Saperstein

We can dream of nonpolluting power plants and fail-safe birth control, but technology alone can't solve today's population and environmental challenges. While some technologies may be inherently better than others, it is their social context that determines whether they advance—or limit—human rights and well-being.

Technology can be a double-edged sword. Its impact depends on the specific cultural, economic, social, and political contexts into which it is introduced. For example, new options for contraception and abortion hold great promise, but they have failed to improve women's lives where underlying health, rights, and poverty issues have not also been addressed. In unjust environments, they may exacerbate rather than alleviate human suffering and undermine rather than advance human rights. Consider, for example, the mass sterilization campaigns in India during Indira Gandhi's emergency regime, or the coercive use of abortion in China today.

Closer to home, women in the United States have often encountered separate and unequal treatment when seeking reproductive health care. Women of color have been steered—and sometimes forced—to use permanent methods of contraception, like sterilization, or "provider-controlled" methods, like Depo-Provera and Norplant. So the same technologies can be experienced very differently: For some, they are liberating; for others, they are a source of oppression.

Power and wealth are not evenly distributed, nor are the costs and benefits of technology. Too often, the benefits accrue to elites, and costs are borne by the marginalized. For example, the Green Revolution was to let us break free of limits to the planet's carrying capacity. Indeed, hybrid plant varieties and fertilizers led to dramatic increases in crop yields. But they also had unintended effects. As agriculture became more profitable, rich farmers and corporations bought up the best, flattest land, forcing poor peasants—who could not afford expensive seeds and fertilizer—onto environmentally fragile marginal lands. The Green Revolution also left a distinctly un-green legacy: depleted soils and aquatic "dead zones" from fertilizer runoff.

Biofuels offers a more recent example. The United States has subsidized production of ethanol, made from corn, as an alternative fuel. But given industrialized agriculture's heavy use of oil, ethanol has proved no greener than conventional fuels. And by forcing the world's poor to compete with automobiles for a share of the corn crop, ethanol contributed to a food crisis in 2007–2008 that drove millions to the brink of starvation.

Of course, new technologies can play an important role in advancing reproductive rights and health and in shrinking our environmental footprint. For example, more-efficient technologies—like those used in Europe—could sharply reduce resource consumption in the United States. The average Italian uses exactly half the energy of the average American.[a] The Washington, DC, area produces 25 percent more CO_2 than all of Sweden, which has nearly twice as many people.[b]

But real sustainability will require sweeping changes in behavior, which will require a new menu of policy choices. On the environmental side, those could include subsidized public transportation, rather than oil and gas exploration. On the reproductive health side, they might include health coverage for contraception, not just for Viagra. Unfortunately, the menu of choices is often set by the powers that be.

Recently, some 12,000 young people gathered in Washington, DC, for PowerShift 2009. They marched in the streets, lobbied in Congress, and engaged in civil disobedience at a coal-fired power plant. Of course, PowerShift attendees want to see coal plants replaced with clean, renewable sources of energy. But the conference's name and agenda show a deeper understanding of the challenges we face: "We are having a broader conversation than just climate change, or climate science," said Marcie Smith,[c] a student at Transylvania University in Kentucky, who attended the meeting. "This is a conversation about justice, equity and opportunity."[d]

Climate change and the other urgent problems we face today call for a true shift in power—a democratization of decision-making about the technological "hardware" we use and the social "software" that drives it. Anything less is a temporary fix, not the durable solution we need.

....

[a] Energy Information Administration, 2008, *International Energy Annual 2006*, table posted December 19, http://www.eia.doe.gov/emeu/international/energyconsumption.html.

[b] Farenthold, David A., 2007, DC area outpaces nations in pollution, *The Washington Post*, September 30.

[c] http://www.transy.edu/news/new_story.htm?id=465&obj=index.

[d] Quoted in Jenkins, Jesse, What the Press Didn't Tell You about the Largest Youth Movement in Decades, http://watthead.blogspot.com/2009/03/what-press-didnt-tell-you-about-largest_4486.html.

reflected—and exacerbated—the low status of women. Women's health and human rights advocates criticized these programs and went on to articulate and implement an alternative paradigm for reproductive health.

Then, in the 1990s, demographers and reproductive health professionals made a game-changing realization: You don't need to control anyone to slow population growth. Given high levels of unwanted fertility in many parts of the world, simply addressing this "unmet need"— by providing reproductive health services that enable women to realize their own fertility goals—would decrease birthrates by as much as, or more than, was called for in most countries' demographic targets.[46]

This shift in thinking enabled feminists, reproductive health advocates, and demographers to make common cause at the 1994 International Conference on Population and Development, in Cairo. That meeting produced a new, rights-based approach to population that embraces voluntary contraception and comprehensive reproductive health services, as well as efforts to empower women and foster development. The Cairo consensus holds that when women have more control over their lives—including their reproductive destinies—they will have healthier, smaller families and invest more in each child. This has immediate benefits for women and families, and those benefits reverberate outward to communities, nations, and the world. The Cairo approach won unprecedented support from NGOs and governments, and it has spurred change in reality as well as rhetoric.

But the revolution remains unfinished. As growth rates fell and the "population bomb" was defused, policymakers moved on to other urgent priorities. The Cairo agenda was eclipsed by the HIV/AIDS crisis, which gathered deadly momentum throughout the 1990s. While funding for HIV/AIDS has increased, funds for family planning and other elements of reproductive health have fallen sharply.[47] At the same time, support for reproductive health declined as the Religious Right rose in power in the United States and elsewhere. As a result, developing countries and donors alike failed to make good on the financial commitments they made in Cairo. Since the mid-1990s, funding for reproductive health services has declined relative to other health spending.

Because of the shortfall in public funding, shocking inequities persist in the reproductive health of rich and poor, both within and between countries. Some 200 million women in developing countries lack access to family planning services.[48] Access can be a matter of life or death: Every year, pregnancy-related complications kill half a million women, one every minute.[49] According to *The Lancet*, a quarter of those lives could be saved if women were able to delay or limit childbearing.[50]

As long as women risk their lives to bring children into the world, the work of the family planning/reproductive health movement is far from complete. As the members of the largest generation in history move into their childbearing years, the need for reproductive health services will grow exponentially. Access to these services is essential for the health and well-being of today's young women and men. And ensuring that they can make real choices about childbearing could help stabilize world population at 8 rather than nearly 11 billion—which will, in turn, make climate change and poverty easier to address. As Timothy E. Wirth argues in his Foreword, "The world has chosen precisely the wrong moment to neglect family planning and reproductive health."

THE RETURN OF "POPULATION CONTROL"

While the neglect of reproductive health continues, population issues are making something of a comeback today, in the context of concern about climate change and other environmental problems. The new interest in population growth presents both an opportunity and a challenge. It could, for example, help mobilize support for reproductive health, gender equity, and other measures to slow population growth— the long-neglected Cairo agenda.

Or—and here's the challenge—it could take us back to the pre-Cairo days of "population control." Unfortunately, some players in the debate take a simplistic view of the relationship between human numbers and environmental harm, and propose solutions that are coercive at best.

A few examples: In a bestselling book called *The World Without Us*, environmental journalist Alan Weisman takes the reader on a tour of the planet after human beings have been wiped out by some hypothetical disaster. He then makes a modest proposal for how we can avoid extinction: by adopting a mandatory "one child per human mother" policy. In a post on the progressive Web site AlterNet, journalist Chris Hedges asserts that "all efforts to stanch the effects of climate change are not going to work if we do not practice vigorous population control."[51] A rather less draconian solution was offered in the *Medical Journal of Australia*: a "carbon tax" for families with more than two children.

Coercive measures to slow population growth are unnecessary, counterproductive, and just plain wrong. First, as we have seen, it is not necessary to control anyone to slow population growth: Birthrates come down where individuals have the means and the power to make their own reproductive choices.

And coercion often backfires. In Chapter 22, Going to Extremes: Population Politics and Reproductive Rights in Peru, Susana Chávez Alvarado offers a cautionary tale about what can happen when demographic goals trump reproductive rights. In the 1990s, Peruvian President Alberto Fujimori sought to alleviate poverty by reducing birthrates among poor women. Targets were set for numbers of women to be sterilized, and coercive tactics—including brute force—were used to meet those targets. The legacy of Peru's forced sterilizations endures, most notably for the women who were its victims. It also produced a backlash by the Religious Right, which launched a successful campaign against sterilization and other forms of contraception. Where once women were sterilized against their will, now women who want permanent contraception cannot obtain it.

In another troubling development, the population-environment banner has lately been seized by groups seeking to sharply limit immigration to the United States. As Priscilla Huang observes in Chapter 28, Over-Breeders and the Population Bomb, a coalition of anti-immigration groups has launched a high-profile advertising campaign that links immigration-fueled population growth to a host of environmental problems, from climate change to suburban sprawl.

Must we limit immigration to save the environment? It is true that immigration contributes to U.S. population growth, as it has throughout our history. But environmental destruction is driven by a wide range of policies and practices, as well as population growth. It is disingenuous to blame burgeoning human numbers for traffic, as the anti-immigration groups do, without mentioning, say, the chronic neglect of public transportation in the United States.

More broadly, the anti-immigrant groups misrepresent the nature of the environmental challenges we face today. They imply that we are in a lifeboat with limited resources, and if too many people get in, we will all sink. But there is a flaw in that thinking: We may be in a lifeboat, but it's not the United States. It's our planet, and we are all in it together.

POPULATION JUSTICE: AN ETHICAL COMPASS

In a world where limits are near and population is still growing, a world riven by inequality and threatened by catastrophic climate change, ethical dilemmas will continue to arise. In that context, how can individuals, nations, and the global community address the complex knot of interrelated issues at the nexus of population and the environment?

Here we propose a "population justice" framework for understanding—and acting upon—these issues. The framework draws inspiration from the reproductive justice and environmental justice movements, both of which grew from the struggle for civil rights in the United States.

Reproductive justice asks us to look at the totality of people's lives, and especially at inequalities of gender, race, and class that shape and constrain women's choices.[52] For example, a legal right to abortion doesn't mean much to a woman who cannot afford one, and reproductive choice remains elusive where families lack health care and other resources they need to raise healthy children. Similarly, the environmental justice movement looks at the inequalities that affect the quality of the air we breathe and the water we drink—for example, by influencing which neighborhood gets the polluting bus depot or waste incinerator.

In this vein, population justice takes a broad view of the population-environment nexus. It calls for a nuanced understanding of the relationship between human numbers and environmental harm, and the inequitable patterns of consumption that mediate that relationship—because a simplistic understanding can lead to simplistic, and even dangerous, solutions. Most importantly, population justice urges attention to the inequalities—both gender and economic—that underlie both rapid population growth and the destruction of the natural environment.

A population justice framework encompasses individual human rights—including the right to bodily integrity and autonomy and free decision-making about sexuality, reproduction, and family. But it's bigger than individual rights. Justice addresses our obligations to one another. If our basic rights are secured (a big "if" for many people in the world), then we have an obligation to ask what impact our choices have on others, including future generations. In the context of an unfolding environmental crisis, the question is an urgent one.

This framework can help prioritize action on population-environment issues. Of each proposed action, we must ask, Does it uphold and enhance established human rights? Does it advance the cause of social justice; will it reduce inequality? Will it promote human well-being and protect the environment?

When viewed through this lens, certain priorities emerge—though they will change in different times and places. For example, forty years ago, when few women in the developing world had access to contraception, ensuring access to voluntary family planning was an obvious way to enhance health and human rights. Today, as Judith Bruce and John Bongaarts observe in The New Population Challenge (Chapter 20), family planning is necessary but not sufficient to achieve those ends.

Current population-environment challenges suggest a range of actions around the globe. In Ghana, a top priority might be to rethink export-led development that is ravaging the resource base, while making sure that poor, marginalized Ghanaians have access to education and health care, including family planning. In south Asia, where almost half of girls are married before their eighteenth birthday, ending child marriage and ensuring equal rights and opportunity for women demands a place at the top of the agenda. And in the United States, the number one priority is to change our systems of production and consumption, so that Americans—with 5 percent of world population—no longer account for one-third of global consumption.[53]

Meeting the ethical criteria outlined above seems a tall order, and indeed, there are several interventions that do *not* fit the bill: government-imposed limits on childbearing, which violate human rights; carbon taxes on children and similar disincentives, which may penalize those who lack reproductive choice; draconian limits on immigration and the police-state tactics needed to enforce them.

But there are many ways to address population-environment issues that meet the highest ethical standards. These include voluntary family planning and other reproductive health services; broad-based "human development" programs that ensure equitable access to education, health care, and sustainable livelihoods; and efforts to improve efficiency and conserve resources.

If our goal is to create a world that is sustainable and just, population-environment policies must serve those ends. "Solutions" that do not meet that ethical test are not really solutions; they are problems in themselves.

THE JOURNEY

Ethical compass in hand, the first steps of the journey become clear. But there is a great distance between where we are now and where we must go, if we are to "not only survive but also demonstrate that we deserve to," as Gordon McGranahan puts it. To get from here to there will require a sea change in consciousness, and a monumental mobilization of political will. Climate change and other environmental crises are truly without precedent in human history. Never before have we held such capacity to transform the thin layer of earth and sky that is home to all known life. Never before have the fates of the world's people been so closely entwined.

As I think about these challenges, I contemplate what my kids call "the extra-stinky jar of nature" sitting on the windowsill of my office. It's an old mason jar filled with treasures found in the small patch of woods behind our house: turquoise robin's egg fragments; a goldfinch feather; monarch wings; a shed snakeskin. The jar is named for the powerful aroma of decomposing organic matter that awaits anyone who makes the mistake of opening it.

I inherited a passionate appreciation for nature from my mother and grandmother, which I've passed on to my snakeskin-collecting children. But what else will my kids inherit? The extra-stinky jar of nature, now a living record of their world, could be a relic by the time they are grown. We have now entered what scientists call the Sixth Great Extinction, which could eliminate one in three species from the planet. Unlike other mass extinctions, like the one that did in the dinosaurs, this one is entirely caused by humans.

Last fall, no acorns fell from the oak trees in our backyard. Nor, for the first time in memory, did they fall throughout much of the eastern United States.[54] The honeybees that pollinate our crops have been decimated by a mysterious "colony collapse disorder."[55] The Chesapeake Bay, ridiculously bountiful back when my great grandfather worked on a crabbing boat there, is in its death throes: Oyster and rockfish populations have collapsed, and the blue crab—the cultural icon of Baltimore, my hometown—is not far behind.[56] Of course, the bay may recover, the bees and acorns may rebound. But there is a growing sense today that the natural world is unraveling, that we have crossed a threshold, or are peering over its edge.

My kids will inherit a natural world that is greatly diminished, but they are the lucky ones. What about the other children—the other members of the largest generation in history?

Among the reports on my desktop, two images call out to me. One shows a boy about the age of my younger son, digging in a hellish open-pit gold mine in the Congo. His filthy shirt is torn, his slender back bent over his work. Another shows an eleven-year-old girl in Bangladesh, dressed in a bright pink sari and an orange head cloth. It is hard to read the expression in her shadowed eyes, but her round cheeks are those of a child. It is her wedding day. Her husband, a man of about twenty-five, stands by her side.

What choices do they have, and what kind of future awaits them?

As we peer over the edge, as we contemplate the damage we have done—and could do—to the habitability of our planet, it has never been more clear that we are in this together. The fate of every living thing on Earth, and of the natural systems that sustain us, is bound up with the hopes and dreams and needs of the young men and women of the largest generation. It is my hope that that recognition will help mobilize the ingenuity, and the resources, and—most important—the compassion needed to create a more sustainable and equitable world for all of us.

REFERENCES

1. Adam, David, 2007, UN scientists warn time is running out to tackle global warming, *The Guardian,* May 5.

2. Changing climate numbers, 2009, editorial, *The New York Times,* February 21.

3. UN, 2007, Department of Economic and Social Affairs, Population Division, World Population Prospects: The 2006 Revision, Highlights, Working Paper Number ESA/P/WP.202.

4. Population Division of the Department of Economic and Social Affairs of the United Nations Secretariat, *World Population Prospects: The 2006 Revision* and *World Urbanization Prospects: The 2005 Revision,* http://esa.un.org/unpp.

5. Diamond, Jared, 2008, What's your consumption factor? *The New York Times,* January 2.

6. AAAS Atlas on Population and the Environment, http://atlas.aaas.org/index.php?part=1.

7. UN Population Division, 1998, *The World at Six Billion,* United Nations Population Division, New York.

8. Millennium Ecosystem Assessment Board of Directors, 2005, *Living Beyond Our Means: Natural Assets and Human Well-being,* Island Press, Washington, DC, http://www.millenniumassessment.org/documents/document.429.aspx.pdf.

9. Vitousek, Peter M., et al., 1997, Human domination of the Earth's ecosystems, *Science* 277, July 25.

10. Hansen, James, 2008, Global Warming Twenty Years Later: Tipping Points Near, briefing to the House Select Committee on Energy Independence and Global Warming, June 23.

11. Intergovernmental Panel on Climate Change, 2007, Fourth Assessment Report, *Climate Change 2007: Synthesis Report, Summary for Policymakers,* IPCC.

12. Gelling, Peter, 2007, Focus of climate talks shifts to helping poor countries cope, *The New York Times,* December 13, http://www.nytimes.com/2007/12/13/world/13climate.html.

13. Watkins, Kevin, et al., 2007, *Human Development Report 2007/8—Fighting Climate Change: Human Solidarity in a Divided World,* UN Development Program.

14. Pachauri, R. K., 2007, Acceptance Speech for the Nobel Prize Awarded to the Intergovernmental Panel on Climate Change, December 10.

15. UNICEF, Millions of people across south Asia affected by monsoonal flooding, press release, http://www.unicef.org/media/media_40495.html.

16. French heat toll almost 15,000, 2003, BBC News, September 25, http://news.bbc.co.uk/2/hi/europe/3139694.stm.

17. Pascual, M., et al., 2006, Malaria resurgence in the east African highlands: Temperature trends revisited, *Proceedings of the National Academy of Sciences* 103(15): 5829–5834, April 11, http://www.pnas.org/content/103/15/5829.full.pdf+html.

18. Millennium Ecosystem Assessment Board of Directors, 2005, *Living Beyond Our Means: Natural Assets and Human Well-being,* Island Press, Washington, DC, accessed online at http://www.millenniumassessment.org/documents/document.429.aspx.pdf.

19. Ibid.

20. Revkin, Andrew, 2008, interviewed by Alex Steffen of Worldchanging, December 1, http://www.worldchanging.com/archives//009111.html.

21. Watkins, Kevin, et al., 2007, *Human Development Report.*

22. World Wildlife Fund, the Zoological Society of London, and Global Footprint Network, 2008, *2008 Living Planet Report* (see box in Chapter 15 in this volume, The Biggest Footprint: Population and Consumption in the United States).

23. BBC News, 2004, Shrimp farms "harm poor nations," May 19, http://news.bbc.co.uk/2/hi/science/nature/3728019.stm.

24. Clark, Duncan, 2009, West blamed for rapid increase in China's CO_2, *The Guardian,* February 23.

25. Population Reference Bureau, 2008, *World Population Data Sheet 2008,* PRB, Washington, DC.

26. UN Environment Programme, 2008/2009, Africa: Atlas of Our Changing Environment, UNEP, Nairobi, http://www.unep.org/pdf/Press Releases/Ghana_Africa_Atlas.pdf.

27. Ibid.

28. Ibid.

29. Glastra, Rob, ed., 1999, *Cut and Run: Illegal Logging and Timber Trade in the Tropics,* International Development Research Center, Ottawa, http://www.idrc.ca/en/ev-28728-201-1-DO_TOPIC.html.

30. UN Environment Programme, 2008/2009, Africa.

31. World Bank, Ghana Country Brief, http://web.worldbank.org/WBSITE/EXTERNAL/COUNTRIES/AFRICAEXT/GHANAEXTN/0,,menuPK:351962~pagePK:141132~piPK:141107~theSitePK:351952,00.html.

32. Stickler, Angus, 2006, Ghana's ruthless corporate gold rush, BBC

News, July 18, http://news.bbc.co.uk/2/hi/programmes/file_on_4/5190588
.stm.

33. UN Development Programme, 2007, *Ghana Human Development Report 2007*, UNDP, Nairobi, http://www.undp-gha.org/docs/Overview.pdf.

34. World Bank Operations Evaluation Department, 2003, Evaluation of the World Bank Group's Activities in the Extractive Industries: Background Paper, Ghana Case Study, October 15.

35. UN Development Programme, 2007, *Ghana Human Development Report 2007*.

36. Population Reference Bureau datafinder, http://www.prb.org/Data finder/Geography/Summary.aspx?region=22&region_type=2.

37. Ibid.

38. Millennium Ecosystem Assessment Board of Directors, 2005, *Living Beyond Our Means: Natural Assets and Human Well-being*, Island Press, Washington, DC, accessed online at http://www.millenniumassessment.org/documents/document.429.aspx.pdf.

39. Center for Global Development, 2007, New Database Ranks CO_2 Emissions from 50,000 Power Plants Worldwide, November 15.

40. World Bank online database, 2004.

41. Reproductive health funding estimates: Speidel, J. Joseph, et al., 2007, *Family Planning and Reproductive Health: The Link to Environmental Preservation*, Bixby Center for Reproductive Health & Policy, UCSF, October; Bankers' bonuses: Stolberg, Sheyl Gay, and Stephen Labaton, 2009, Obama calls Wall Street bonuses "shameful," *The New York Times*, January 29, http://www.nytimes.com/2009/01/30/business/30obama.html.

42. See Judith Bruce and John Bongaarts, Chapter 20, The New Population Challenge, in this volume.

43. UN Millennium Campaign, http://www.un.org/millenniumgoals/poverty.shtml.

44. Sinding, Steven W., 2007, Overview and Perspective, *The Global Family Planning Revolution: Three Decades of Population Policies and Programs*, Warren C. Robinson and John A. Ross, eds., The World Bank, Washington, DC.

45. Ibid.

46. Sinding, Steven W., John Ross, and Allan Rosenfield, 1994, Seeking Common Ground: Unmet Needs and Demographic Goals, *Beyond the Numbers: A Reader on Population, Consumption and the Environment*, Laurie Ann Mazur, ed., Island Press, Washington, DC.

47. UN Millennium Project, 2005, *Public Choices, Private Decisions: Sexual and Reproductive Health and the Millennium Development Goals*, Global Health Council.

48. UNFPA Web site, http://www.unfpa.org/rh/planning.htm.

49. Ibid.

50. Cleland, J., S. Bernstein, et al., 2006, Family planning: The unfinished agenda, *The Lancet* 368(9549): 1810–1827.

51. Hedges, Chris, 2009, Are we breeding ourselves to extinction? AlterNet, March 11, http://www.alternet.org/environment/130843/are_we_breeding_ourselves_to_extinction/?page=1.

52. Asian Communities for Reproductive Justice, 2005, *A New Vision for Advancing Our Movement for Reproductive Health, Reproductive Rights and Reproductive Justice,* ACRJ, Oakland, CA.

53. Cassara, Amy, How Much of the World's Resource Consumption Occurs in Rich Countries? World Resources Institute, Earthtrends database, http://earthtrends.wri.org/updates/node/236.

54. Walton, Marsha, 2008, Scientists baffled by mysterious acorn shortage, CNN, December 12, http://www.cnn.com/2008/TECH/science/12/12/acorn.shortage/index.html.

55. Barrionuevo, Alexei, 2007, Honeybees vanish, leaving keepers in peril, *The New York Times,* February 27.

56. Farenthold, David, 2008, Maryland proposes restrictions on blue crab catch, *The Washington Post,* April 10.

PART I

The Numbers

CHAPTER I

....

Human Population Grows Up

JOEL E. COHEN

In the next half century, humanity will undergo historic changes in the balance between young and old, rich and poor, urban and rural. Our collective and individual choices now and in the years ahead will determine how well humankind copes with its coming of age.

The current decade spans three unique, important transitions in the history of humankind. Before 2000, young people always outnumbered old people. From 2000 forward, old people will outnumber young people. Until approximately 2007, rural people always outnumbered urban people. From 2008 forward, urban people will outnumber rural people. From 2003 on, the median woman worldwide had, and will continue to have, too few or just enough children during her lifetime to replace herself and the father in the following generation.[1]

The century with 2000 as its midpoint marks three additional unique, important transitions in human history. First, no person who died before 1930 had lived through a doubling of the human population. Nor is any person born in 2050 or later likely to do so. In contrast, everyone born in 1965 or earlier and still alive has seen human numbers more than double from 3.3 billion in 1965 to 6.8 billion in 2009. The fastest population growth rate ever reached, about 2.1 percent a

Joel E. Cohen is Abby Rockefeller Mauzé Professor of Populations and head of the Laboratory of Populations at the Rockefeller University and Columbia University.

year, occurred between 1965 and 1970.[2] Human population never grew with such speed before the twentieth century and is never again likely to grow with such speed. Our descendants will look back on the late 1960s peak as the most significant demographic event in history, even though those of us who lived through it did not recognize it at the time.

Second, the dramatic fall since 1970 of the global population growth rate to 1.1 or 1.2 percent a year today resulted primarily from choices by billions of couples around the world to limit the number of children born.[3] Global human population growth rates have probably risen and fallen numerous times in the past. The great plagues and wars of the fourteenth century, for example, reduced not only the growth rate but also the absolute size of global population, both largely involuntary changes. Never before the twentieth century has a fall in the global population growth rate been voluntary.

Finally, the last half century saw, and the next half century will see, an enormous shift in the demographic balance between the more developed regions of the world and the less developed ones. Whereas in 1950 the less developed regions had roughly twice the population of the more developed ones, by 2050 the ratio will exceed six to one.[4] These colossal changes in the composition and dynamics of the human population by and large escape public notice.

Here, I will focus on the four major underlying trends expected to dominate changes in the human population in the coming half century. The population will be bigger, slower-growing, more urban, and older than in the twentieth century. Of course, precise projections remain highly uncertain. Small changes in assumed fertility rates have enormous effects on the projected total numbers of people, for example. Despite such caveats, the projections do suggest some of the challenges humanity will face over the next fifty years.

RAPID BUT SLOWING GROWTH

Although the rate of population growth has fallen since the 1970s, current rates (as a percentage) and absolute numbers of global population growth are still greater than any experienced prior to World War II. Whereas the first absolute increase in population by a billion people took from the beginning of time until the early nineteenth century, a billion people will be added to today's population in only thirteen to fourteen years. By 2050 the world's population is projected to reach 9.2 billion, depending on future birth and death rates.[5] This anticipated

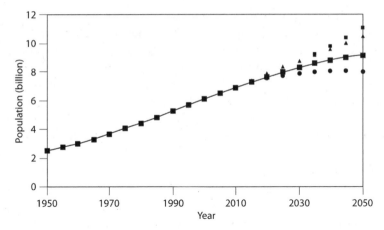

- ■ Constant-fertility variant
- ▬■▬ Medium variant
- ● Low variant
- ▲ High variant

FIGURE 1.1. Population projections to 2050.
Source: UN World Population Prospects, 2006 Revision.

increase from 2009 to 2050 exceeds the total population of the world in 1930, which was around 2 billion.

Childbearing choices made today and tomorrow will help determine the future size of the human population. In the unlikely event that fertility did not decline at all from today's levels, population would grow to 11.9 billion by 2050, nearly doubling from 6 billion in 1999. The 9.2 billion projection above assumes that family planning will be more widely practiced and the trend toward smaller families will continue. If, instead, women average just one more child for every two women, world population could reach 10.8 billion by 2050; if women have one fewer child for every two women, world population could be 7.8 billion by 2050.[6] A difference in fertility of a single child per woman's lifetime between now and 2050 alters the projection by 3 billion, a difference equal to the entire world population in 1960.

> In short, rapid population growth has not ended. Human numbers currently increase by 75 million to 80 million people annually, the equivalent of adding another United States to the world every four years or so.

In short, rapid population growth has not ended. Human numbers currently increase by 75 million to 80 million people annually, the equivalent of adding another United States to the world every four years or so.[7] But most of the increases are not occurring in countries with the wealth of the United States. Between 2005 and 2050, population will at least triple in Afghanistan, Burkina Faso, Burundi, Chad, Congo, Democratic Republic of the Congo, East Timor, Guinea-Bissau, Liberia, Mali, Niger, and Uganda.[8] These countries are among the poorest on Earth.

Virtually all population growth in the next forty-five years is expected to happen in today's economically less developed regions. Despite higher death rates at every age, poor countries' populations grow faster than rich countries' populations, because birthrates in poor countries are much higher. At present, the average woman bears nearly twice as many children (2.8) in the poor countries as in the rich countries (1.6 children per woman).[9]

Half the global increase will be accounted for by just nine nations. Listed in order of their anticipated contribution, they are India, Pakistan, Nigeria, Democratic Republic of the Congo, Bangladesh, Uganda, the United States, Ethiopia, and China.[10]

In contrast, fifty-one countries or areas, most of them economically more developed, will lose population between now and 2050. Germany is expected to drop from 83 million to 79 million people, Italy from 58 million to 51 million, Japan from 128 million to 112 million, and most dramatically, the Russian Federation from 143 million to 112 million. Thereafter Russia will be slightly smaller in population than Japan.[11]

> The poor countries will have to build the equivalent of a city to accommodate a million people every five days for the next forty to forty-five years.

Slowing population growth everywhere means that the twentieth century was probably the last in human history in which younger people outnumbered older ones. The proportion of all people who were children aged four years and younger peaked in 1955 at 14.5 percent and gradually declined to 9.5 percent by 2005, whereas the fraction of people aged sixty years and older increased from a low of 8.1 percent in 1960 to 10.4 percent in 2005.[12] Around 2000, each group constituted about 10 percent of humanity. Now and henceforth, the elderly have the numerical upper hand. (Yet while young people comprise a smaller percentage of the world's population, the current gen-

BOX 1.1

Crossroads for Population

The Challenge

Rapid population growth will boost human numbers by nearly 50 percent, from 6.8 billion now to 9.2 billion in 2050. Virtually all this growth will happen in existing or new cities in developing countries. During the same period, many richer nations will lose population. Falling fertility and increasing longevity worldwide will expand the proportion of potentially dependent elderly people.

The Solutions

Create a bigger pie, and fewer forks, and better manners: Intensify human productive capacity through investment in education, health, and technology. Increase access to reproductive health care and contraception to slow population growth voluntarily. Improve the terms of people's interactions by reforming economic, political, civil, and social institutions, policies, and practices and achieving greater social and legal equity.

eration of people under the age of twenty-five is the largest ever—see Chapter 2.)

This crossover in the proportions of young and old reflects both improved survival and reduced fertility. The average life span grew from perhaps thirty years at the beginning of the twentieth century to more than sixty-five years at the beginning of the twenty-first century.[13] The more powerful influence, however, is reduced fertility, adding smaller numbers to the younger age groups.

The graying of the population is not proceeding uniformly around the globe. In 2050 nearly one person in three will be sixty years or older in the more developed regions, and one person in five in the less developed zones. But in eleven of the least developed countries—Afghanistan, Angola, Burundi, Chad, Democratic Republic of the Congo, Equatorial Guinea, Guinea-Bissau, Liberia, Mali, Niger, and Uganda—half the population will be aged twenty-three years or younger.[14]

If recent trends continue as projected to 2050, virtually all of the world's population growth will be in urban areas. In effect, the poor countries will have to build the equivalent of a city to accommodate a million people every five days for the next forty to forty-five years.[15]

Projections of billions more people in developing countries and

BOX 1.2

The Migration Wild Card

Migration has little immediate effect on global population size but may accelerate the slowing of population growth. Migrants who move from high-fertility to low-fertility regions, or their descendants, often adopt the reduced-fertility patterns of their new home, with some time delay.

From 2005 to 2050, the more developed regions are projected to have about 2.2 million more immigrants than emigrants a year, and the United States is expected to receive about half of these.[a] More than most demographic variables, future international migration is subject to intentional policy choices by national governments, making it difficult to predict. Assuming that recent levels of migration continue, the 98 million net migrants expected to move to the developed regions during 2005–2050 would more than offset the projected loss of 73 million people in those countries from an excess of deaths over births.[b]

Different international migration scenarios would not greatly affect the sharp rise in the rich countries' proportion of dependent elderly projected for the coming century, although they could dramatically affect population size. In 2000, for example, the U.S. Census Bureau projected the nation's numbers in 2050 with different levels of immigration. Results ranged from 328 million, representing a 20 percent population increase with zero immigration, to 553 million, representing an 80 percent increase with the highest level of immigration—hypothetical net annual immigration rising to 2.8 million by 2050.[c] Regardless of migration, though, the U.S. ratio of elderly to working-age people will rise steeply from 2010 until around 2035 and will gradually increase thereafter.[d] By 2050 it is projected to reach 39 percent with zero immigration, and 30 percent with the highest immigration.[e]

....

[a]United Nations, Department of Economic and Social Affairs, Population Division, International Migration 2006, http://www.un.org/esa/population/publications/2006Migration_Chart/Migration2006.pdf.

[b]Ibid.

[c]U.S. Census Bureau, Annual projections of the total resident population as of July 1: middle, lowest, highest, and zero international migration series, 1999 to 2100, (NP-T1). Internet release date: January 13, 2000. Highest date: February 14, 2000. http://www.census.gov/population/projections/nation/summary/np-t1.txt.

[d]Ibid.

[e]Ibid.

more elderly people everywhere, coupled with hopes of economic growth especially for the world's poor, raise concerns in some quarters about the sustainability of present and future populations.

BEYOND CARRYING CAPACITY

In the short term, our planet can provide room and food, at least at a subsistence level, for 50 percent more people than are alive now. The estimated cereal production in the 2007–2008 crop year of over 2.1 billion metric tons of cereal grains[16] was enough to feed more than 10 billion people a vegetarian diet, while the number of undernourished people rose by 75 million in 2007, bringing the estimated world total to 923 million, with roughly one person in seven undernourished.[17] But as demographer-sociologist Kingsley Davis observed in 1991, "there is no country in the world in which people are satisfied with having barely enough to eat."[18] The question is whether 2050's billions of people can live with freedom of choice and material prosperity, however freedom and prosperity may then be defined, and whether their children and their children's offspring will be able to continue to live with freedom and prosperity, however they may define them in the future. That is the question of sustainability.

This worry is as old as recorded history. Cuneiform tablets from 1600 BC showed that the Babylonians feared the world was already too full of people. In 1798 Thomas Malthus renewed these concerns,[19] as did Donella Meadows and her coauthors in their 1972 book *The Limits to Growth*.[20] While some people have fretted about too many people, optimists have offered reassurance that deities or technology will provide for humankind's well-being.

Attempts to quantify Earth's human carrying capacity or a sustainable human population size face the challenge of understanding the constraints imposed by nature, the choices faced by people, and the interactions between them.[21] For example, what will humans desire and what will they accept as the average level and distribution of material well-being in 2050 and beyond? What technologies will be used? What domestic and international political institutions will be used to resolve conflicts? What economic arrangements will provide credit, regulate trade, set standards, and fund investments? What social and demographic arrangements will influence birth, health, education, marriage, migration, and death? What physical, chemical, and biological environments will people want to live in? What level of variability will people be willing to live with? (If people do not mind seeing human population size drop by billions when the climate becomes unfavorable, they may regard a much larger population as sustainable when the climate is favorable.) What level of risk are people willing to live with? (Are mud slides, hurricanes, or floods acceptable risks, or not? The answer will influence the area of land viewed as habitable.) What time horizon is assumed? Finally, and significantly, what will people's values and tastes be in the future? As anthropologist Donald L. Hardesty noted in 1977, "a plot of land may have a low carrying capacity, not because of low soil fertility but because it is sacred or inhabited by ghosts."[22]

Most published estimates of Earth's human carrying capacity have uncritically assumed answers to one or more of these questions. In my book *How Many People Can the Earth Support?* I collected and analyzed more than five dozen of these estimates published from 1679 onward. Those made in just the past half century ranged from less than a billion to more than 1,000 billion. These estimates are political numbers, intended to persuade people, one way or another: either that too many humans are already on Earth or that there is no problem with continuing rapid population growth. Scientific numbers are intended to describe reality. Because no estimates of human carrying capacity have explicitly addressed the questions raised above, taking into account the diversity of views about their answers in different societies and cultures, no scientific estimates of sustainable human population size can be said to exist. Too often, attention to long-term sustainability is a diversion from the immediate problem of making tomorrow better than today, a task that does offer much room for science and constructive action. Let us therefore briefly consider two major demographic trends, urbanization and aging, and some of the choices they present.

BOOM OR BOMB?

Many major cities were established in regions of exceptional agricultural productivity, typically the floodplains of rivers, or in coastal zones and islands with favorable access to marine food resources and maritime commerce. If the world's urban population roughly doubles in the next half century, from 3 billion to 6 billion, while the world's rural population remains roughly constant at 3 billion, and if many cities expand in area rather than increasing in density, fertile agricultural lands around those cities could be removed from production, and the waters around coastal or island cities could face a growing challenge from urban waste. Right now the most densely settled half of the planet's population lives on 2 to 3 percent of all ice-free land. If cities double in area as well as population by 2050, urban areas could grow to occupy 6 percent of the land. Withdrawing that amount mostly from the 10 to 15 percent of land considered arable could have a notable impact on agricultural production. Planning cities to avoid consuming arable land would greatly reduce the effect of their population growth on food production, a goal very much in the urbanites' interest because the cities will need to be provisioned.

Unless urban food gardening surges, on average each rural person will have to shift from feeding herself (most of the world's agricultural workers are women) and one city dweller today to feeding herself and two urbanites in less than a half century. If the intensity of rural agricultural production increases, the demand for food, along with the technology supplied by the growing cities to the rural regions, may ultimately lift the rural agrarian population from poverty, as has happened in many rich countries. On the other hand, if more chemical fertilizers and biocides are applied to raise yields, the rise in food production could put huge strains on the environment.

For city dwellers, the threats of urbanization include frightening hazards from infectious disease unless adequate sanitation measures supply clean water and remove wastes. Yet cities also concentrate opportunities for educational and cultural enrichment, access to health care, and diverse employment. Therefore, if half the urban infrastructure that will exist in the world of 2050 must be built in the next forty to forty-five years, the opportunity to design, construct, operate, and maintain new cities better than old ones is enormous, exciting, and challenging.

After 2010, most countries will experience a sharp acceleration in the rate of increase of the elderly-dependency ratio—the ratio of the number of people aged sixty-five and older to the number aged fifteen to sixty-four. The shift will come first and most acutely in the more developed countries, whereas the least developed countries will experience a slow increase in elderly dependency after 2020. By 2050 the elderly-dependency ratio of the least developed countries will approach that of the more developed countries in 1950.

Extrapolating directly from age to economic and social burdens is unreliable, however. The economic burden imposed by elderly people will depend on their health, on the economic institutions available to offer them work, and on the social (including familial) institutions on hand to support their care.

The sustainability of the elderly population depends in complex ways not only on age, gender, and marital status but also on the availability of supportive offspring and on socioeconomic status—notably educational attainment. Better education in youth is associated with better health in old age. Consequently, one obvious strategy to improve the sustainability of the coming wave of older people is to invest in educating youth today, including education in those behaviors that preserve health and promote the stability of marriage.[23] Another obvious strategy is to invest in the economic and social institutions that facilitate economic productivity and social engagement among elderly people.

> "Virtually everything that needs doing from a population point of view needs doing anyway."

No one knows the path to sustainability because no one knows the destination, if there is one. But we do know of many actions we could take today to make tomorrow better than it would be if we do not put our knowledge to work. These include investments in education, health, and technology; better access to reproductive health care and contraception to slow population growth voluntarily; and reforms that promote greater social and legal equity. As economist Robert Cassen remarked,[24] "virtually everything that needs doing from a population point of view needs doing anyway."

REFERENCES

1. Wilson, Chris, and Gilles Pison, 2004, La majorité de l'humanité vit dans un pays où la fécondité est basse, *Population et Sociétés*, Number 405 (October), ISSN 0184 77 83.

2. UN, Department of Economic and Social Affairs, Population Division, *World Population Prospects: The 2006 Revision and World Urbanization Prospects*, http://esa.un.org/unpp.

3. Ibid.

4. Ibid.

5. Ibid.

6. Ibid.

7. UN, Department of Economic and Social Affairs, Population Division, *World Population Prospects: The 2006 Revision, Highlights*, Working Paper Number ESA/P/WP.202, 2007.

8. UN, Department of Economic and Social Affairs, Population Division,

World Population Prospects: The 2006 Revision and World Urbanization Prospects, http://esa.un.org/unpp.

9. Ibid.

10. Ibid.

11. Ibid.

12. Ibid.

13. Ibid.

14. Ibid.

15. Cohen, Joel E., 2008, Sustainable cities, *Bulletin of the American Academy of Arts and Sciences* 61(4): 6–8, summer.

16. Food and Agricultural Organization, 2008, *Food Outlook & Global Market Analysis*, November, p. 2, http://www.fao.org/docrep/011/ai474e/ai474 e00.HTM.

17. Ibid., p. 1.

18. Davis, Kingsley, and Mikhail S. Bernstam, eds., 1991, *Resources, environment and population: Present knowledge and future options,* Supplement to volume 16 of *Population and Development Review, 1990,* Population Council, New York, and Oxford University Press, New York/Oxford.

19. Malthus, T. R., 1798, *An Essay on the Principle of Population . . .* , complete first edition (1798) and partial seventh edition (1872) reprinted in *On Population,* Gertrude Himmelfarb, ed., Modern Library, New York, 1960. Malthus, T. R., 1798/1970, *An Essay on the Principle of Population,* A. Flew, ed., Penguin, London. Malthus, Thomas R., 1970, *An Essay on the Principle of Population and a Summary View of the Principle of Population,* third edition, Penguin Books, Middlesex, UK. Malthus, Thomas, Julian Huxley, and Frederick Osborn, 1960, *Three Essays on Population,* New American Library of World Literature, New York.

20. Meadows, Donella H., Dennis L. Meadows, Jírgen Randers, and William W. Behrens III, 1972, *The Limits to Growth: A Report for the Club of Rome's Project on the Predicament of Mankind,* 1974, second edition, Signet, New American Library, New York.

21. Cohen, Joel E., 1995, *How Many People Can the Earth Support?* W. W. Norton, New York.

22. Hardesty, Donald L., 1977, *Ecological Anthropology,* John Wiley, New York.

23. Cohen, Joel E., 2008, Make secondary education universal, *Nature* 456 (December).

24. Cassen, Robert, et al., 1994, *Population and Development: Old Debates, New Conclusions,* Transaction Publishers, New Brunswick, NJ, Oxford, UK.

CHAPTER 2

The Largest Generation Comes of Age

MARTHA FARNSWORTH RICHE

More than a half century ago, what some now call the "greatest generation" came of age. Those men and women fought in World War II and returned home to build the infrastructure of the world as we've known it. Now a potentially even more influential generation stands on the brink of adulthood. Today, the largest generation ever—3 billion strong—is under the age of twenty-five. Choices made by these young people will shape the future—politically, economically, and demographically. The choices they make will be determined, in turn, by the opportunities available to them. And those opportunities are starkly different for young people in different regions of the world.

> Today, the largest generation ever— 3 billion strong—is under the age of twenty-five.

It is difficult to predict the ways in which the largest generation will remake the world. But its influence on the size of world population is clear. Collectively, young people's choices about childbearing will

Martha Farnsworth Riche served as Director of the U.S. Census Bureau between 1994 and 1998 and is now a fellow at the Center for the Study of Economy and Society, Cornell University.

determine whether the world will contain 7.8 billion, 9.2 billion, or 11.9 billion people in 2050. (These are the low, medium, and high projections envisaged by the United Nations.)[1]

AN INCOMPLETE TRANSITION

To understand the demographic impact of the largest generation, it is helpful to look at the trajectory of past growth. Population growth is new, historically speaking. Until 1800, as far as scientists can tell, the world had fewer than a billion people. Many were born but few survived infancy and childhood, and famines and epidemics regularly reduced population sizes attained during good times. Then, thanks to more-stable food supplies, better housing and sanitation, and medical advances in more-developed countries, world population grew slowly but steadily throughout the nineteenth century. However, there were still fewer than 2 billion people worldwide when the twentieth century began.

The twentieth century was the century of "demographic transition." This is the theoretical construct that explains the surge of population growth that took place largely in the last half of the century, when human numbers increased from 2.5 billion people to over 6 billion (Figure 2.1).

Here's how demographic transition works: First, improvements in health and living conditions improve child survival rates. Some time later—often a few decades—fertility rates fall when people realize that they don't need to have so many babies to make sure their children will survive them. Until they make that adjustment, the children who would have died at birth or in childhood under earlier conditions survive and have children of their own. These additional children make population grow rapidly, even though they have fewer children than their parents did, because the sheer number of new parents is so much larger than in previous generations. This is known as "population momentum." The math of momentum is easy: Say two couples have six children each, who all marry; if these six couples have only three children each, they produce eighteen in all—50 percent more children despite a halving of fertility.

In 1930 the world was home to 2 billion people; by 1960, 3 billion. Then population momentum kicked in, and the population was at 4 billion in 14 years, 5 billion in 13 more years, and 6 billion in 12 more years, by 1999. Subsequent billions are expected to arrive more slowly, as the transition reaches its presumed conclusion. Ultimately, popula-

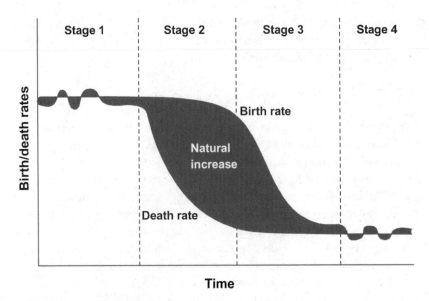

FIGURE 2.1 The classic stages of demographic transition.
Note: Natural increase results when births outnumber deaths.

tion is expected to stabilize, as fertility falls in line with death rates worldwide.

The demographic transition was first seen in Europe, where it took place over nearly two centuries as Europeans developed and implemented the innovations that we call "modernization." Then, during the latter half of the twentieth century, modern public health practices spread mortality improvements around the world, and rapid growth in several regions contributed to a population explosion worldwide. Instead of the slow transition that accompanied modernization in the "old" world, an accelerated transition took place in the "new" one, as modern public health practices were implanted in countries that were, socioeconomically speaking, premodern. In these countries, the as-yet-incomplete transition is taking place over decades, not centuries. And fertility was so high at the beginning that the younger generations vastly outnumber the older ones in Asia, Africa, and Latin America.

The twenty-first century marks the end of this transition, in global terms, as most countries have now completed the demographic transition. It looks like 1989 was the peak year for world population growth, with almost 90 million more people at the end of the year than at the beginning. But, in some regions, the demographic transition is not far advanced. These include sub-Saharan Africa and several countries, largely in the Middle East but also elsewhere, such as Guatemala.

COMING-OF-AGE SCENARIOS

The young men and women of the largest generation will shape the demographics of the future, but their opportunities and challenges will be determined, in part, by the demographic landscape of today. And that landscape varies greatly from one region to another. Put simply, from a global perspective, youth are relatively scarce in some regions while in others they are very numerous. This reflects a stark contrast between "more" and "less" developed regions.

Half the population of less-developed regions was under age twenty-five in 2005, compared with little more than 30 percent in the more-developed regions.[2] Little more than 10 percent of youth under the age of fifteen are in more-developed countries, and nearly 90 percent are in less-developed countries—three in five are Asian, one in five African.[3] These proportions essentially dictate the locus of tomorrow's global labor force, absent unprecedented immigration to more-developed regions. They will also profoundly affect the world's economy, as they put the bulk of potential consumer growth in regions that are relatively poor. Given today's population dynamics, the United Nations expects the youth imbalance between regions to continue and even widen slightly during this half century. In 2050, the UN projects that half the world's youth under age fifteen will be in Asia, and 30 percent in Africa. Latin America will also continue to have a youthful population.

OUTLOOK FOR YOUTH I: THE AFFLUENT COUNTRIES

Opportunities look very different for the relatively small share of young people coming of age in wealthy countries than for the large numbers of youth in the rest of the world. While there are vast disparities *within* these countries, some generalizations can be made. The majority of young people in developed countries are in families with only one or two children, which means that, on a per person basis, they have received more attention from their parents than previous generations did, and had more resources invested in them.[4] Indeed, in many developed countries, the direct and indirect (as in mother's forgone income) costs of meeting high expectations for parental investment in children are influencing childbearing decisions. Societies that allocate these costs narrowly, rather than broadly, tend to induce many women to limit or even forgo childbearing, such that some countries are worrying about population decline.

Japan is the extreme example. In Japan, adherence to traditional gender roles accompanies high and costly expectations for children. The former makes women choose between work and family, leading many young women to postpone marriage and thus childbearing, while the latter limits family size for those who do marry. A recent survey of today's young Japanese adults (aged eighteen to thirty-four) found that nine out of ten want to marry, and many say they want two children.[5] But with a total fertility rate of 1.3 children per woman in 2005, the Japanese population has begun a decline in absolute numbers that is projected to amount to 20 percent between 2006 and 2050.[6] That decline would be even steeper if life expectancy were not continuing to rise, delaying older people's "exit" from the population.

As a relatively small segment of the Japanese population, tomorrow's young adults will be sought after by employers as the large post–World War II generation retires. However, Japanese life expectancy is the world's highest, suggesting a large support burden, both familial and economic, for the smaller generations that succeed more-numerous ones. In 2006, the 14 million Japanese under age fifteen were vastly outnumbered by the 20 million aged sixty-five and older. Meanwhile, a lack of family-friendly policies by government, employers, and society as a whole has resulted in many young people's pioneering new, non-marital lifestyles, while delaying or forgoing childbearing. There is no evidence that a declining population is necessarily a bad thing, especially in a country as densely populated and resource-dependent as Japan. Still, there may be considerable impact on individual lives, if young men and women are unable to marry or have the children they want. And the failure of large numbers of young adults to pay into the government retirement program, because they doubt it will pay off when they retire, suggests that their worries may continue into old age.[7]

Population numbers can grow in three ways: if births outnumber deaths ("natural increase"), if people move in from other countries ("net migration"), or if life expectancy increases ("mortality improvement"). Reversing any of these processes can cause the numbers to decline: fewer births, more people moving to other countries, and people dying sooner than usual, as with the AIDS epidemic. Japan's numbers are declining because it is having considerably fewer births, and the longer life expectancy of its people does not make up for the loss of youth.

In most European countries, population is growing, in part, because of improved life expectancy. This means that young Europeans, too, will eventually support and care for a large elderly population. Immigration also contributes to Europe's population growth. In Austria, for example, births and deaths are roughly equal, meaning no population growth from "natural increase." However, Austria is a net gainer from migration, so its population is projected to grow, albeit slowly. Immigration can ease the economic adjustment to a new age structure, by providing more workers to support growing older generations. At the same time, immigration poses social challenges of its own as Europeans adapt to greater cultural heterogeneity. There is also migration within Europe; the broadening of the European Union gives young eastern Europeans access to better opportunities in the west, but at some social cost to them and to their families—and to their increasingly elderly countries of origin.

Births in northwestern Europe are numerous enough to keep younger generations growing in line with older ones; whether those countries' populations stabilize or grow reflects to a large extent their acceptance of immigrants. In these countries, family-friendly policies are making it easier to both work and raise children, keeping fertility levels high relative to other industrialized countries. Some countries focus on gender equality (e.g., Sweden); others focus on flexibility at work and in education (e.g., Britain); still others focus on help with children (e.g., France). Although these changes were not made to increase fertility but to rationalize labor markets and equilibrate rights, they are having that effect: Births per 1,000 people slightly outnumber deaths. As a result, young adults in these countries face a future where they will be valuable additions to the labor force, and they can expect broad societal support in blending work and family.[8]

In wealthy countries that have been slow to adopt policies that are friendly to working families—such as Germany, Italy, Greece, and Spain—older people outnumber younger ones. Some say that these countries are now in a "fertility trap." Countries fall into the "trap" when women defer childbearing over long periods and new social norms such as childlessness or the one-child family replace the old two-child norm. But the word *trap* implies that women have been ensnared against their will, and most young adults in these countries say they are satisfied with their lower fertility.

The United States, in contrast, is the only large industrial country where domestic population is growing significantly. Because of relatively high fertility (only Iceland has a higher fertility rate among the world's more developed countries), births outnumber deaths in the United States even as rising life expectancy increases the numbers of elderly. With relatively high net migration rates, the U.S. population is projected to grow by 40 percent between 2006 and 2050. The racial and ethnic composition of both immigrant and high-fertility groups is also increasing the nation's nonwhite population, continuing a trend toward greater racial and ethnic diversity.

Wherever they are, and whatever the dynamics that shape their outlook, young adults from industrialized countries are a decreasing share of the global labor force. In 2005, seven out of eight people under age twenty-five were in "less" developed countries; by 2050, a projected nine out of ten young people will be in those countries.[9] In today's global economy, youth in higher-wage countries face growing competition from their increasingly educated counterparts in the developing world.

OUTLOOK FOR YOUTH II: THE DEVELOPING COUNTRIES

The rapid demographic transition of recent generations has placed most of the world's largest generation in its less economically developed regions. If current trends in education, reproductive health, and economic development continue, this transition may reach completion. Between 2005 and 2050, the UN projects only a 3.5 percent increase in the numbers of people under age twenty-five in today's less developed countries. Combined with a 14 percent decrease in today's more developed countries, that adds up to little more than a 1 percent increase in the numbers of youth overall.

> The outlook for youth in less developed countries reflects how far the demographic transition has advanced in their countries, along with the level of national investment in education beyond the primary level.

To a large extent, the outlook for youth in less developed countries reflects how far the demographic transition has advanced in their countries, along with the level of national investment in education beyond the primary level. Thus, the outlook for young adults in less developed countries can be generalized in two ways. Youth in countries that have invested in education are seeing increasing economic opportunity. In turn, these opportunities are constraining fertility by raising the cost of having children, so population growth is slowing. Meanwhile, countries that put tradition before modernization, especially by limiting women's roles outside the home, continue to experience high fertility and thus rapid population growth—with many more youths seeking employment than those countries' economies can easily absorb.

In many developing countries, UN projections suggest, the numbers of youth may be peaking. If the trends that underlie these projections can be sustained, these countries will see the "demographic window," a period in which there are large numbers of young workers with relatively few dependents—children and the elderly. With a smaller number of nonworkers to care for, societies that have entered the demographic window

> Most demographic projections assume that fertility rates will continue their orderly decline. But that is up to the men and women of the largest generation.

can make investments in human and other productive capital, which can boost economic growth and human well-being. (See Chapter 11.)

However, it is not a given that all developing countries will see a demographic window. Most demographic projections assume that fertility rates will continue their orderly decline. But that is up to the men and women of the largest generation, whose fertility choices will determine the size of the next generation. Recall that three out of five people under age fifteen are in Asia, where, outside of China, they average a third or more of the population—yet to have their children.[10] The fertility rate for the region (again excluding China) is 2.8, and the population (excluding China) is projected to grow by 45 percent between 2005 and 2050.

China and a few other Asian countries (e.g., Thailand, Vietnam, and Iran, among populous countries) have fertility rates that are akin to the more developed countries: below the "replacement level" of approximately two children per couple. However, their populations will still grow because they have so many young people. That is, even with low fertility rates, demographic momentum will assure growth when the large generations of youth become parents in their turn.

Momentum explains why countries in this region that have fertility rates between two and three children per woman will experience much more population growth. India, for example, has a projected population growth of 45 percent, even with its current low fertility rate of 2.9

children per woman. Ditto Bangladesh, with a projected growth of nearly 60 percent. Population momentum is especially visible in Indonesia and Myanmar. Both countries have a relatively low fertility rate—2.4 children per woman—but extremely youthful populations. With about six times as many people under age fifteen as over age sixty-five, both countries will grow by 25 percent by 2050 on the basis of current trends.

Some Asian countries have much higher fertility rates and thus much higher projected population growth. Like the very low-fertility European countries, these countries tend to prioritize customs and tradition. The common denominator is resistance to wider roles for women. In industrialized countries, reluctance to ease women's family responsibilities leads women to reduce their family size; in developing countries, women's lack of options outside the home leads them to maximize family size—or denies them the ability to control their fertility.

As a result, high fertility rates in Asia tend to be in very traditional societies, such as Muslim countries in western Asia like Saudi Arabia, Pakistan, and Yemen. In these countries, resources such as water and arable land are already constrained, and economies are struggling to produce jobs for the ever-growing numbers of youth. Yet with 40 percent or more of the population under age fifteen, these countries' populations are expected to double or even triple by 2050.

Although this situation represents a minority of Asian countries, it is typical of most countries in Africa, where population is expected to more than double during this half century. As a result, Africans will account for 22 percent of the world's population in 2050, up from 14 percent in 2006. As in Asia, rapid population growth is occurring in some of the world's most environmentally ravaged, economically challenged, and politically unstable countries. But in these countries, the peak of the youth population is not in sight, and there is no demographic window to help build resources for investing beyond simple subsistence.

The outlook for African youth depends on where they are coming of age. In northern and southern Africa, except Sudan, youth are relatively less numerous than they are elsewhere on the continent, because of fertility rates that currently average around 3.0. However, because of higher fertility among previous generations, youth make up such a large share of the population that substantial population growth is ensured. Based on current trends, UN demographers project 55 percent population growth in southern Africa and 66 percent in northern Africa by 2050.

In contrast, population in western, eastern, and middle Africa will more than double during the first half of the twenty-first century. In these regions, fertility rates of five, six, or more children per woman are common. Early and continued childbearing is directly related to limited schooling, especially for girls, and to early marriage where girls lack alternative roles. With available resources largely devoted to subsistence, economic opportunities for youth of both sexes are limited in these countries. The lack of opportunity fuels emigration, without which population growth would be even greater. The World Bank reckons that such migration is likely to intensify in response to the opportunities gap, especially as the growth of migrant networks and the advanced information and communications technology favored by youth support it.[11] (The Bank also notes the positive development effects of return migration, as youth return home with new skills.)

Sub-Saharan Africa is also the region that has been most hard-hit by the HIV/AIDS epidemic. With just over 10 percent of the world's population, sub-Saharan Africa is home to more than two out of three adults (68 percent) and nearly 90 percent of children infected with HIV. And this region has the highest rate of mortality from the disease: In 2007, more than three in four deaths (76 percent) from AIDS-related illnesses occurred in sub-Saharan Africa.[12] Here, the face of the epidemic is often female; in sub-Saharan Africa, six in ten adults living with HIV/AIDS are women.[13] The devastating epidemic has taken a toll on every aspect of life in this region. It has also changed its growth trajectory: Although continued high fertility means that Africa's population is projected to reach a billion by 2050, that number is about 350,000 less than projected without AIDS. In a few countries where fertility is relatively low, the epidemic is expected to bring about population decline.

For the continent's youth, the epidemic darkens an already bleak outlook. First, they are directly affected by the ravages of the disease, which disproportionately afflicts the sexually active, who tend to be young adults. Second, youth suffer from the diminished presence of teachers, health workers, and others who care for the young, while those who prematurely lose parents and other adult family members face the double burden of caregiving and loss of family support. Youth who survive to adulthood without the investments in their education that would otherwise have been made, whether by their families or their societies, face reduced economic opportunities throughout their lives. In Kenya, for example, the epidemic's effects on secondary educa-

BOX 2.1

Peak Population and Generation X

Alex Steffen

The babies born in the late 1960s were historic. They were part of the highest global population growth rate ever achieved, 2.1 percent a year. You might say that the birth of Generation X (which more or less bookends those years) was the beginning of our planet's era of peak human population.

It's easy to get blasé about demographics; big, abstract numbers thought about over numbing time periods, and recounted by people who love statistics. It would be a mistake, however, to fail to see peak population as a hugely important insight, because when we know that we are riding a wave of increasing numbers (and increasing longevity) that will crest some time after the middle of this century, we can also see this:

- The longer population growth rates remain high, the more total people there will be on the planet when we reach peak population, so we ought to see to it by every ethical means possible that the wave of population growth crests sooner rather than later.
- If we reach peak population sooner, at a lower number of people, we will be much more able to confront the myriad interlocking crises we face.
- We know the single best way to bring down high birthrates is to empower women by giving them access to reproductive health choices (including contraception and abortion), education, economic opportunities, and legal protection of their rights—so this ought to be one of our highest priorities.
- Our other main task is to preserve natural systems and transform human economies in order to best withstand this wave of human beings, avoid catastrophe, and leave behind as intact a world as we can—to save not just biodiversity but also the diversity of human cultures and histories—so future generations have as many options as possible.
- Our best hopes for both avoiding catastrophe and preserving our heritage all hinge on our actions over roughly the next two decades. After that time, all of these jobs will grow progressively harder, trending quickly toward impossibility.

tion have lowered human capital and per capita income so much that they will not return to their 1990 levels until 2030.[14]

In Latin America, the other large region in terms of population, the numbers of young people are also relatively large, and the age structure is a rough match of Asia's. But a higher rate of natural increase

Add this information together, and a generational imperative emerges. Generation X can be seen as the beginning of peak population; many of us (born between roughly 1960 and 1980) will live to see population peak in the middle of this century, and much of the most important work to see us through to the other side of that watershed will need to be done when Generation Xers are in their professional prime. We did not cause the crisis we face—unless you count us guilty at birth—but if the crisis is solved, it'll have to be in large part through the leadership of people born in my generation. Our historic call is to save the planet during peak population.

I am optimistic that we can do this. We have a rising network of brilliant and dedicated world-changing leaders. We live, despite the financial crisis, at a moment of great wealth. We have the motive, means, and opportunity.

None of this is to say that Gen X will do it alone. In particular, if you're young today, you have a huge choice to make: This transition will be unfolding during your entire career, and the role you choose to play in making it happen will be vitally important to your life, the planet, and the future. You too are called. At the same time, few eighteen-year-olds have the mix of experience, energy, and resources for changing the world that, say, a thirty-five-year-old has. Since the moment is now, it's those of us at the height of our powers who will have to lead the way.

Contemplating the journey beyond peak population, and the duty we have to lead it—well, it can weigh on you. I find it useful to remember that by changing the world today, we're building a better future beyond the crisis, that we work not only on our own behalf but for children who will not be born within our lifetimes, and their children, and theirs—that we'll make great ancestors.

But I also find it helpful to remember that these are our lives, and this is our adventure; and though times are tough and the planet demands our hard work, it also needs people who are happy, healthy, and creatively energetic. The world needs our best-lived lives, not our martyrdom.

....

Alex Steffen is the Executive Editor of worldchanging.com, from which this blog post is adapted.

portends faster population growth in this region, even with the tempering impact of emigration. Latin American governments' success in increasing economic opportunities will determine the extent to which young people perceive their future at home, or as immigrants to industrialized countries.

FAMILY PLANNING AND EDUCATION ARE KEY

In the best-case scenario, the developing countries will complete the demographic transition and take advantage of the demographic window to invest in human and productive capital. But that scenario rests on assumptions and projections—estimates of what will happen tomorrow based on what happened today. Thus, the projections cited so far are only the most likely of a range of scenarios, because they assume a continuation of current trends.

In particular, these projections assume that fertility will continue to decline in developing countries. In the fifty least-developed countries, largely in sub-Saharan Africa and the Middle East, they assume a sharp decline—from over 4.6 children per woman now to 2.5 children per woman in 2050. This in turn is based on an assumption that access to family planning will improve in these countries. But if this does not occur and fertility levels stay where they are now, the population of the less developed regions will reach 10.6 billion, not the 7.9 currently forecast (while the population of the developed regions remains stable, at 1.2 billion). Worldwide, that would mean a doubling of the population just since 1990. Access to family planning is thus a variable that will help determine the choices made by the largest generation, which will, in turn, determine the size of world population by mid-century.

Education is another crucial variable. Economic productivity is closely tied to education, and for the bulk of the world's youth in developing countries, higher population growth rates would make it harder for governments to provide them with more years of schooling. Education has a strong correlation with personal health as well as with better economic opportunities, especially for girls and women. Thus, it also determines fertility levels, especially in developing countries. For instance, in Brazil, women with no formal education have 5 children, on average, compared with 1.5 for women with some college education.[15] Education, then, is both a cause and a result of lower fertility rates—and part of a "virtuous circle" of mutually reinforcing improvements in health and productivity.

CONCLUSION

The largest generation—like the "greatest generation" before it—will leave a lasting imprint on the world. Just as their differing opportunities reflect choices made by previous generations, so, too, will the choices they make—especially in the areas of sexuality and childbearing—shape the course of their lives and the well-being of their families. Collectively, their choices will help determine the ultimate size of the human population and, to an extent, its impact on the natural systems that sustain all life. Decisions made today about investments in education and family planning will determine how much leeway today's youth will have to shape, rather than endure, the world their children and grandchildren will be born into.

REFERENCES

1. Population Division of the Department of Economic and Social Affairs of the UN Secretariat, 2007, World Population Prospects: The 2006 Revision, Highlights: Table 1.1, UN, New York.

2. Population Division of the Department of Economic and Social Affairs of the UN Secretariat World Population Prospects Database, http://esa.un .org/unpp/index.asp?panel=2, consulted July 26, 2007.

3. Population Division of the Department of Economic and Social Affairs of the UN Secretariat, 2007, World Population Prospects: The 2006 Revision, Highlights: Table 1.5, UN, New York.

4. Bianchi, for example, found that despite the greatly increased presence of mothers in the workforce compared with the past, U.S. children received slightly more maternal attention per week from mothers (and much more paternal attention) once that attention was spread over fewer children, Bianchi, Suzanne, 2000, Maternal employment and time with children, *Demography* 37: 4, 401–414.

Gauthier, Smeeding, and Furstenberg found that parental time invested in children has increased over the past forty years in sixteen industrialized countries, and the consistency of their results suggests a global trend, Gauthier, Anne H., Timothy M. Smeeding, and Frank F. Furstenberg Jr., 2004, Are parents in industrialized countries investing less time in children? *Population and Development Review* 30(4): 647–672.

5. *The Economist*, 2007, July 28, p. 26.

6. Unless otherwise noted, all numbers are derived from the Population Reference Bureau's 2006 World Population Data Sheet. This invaluable compendium is revised annually and is available at http://www.prb.org/.

7. *The Economist*, 2007, July 28, p. 26.

8. These societies also tend to tolerate nonmarital childbearing, as do the United States and other English-speaking industrialized countries.

9. Ibid.

10. China is often excluded from demographic projections because its sheer size—a population of 1.3 billion—would distort the data for other countries.

11. World Bank, World Development Report 2007, Chapter 8.

12. UNAIDS 2007 AIDS epidemic update—regional summary, accessed April 30, 2008 at http://www.unaids.org/en/CountryResponses/Regions/SubSaharanAfrica.asp.

13. Ibid.

14. Bell, Clive, Romona Bruhns, and Hans Gersbach, 2006, Economic Growth, Education, and AIDS in Kenya Model: A Long-run Analysis, prepared for World Bank, World Development Report 2007, and cited on p. 5.

15. DHS data analyzed by Wolfgang Lutz and Anne Goujon, in The world's changing human capital stock, *Population and Development Review* 27: 2, June 2001.

CHAPTER 3

....

People on the Move

Population, Migration, and the Environment

SUSAN GIBBS

The growing scope and scale of global migration illustrates the inter-connections among population growth, economic inequality, and environmental degradation. People choose to migrate across national borders for many reasons: Some set out in search of economic opportunity or a new beginning; others migrate because they are forced from their homes by violence or disaster and their lives are in grave danger. Migration issues have become increasingly heated in many countries and regions as new population influxes have challenged and at times threatened their new communities and nations.

This chapter will provide a brief overview of global migration trends and argue that this issue cannot be understood—or effectively managed—without employing a global perspective. Migration pressures are a symptom of deeper demographic, environmental, and economic challenges, and they cannot be alleviated solely by bolstering border patrols and resorting to increasingly draconian enforcement mechanisms. Global migration will continue to accelerate unless and until we more fully embrace the challenges of global stewardship and sustainability.

The number of migrants crossing national borders has reached record levels. An estimated 200 million people—about 3 percent of the

Susan Gibbs is a consulting program officer for the Wallace Global Fund responsible for the Fund's grantmaking in global population and women's empowerment.

world's population—now live outside their countries of birth. Nearly half of the world's migrants move from one developing country to another. In other words, south-south migration is almost as common as south-north migration.[1] Women now comprise half the world's migrant population: The percentage of women migrants increased from 46.7 percent in 1960 to 49.6 percent in 2005.[2]

> An estimated 200 million people—about 3 percent of the world's population—now live outside their countries of birth.

Why do people migrate? Many potent "push" and "pull" factors have contributed to what one author has coined the "age of migration."[3] Advances in communications and technology have made economic opportunities in distant lands more accessible, and transportation advances have reduced the costs and eased the logistics of international travel. Past migration patterns have resulted in well-established family and community networks that facilitate continued migrant flows. Finally, migration is the inevitable response to global economic, environmental, and demographic trends and inequities that are explored more fully below.

MIGRATION, POPULATION CHANGE, AND GLOBALIZATION

Global demographic and economic disparities create powerful catalysts for continued migration. The world's population continues to grow by 70 to 80 million people per year, with 97 percent of this growth in developing countries. As detailed in the previous chapters, the populations of many developed countries are aging and shrinking while the populations of many developing countries are youthful and growing. Overall, developed and developing countries will confront starkly contrasting demographic futures: High-income countries will face a "youth dearth" in contrast to low-income countries' "youth bulge."[4]

These demographic trends will amplify the unequal distribution of the world's workforce, which in turn will contribute to global migration pressures: The world's total labor force of 3.1 billion in 2005 included 600 million workers in developed countries and 2.4 billion in developing countries.[5] As American, European, and Japanese baby boomers retire in large numbers, young migrants can bolster the workforce and tax base.[6] In this formulation, migration from low-income to high-income nations is a win-win strategy, summarized in a recent article in *Newsweek International*: In an ideal world, "labor would flow from the overpopulated, resource-poor south to the depopulating north, where jobs would continue to be plentiful. Capital and remittance income from the rich nations would flow along the reverse path, benefiting all."[7]

The actual politics and policies of international migration bear little resemblance to this "ideal world." Developed nations strictly regulate migration flows, largely blocking entry to unskilled migrants while selectively admitting the highly skilled, thereby reinforcing the flight, or "brain drain," of higher-skilled workers from poor to rich countries.

Restrictionist policies and enforcement mechanisms have consumed enormous resources but have met with mixed success because global economic inequities are a key factor propelling global migration, and these disparities are growing. The United States, Europe, and Japan are now a hundred times richer on average than Ethiopia, Haiti, and Nepal. Globalization has rendered markets "inherently disequalizing, making rising inequality in developing countries more rather than less likely."[8] These economic inequities *within* nations can also stimulate migration: In developing countries, some 40 percent of workers are employed in the rural agricultural sector where comparatively low wages have encouraged millions of workers to leave the land and head to cities.[9] Unequal land distribution and tenure policies in which a small elite control most of the agriculturally productive land are also associated with high levels of rural out-migration.[10]

The acceleration of globalization, and related trends of privatization, deregulation, and trade liberalization, have recalibrated the "push" and "pull" dynamics of migration. In some cases, globalization has reduced migration pressures, at least for the highly skilled. As Thomas Friedman observed in *The World Is Flat*, foreign investment has stimulated job creation in some developing countries. In Friedman's optimistic formulation, "When the world is flat you can innovate without having to emigrate." However, this new economic paradigm has not distributed benefits evenly: While some highly skilled workers may experience new options and employment opportunities in their home countries, unskilled laborers are often left behind as firms chase lower labor costs and higher profit margins elsewhere.

> Global economic inequities are a key factor propelling global migration. The United States, Europe, and Japan are now a hundred times richer on average than Ethiopia, Haiti, and Nepal.

Globalization has broken down barriers impeding the flow of trade and investment while leaving regulations against the transnational movements of labor largely in place. In other words, jobs can easily move across national borders, but workers can't. That movement of jobs—especially the export of manufacturing jobs from the United States—has helped stoke popular resentment against immigrants, who are perceived as competing for jobs in an unstable economy and contributing to downward pressure on wages and benefits. And while nations have resorted to increasingly aggressive enforcement measures, including erecting walls—in some cases, literally—to staunch the flow of unauthorized migration, the economic and social forces that propel these movements of people have overtaken attempts by national governments to staunch the flow.

The effects of the 1994 North American Free Trade Agreement (NAFTA) on Mexican out-migration illustrate the links between migration, free trade, and globalization. Early NAFTA proponents argued that the benefits of free trade would stimulate economic growth and job creation in Mexico, thus reducing Mexican out-migration. In fact, NAFTA has had the opposite effect. The National Campesino Front estimates that NAFTA displaced 2 million Mexican farmers.[11] Why? While other political and economic factors played a role in stimulating Mexican out-migration, such as the nation's 1994 peso devaluation and subsequent economic crisis, NAFTA succeeded in opening Mexico's agricultural and manufacturing sectors to foreign investment and agricultural imports while weakening public sector supports. Mexican farms and firms were unable to compete with subsidized U.S. corn and other agricultural products flooding the Mexican market, and the result was that thousands of poor Mexican farmers joined the ranks of migrants to the United States, where many contributed "their poorly paid labor to the same agricultural sector that displaced them."[12]

While economic inequality is one root cause of global migration, economists continue to debate migration's effects on income inequality. Remittances sent home by migrants abroad to their countries of origin are among the fastest-growing international financial flows and contribute to economic development, poverty reduction, and the maintenance of household consumption levels during times of economic adversity. Migrants sent home $67 billion in 1990, and this total grew to some $300 billion in 2006—nearly three times the world's foreign-aid budgets combined.[13] Egypt receives more money from remittances than it does from transit fees from the Suez Canal.[14] The state of Kerala in India has long been touted as a development success story, lauded for its farsighted investments in education and social services, including girls' education and maternal and child health. However, the $5 billion in migrant remittances now flowing into Kerala each year are so significant that, according to leading demographer Irudaya Rajan, "remittances from global capitalism are carrying the whole Kerala economy."[15] Still, while remittances infuse many developing-country economies with needed funds, they cannot on their own address public infrastructure and development needs.

> Migrants sent home $300 billion in 2006—nearly three times the world's foreign-aid budgets combined.

Attempts to address and regularize migration at the global policy level have not advanced far. Some have urged the establishment of a world migration organization as a first step toward institutionalizing a global response to migration, similar in scope to the World Trade Organization.[16] Another proposal centers on creating an international remittances institute to streamline and reduce the costs of remittance transfers.[17] The Mode IV commitments outlined in the General Agreement on Agriculture and Trade in Services (GATS) offer a potential framework for temporary labor migration, although this would likely only benefit highly skilled workers. Fordham Law School Professor Jennifer Gordon has advanced the concept of a "Transnational Labor Citizenship Visa" for members of cross-border worker organizations that would facilitate and oversee temporary labor migration, as well as uphold workplace standards.[18] Gordon notes that some 200 bilateral labor migration agreements have already been negotiated between individual sending and receiving countries, and new transnational civil society networks are proliferating, but transnational labor experiments that emphasize workers' rights are still "embryonic, fragile and often distorted by the flawed guest worker programs around which they have been shaped." However, she is optimistic that each new initiative will bring us "closer to a global perspective on labor migration."[19]

MIGRATION, CLIMATE CHANGE, AND ENVIRONMENTAL DEGRADATION

Some migrants cross international borders not in pursuit of economic well-being but to flee political repression and warfare. The 1951 United Nations Convention Relating to the Status of Refugees stipulates that only those with a "well-founded fear of persecution on account of race, religion, nationality, membership in a particular social group, or political opinion" are eligible for international protection. These strict criteria for obtaining refugee status mean that only a small percentage of forced migrants receive international care and protection. Forced migrants who do not cross national borders are classified as internally displaced persons (IDPs). In 2007, there were over 14 million refugees and asylum seekers worldwide and 24.5 million internally displaced persons.

Growing numbers of migrants are likely to be forced from their homes by threats unanticipated by the framers of the 1951 Geneva Convention. Human-induced climate and hydrologic changes such as temperature and sea level rise will likely imperil local livelihoods and increasingly stimulate migration.[20] Low-elevation coastal areas less than 10 meters above sea level contain 2 percent of the world's land but 10 percent of the world's population.[21] A 1-meter sea level rise will likely displace 56 million people in eighty-four developing countries.[22] Falling water tables due to human-induced depletion of groundwater sources, salinization of the groundwater supply, flooding, and severe weather events—all phenomena associated with climate change—are forecast to swell the ranks of "environmental refugees" to some 200 million by 2050—almost a tenfold increase over the world's current total of internally displaced persons and refugees.[23] The relief organization Christian Aid estimates that environmental migrants will eventually number a billion.[24] Estimates of the current and potential numbers of environmental migrants vary widely, and one leading researcher in the field has admitted that his widely cited global estimate was based on "heroic extrapolation."[25] Nonetheless, these projections have led some to frame the plight of environmental migrants "as one of the foremost human crises of our times."[26]

Climate change in and of itself does not precipitate population movements. Rather, climate change contributes to environmental and economic effects that can lead to migration. In addition to being associated with sudden-onset natural disasters and long-term environmental effects such as desertification, climate change may also lead to conflicts over scarce resources, repressive and ineffective government responses, and widespread human rights violations, all of which can

> A 1-meter sea level rise will likely displace 56 million people in eighty-four developing countries.

lead to displacement.[27] Large-scale development projects such as dams, airports, and roads also contribute to displacement, estimated at up to 10 million people each year.[28] These projects are often erected in remote areas inhabited by indigenous or rural populations already subject to social and economic marginalization.

Climate change tends to magnify existing societal inequalities, and gender inequality is one of society's most pronounced fault lines. Women are often the most vulnerable to the effects of environmental degradation and climate change, because of negative impacts on their work burdens and livelihoods, and their domestic obligations can limit their mobility, placing them at greater risk. In the words of Wangari Maathai, founder of Kenya's Greenbelt Movement, "men can trek and go looking for greener pastures in other areas in other countries . . . but for women, they're usually left on site to face the consequences . . . so when there is deforestation, when there is drought, when there is crop failure, it is the women and children who are the most adversely affected."[29] However, while women can be burdened with "differential vulnerability," they also enjoy "differential adaptive capacity," and their key roles in protecting and managing their households' livelihood strategies can mitigate against climate change's negative effects.[30]

The concept of "environmental refugees" has been criticized for oversimplifying complex dynamics. There is no base of research documenting that migration increases proportionately to worsening environmental degradation, in part because of the resilience and varied adaptations of local communities.[31] It is difficult to isolate climate change as a migration driver, because environmental degradation typically goes hand in hand with political instability and economic insecurity, and all of these factors contribute to population displacement. For example, the destruction and displacement that Cyclone Nargis wreaked upon Myanmar represented the kind of natural disaster that is projected to increase in number and severity because of climate change. However, the cyclone's devastating effects were exacerbated by political repression and unsustainable natural resource management practices, including the destruction of Myanmar's mangrove forests for fuelwood, shrimp farming, and rice cultivation. Researchers estimate that some 83 percent of the mangroves in Myanmar's Irrawaddy Delta were destroyed between 1924 and 1999.[32] Similarly, attributing the mass displacement of residents of New Orleans after Hurricane Katrina solely to climate change oversimplifies the causes and effects of the crisis and its aftermath.

Global policy efforts to address the nexus between migration, development, and the environment have been piecemeal and limited. Some advocates argue that the United Nations Convention Relating to the Status of Refugees should be expanded to include a new category of "environmental persecution," maintaining that the environment can be used as an instrument of harm and that "harm is intentional when a set of policies is pursued in full knowledge of its damaging consequences," such as current U.S. energy policies that increase greenhouse gas emissions.[33] Others caution that expanding the legal definition of refugees eligible for international protection would weaken the existing international refugee regime.[34] The Climate Change, Environment and Migration Alliance (CCEMA), established in 2008 by the International Organization for Migration, the United Nations Environment Programme, and other partners, offers a promising mechanism for initiating global dialogue on this important issue.

In sum, environmental degradation, unsustainable exploitation of natural resources, and chronic poverty are among interrelated "push" factors contributing to displacement and migration, and these pressures show no signs of easing. The international community is still struggling to even define "environmental refugees" and has not even begun to formulate effective policy responses to this growing global challenge.

MIGRATION, JUSTICE, AND THE ENVIRONMENTAL CHALLENGE

While environmental degradation and climate change can stimulate migration, migration itself can be associated with negative environmental impacts. As detailed in a recent World Wildlife Fund and Conservation International Report, biodiversity-rich frontier zones are often viewed as "open access land systems where land and resources are free for the taking" and migrants can accelerate agricultural clearing, habitat destruction, loss of ecological connectivity, and biodiversity loss. Migration may also threaten long-standing residents: Migrants into biodiversity-rich frontier settings may use land and natural resources unsustainably, introduce exotic and invasive plant and other species, and crowd out local residents. A rapid influx of migrants can alter local politics and "weaken the social bonds of reciprocity and trust often required for land and resource management."[35]

These tensions among migration, natural resource management, and environmental sustainability are not restricted to remote ecoregions in Africa, Asia, and Latin America. Because immigration is a significant component of U.S. population growth, some environmental advocates urge sharp restrictions on immigration to reduce the nation's "ecological footprint," and they lament the reluctance of policymakers and environmental advocates to address the pressures of immigration and population growth openly and directly.[36] Their alarm stems from a conviction that the nation cannot absorb an ever-expanding population without compromising its environmental well-being and sustainability. At the same time, environmentalists note an increase in net impact when migrants move from low-consuming to high-consuming societies. While most environmentalists are deeply committed to addressing unsustainable patterns of resource production and consumption, they also argue that continued population growth will make these difficult and complex problems much harder to solve.

Constructive dialogue about the role of population growth and immigration as factors in environmental sustainability has been difficult to achieve in the United States because of the hot-button politics around immigration issues. Environmentalists tend to focus on migration in the aggregate and worry about broad and long-term demographic impacts, whereas the immigrant rights movement is focused on defending individuals in light of the long history of racism, xenophobia, and abuse targeting immigrant communities. From the Chinese Exclusion Act in 1882 to early nineteenth-century bans on paupers, prostitutes, and the mentally ill, the United States has long exercised its prerogative to selectively admit newcomers, and anti-immigrant discrimination and repression has a long and insidious history. Today, the specter of terrorism has exposed Middle Eastern and south Asian immigrants to particularly vitriolic ethnic profiling, harassment, and abuse. Hate crimes against Latinos rose by almost 35 percent between 2003 and 2006.[37] Migrant women often encounter psychological, physical, and verbal violence by smugglers, traffickers, employers, and even government officials.[38] In times of economic stress and rapid social change, immigrants are often demonized and blamed for a range of ills, including congestion and pollution, crime, unemployment, and even a perceived decline of "traditional" American values and cultural traditions.

However, blaming immigrants fails to address the root causes of these complex problems. Managing the challenge of global migration requires a global perspective: Globalization has promoted and accelerated the flows of goods, capital, and ideas across national borders while simultaneously attempting—with limited success—to stymie the global movement of labor. This discrepancy has been criticized as unworkable, unjust, and economically unsound, branded by one economist as "apartheid on a global scale."[39] Addressing migration in a global context also requires shining a spotlight on the United States' own profligate patterns of unsustainable production and consumption, which are closely linked to the growing migration pressures associated with globalization. As one advocate complains, "while right-wing groups want to close the border and drastically limit immigration they have no qualms about importing natural resources and exporting pollution across borders."[40]

> Globalization has promoted and accelerated the flows of goods, capital, and ideas across national borders while simultaneously attempting—with limited success—to stymie the global movement of labor.

The nexus among migration, development, and the environment remains fiercely contested. As framed by one anti-immigration U.S. advocacy group, "focusing solely on the interests of 'justice' for immigrants—both legal and illegal—causes us irresponsibly to ignore the unsustainable society we are creating for future generations."[41] Are the assumptions underlying this claim true? Must there be a zero-sum trade-off between protecting the rights of today's immigrants and preserving the planet? For the sake of our collective future, we can only hope that we can do both. We can only hope that environmentalists, immigrant rights advocates, and others will join forces to address and redirect the global economic and environmental forces that propel so many to migrate in the first place.

In identifying one of the major challenges of the twenty-first century, Nancy Birdsall was not focusing on migration when she cited the urgent need to "strengthen and reform the institutions, rules and customs by which nations and peoples manage the fundamentally political challenge of complementing the benefits of the global market with collective management of the problems, including persistent and unjust inequality, that global markets alone will not resolve."[42] Unfortunately, the world's growing environmental problems not only exacerbate injustice and inequity, they also pose significant challenges to governance and problem-solving—which the world now so desperately needs.

Arnoldo Garcia has framed the goal of the immigrant rights movement as "securing sustainable community development and human rights, including labor, cultural, civil, social, economic, and environmental rights for everyone."[43] But this goal is easier to articulate than to implement—especially when population growth and consumption are outstripping the planet's resource base and the mythical frontier, long a beacon for migrants, is disappearing for good. The challenges and abuses faced by many migrants, including forced migrants and refugees as well as economic migrants simply searching for a means of survival, are pernicious, and their needs for support and protection are real. However, global migration must ultimately be addressed further "upstream": The root causes propelling migration, such as global economic inequality and environmental degradation, must be confronted and addressed. All of the planet's citizens—regardless of migration status—have equal stakes in ensuring a sustainable future.

REFERENCES

1. Ratha, Dilip, and William Shaw, 2007, South-South Migration and Remittances, World Bank, Development Prospects Group, Washington, DC, January.

2. Morrison, Andrew R., et al., 2007, *The International Migration of Women*, Palgrave MacMillan, Houndmills, UK.

3. Castles, Stephen, and Mark Miller, 2003, *The Age of Migration: Population Movements in the Modern World*, Guilford Press, New York.

4. Pritchett, Lant, 2006, *Let Their People Come*, Center for Global Development, Washington, DC, September.

5. Martin, Philip, and Gottfried Zurcher, 2008, Managing migration: The global challenge, *Population Bulletin*, Volume 63, Number 1, Population Reference Bureau, Washington, DC, March.

6. Meyers, Dowell, 2007, *Immigrants and Boomers: Forging a New Social Contract for the Future of America*, Russell Sage Foundation Publications, New York.

7. Meyer, Michael, 2004, Birth dearth, *Newsweek International*, September 27.

8. Birdsall, Nancy, 2005, Rising inequality in the new global economy, *Wider Angle*, World Institute for Development Economics Research, Number 2.

9. Martin, Philip, and Gottfried Zurcher, 2008, Managing migration: The global challenge, *Population Bulletin*, Volume 63, Number 1, Population Reference Bureau, Washington, DC, March.

10. Oglethorpe, Judy, et al., 2007, *People of the Move: Reducing the Impact of Human Migration on Biodiversity*, World Wildlife Fund and Conservation International, Washington, DC.

11. Carlsen, Laura, 2007, Migrants: Globalization's Junk Mail? Institute for Policy Studies Foreign Policy in Focus, February 23, accessed at http://www.fpif.org/fpiftxt/4022.

12. Carlsen, Laura, 2007, NAFTA Free Trade Myths Lead to Farm Failure in Mexico, Center for International Policy, Americas Program Policy Report, December 5.

13. DeParle, Jason, 2007, In a world on the move, a tiny land strains to cope, *New York Times,* June 24.

14. DeParle, Jason, 2008, World banker and his cash return home, *New York Times*, March 17.

15. DeParle, Jason, 2007, Jobs abroad support "model" state in India, *New York Times*, September 7.

16. Bhagwati, Jagdish, 2003, Borders beyond control, *Foreign Affairs*, February.

17. DeParle, Jason, 2008, World banker and his cash return home, *New York Times*, March 17.

18. Gordon, Jennifer, 2008, Toward a freer flow of labor (with rights), *Americas Quarterly*, summer.

19. Ibid.

20. Sachs, Jeffrey D., 2007, Climate change refugees: As global warming tightens the availability of water, prepare for a torrent of forced migrations, *Scientific American*, June.

21. McGranahan, Gordon, et al., The rising tide: Assessing the risks of climate change and human settlements in low elevation coastal zones, *Environment & Urbanization*, 19(10): 17.

22. Brown, Lester, 2004, Environmental refugees: When the soil dies and the well dries, *International Herald Tribune*, February 14.

23. Ibid.

24. Christian Aid, 2007, *Human Tide: The Real Migration Crisis*, London.

25. Brown, Oli, 2008, The numbers game, *Forced Migration Review*, October.

26. Meyers, Norman, 2005, Environmental Refugees: An Emergent Security Issue, Thirteenth Economic Forum, May.

27. Ferris, Elizabeth, 2007, Making Sense of Climate Change, Natural Disasters, and Displacement: A Work in Progress, Calcutta Research Group Winter Course, December 14.

28. Castles, Stephen, 2004, Confronting the realities of forced migration, *Migration Information Source,* May, p. 4.

29. Zabarenko, Deborah, 2008, Women face tough impact from climate change, Reuters, May 7.

30. Women's Environment and Development Organization, ABANTU for Development in Ghana, ActionAid Bangladesh, and ENDA in Senegal, 2008, Gender, Climate Change and Human Security: Lessons from Ghana, Bangladesh and Senegal, May.

31. Black, Richard, Claudia Natali and Jessica Skinner, 2005, Migration and Inequality, The World Bank, Background Paper for 2006 *World Development Report*, January.

32. Spencer, Jane, 2008, Forest clearing may have worsened toll, *Wall Street Journal*, May 9.

33. Conisbee, Molly, and Andrew Simms, 2003, *Environmental Refugees: The Case for Recognition*, New Economics Foundation, London, p. 9.

34. Ferris, Elizabeth, 2007, Making Sense of Climate Change, Natural Disasters, and Displacement: A Work in Progress, Calcutta Research Group Winter Course, December 14.

35. Oglethorpe, Judy, et al., 2007, *People of the Move: Reducing the Impact of Human Migration on Biodiversity*, World Wildlife Fund and Conservation International, Washington, DC.

36. Kolankiewicz, Leon, and Roy Beck, 2001, Forsaking Fundamentals: The Environmental Establishment Abandons U.S. Population Stabilization, Center for Immigration Studies, March.

37. Mock, Brentin, 2007, Immigration backlash: Hate crimes against Latinos flourish, *Southern Poverty Law Center Intelligence Report*, winter.

38. Sachs, Susan, 2000, Sexual abuse reported at an immigration center, *New York Times*, October 5.

39. DeParle, 2007, Should we globalize labor too? *New York Times Magazine*, June 10.

40. Garcia, Arnoldo, 2003, Immigration. Population and environmental justice, *Urban Habitat*, summer.

41. Elbel, Fred, Intergenerational Justice, Colorado Alliance for Immigration Reform, http://www.cairco.org/ethics/intergen_justice.html.

42. Birdsall, Nancy, 2005, Rising inequality in the new global economy, *Wider Angle*, World Institute for Development Economics Research, Number 2.

43. Garcia, Arnoldo, 2003, Immigration. Population and environmental justice, *Urban Habitat*, summer.

CHAPTER 4

....

The Urban Millennium

In 2008, the world reached an invisible but momentous milestone: For the first time in history, more than half its human population, 3.3 billion people, lived in urban areas. By 2030, this is expected to swell to almost 5 billion. Many of the new urbanites will be poor. Their future, the future of cities in developing countries, and the future of humanity itself all depend very much on decisions made now in preparation for this growth.

While the world's urban population grew very rapidly (from 220 million to 2.8 billion) over the twentieth century, the next few decades will see an unprecedented scale of urban growth in the developing world. This will be particularly notable in Africa and Asia, where the urban population will double between 2000 and 2030. That is, the accumulated urban growth of these two regions during the whole span of history will be duplicated in a single generation. By 2030, the towns and cities of the developing world will make up 81 percent of urban humanity.

Urbanization—the increase in the urban share of total population— is inevitable. And while the current concentration of poverty, slum growth, and social disruption in cities does paint a threatening picture,

....

Adapted from The State of World Population 2007: Unleashing the Potential of Urban Growth, *published by the United Nations Population Fund.*

urbanization is also positive. No country in the industrial age has ever achieved significant economic growth without urbanization. Cities concentrate poverty, but they also represent the best hope of escaping it.

Cities embody the environmental damage done by modern civilization, yet experts and policymakers increasingly recognize the potential value of cities to long-term sustainability. If cities create environmental problems, they also contain the solutions. The potential benefits of urbanization far outweigh the disadvantages; the challenge is in learning how to exploit its possibilities.

> If cities create environmental problems, they also contain the solutions.

URBANIZATION'S SECOND WAVE: A DIFFERENCE OF SCALE

Comparing future to past trends helps put current urban growth trends into perspective. The *scale* of current change is unprecedented—though *rates* of urban growth in most regions have slowed.

The first urbanization wave took place in North America and Europe over two centuries, from 1750 to 1950: an increase from 10 to 52 percent urban and from 15 to 423 million urbanites. In the second wave of urbanization, in the less developed regions, the number of urbanites will go from 309 million in 1950 to 3.9 billion in 2030. In those eighty years, these countries will change from 18 percent to some 56 percent urban.

At the beginning of the twentieth century, the now developed regions had more than twice as many urban dwellers as the less developed (150 million to 70 million). Despite much lower levels of urbanization, the developing countries now have 2.6 times as many urban dwellers as the developed regions (2.3 billion to 0.9 billion). This gap will widen quickly in the next few decades (see Figure 4.1).

At the world level, the twentieth century saw an increase from 220 million urbanites in 1900 to 2.84 billion in 2000.[1] The present century will match this absolute increase in about four decades. Developing regions as a whole will account for 93 percent of this growth, Asia and Africa for over 80 percent.

The impact of globalization on city growth patterns marks a critical difference between past and present transitions.[2] Cities are the main beneficiaries of globalization, the increasing integration of the world's economies. People follow jobs, which follow investment and economic

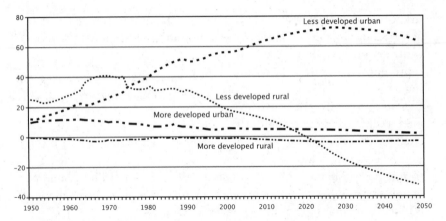

FIGURE 4.1 Annual growth of the urban and rural populations of the more developed and the less developed regions, 1950–2050. Population growth in millions.

Source: UN Population Division.

activities. Most are increasingly concentrated in and around dynamic urban areas, large and small.

However, very few developing-country cities generate enough jobs to meet the demands of their growing populations. Moreover, the benefits of urbanization are not equally enjoyed by all segments of the population; left out are those who traditionally face social and economic exclusion—women and ethnic minorities, for example. The massive increase in numbers of urbanites, coupled with persistent underdevelopment and the shortage of urban jobs, is responsible for conditions that can outmatch the Dickensian squalor of the Industrial Revolution. Nevertheless, like Adegoke Taylor (see Box 4.1), rural-urban migrants generally prefer their new life to the one they left behind.

CITIES: ENVIRONMENTAL BURDEN OR BLESSING?

The battle for a sustainable environmental future is being waged primarily in the world's cities. Right now, cities draw together many of Earth's major environmental problems: population growth, pollution, resource degradation, and waste generation. Paradoxically, cities also hold our best chance for a sustainable future.

Urban concentration need not aggravate environmental problems. These are due primarily to unsustainable patterns of production and consumption and to inadequate urban management. Urban localities

BOX 4.1

Adegoke Taylor

George Packer

A skinny, solemn thirty-two-year-old itinerant trader with anxious eyes, Adegoke Taylor shares an eight-by-ten-foot room with three other young men, on an alley in Isale Eko several hundred feet from the Third Mainland Bridge. In 1999, Taylor came to Lagos from Ile-Oluji, a Yoruba town a hundred and thirty miles to the northeast. He had a degree in mining from a polytechnic school and the goal of establishing a professional career. Upon arriving in the city, he went to a club that played juju—pop music infused with Yoruba rhythms—and stayed out until two in the morning. "This experience alone makes me believe I have a new life living now," he said, in English, the lingua franca of Lagos. "All the time, you see crowds everywhere. I was motivated by that. In the village, you're not free at all, and whatever you're going to do today you'll do tomorrow."

Taylor soon found that none of the few mining positions being advertised in Lagos newspapers were open to him. "If you are not connected, it's not easy, because there are many more applications than jobs," he said. "The moment you don't have a recognized person saying, 'This is my boy, give him a job,' it's very hard. In this country, if you don't belong to the elite you will find things very, very hard."

Taylor fell into a series of odd jobs: changing money, peddling stationery and hair plaits, and moving heavy loads in a warehouse for a daily wage of four hundred naira—about three dollars. Occasionally, he worked for West African traders who came to the markets near the port and needed middlemen to locate goods. At first, he stayed with the sister of a childhood friend in Mushin, then found cheap lodging there in a shared room for seven dollars a month, until the building was burned down during ethnic riots. Taylor lost everything. He decided to move to Lagos Island, where he pays a higher rent, twenty dollars a month.

Taylor had tried to leave Africa but was turned down for a visa by the American and British Embassies. At times, he longed for the calm of his home town, but there was never any question of returning to Ile-Oluji, with its early nights and monotonous days and the prospect of a lifetime of manual labor. His future was in Lagos. . . .

"There's no escape, except to make it," Taylor said.[a]

---

[a] Excerpt courtesy of Anderson Literary Management, Inc., November 13, 2006, "The Megacity," copyright 2006 © by George Packer, *The New Yorker* 82(37): 64.

actually offer better chances for long-term sustainability, starting with the fact that they concentrate half the Earth's population on less than 3 percent of its land area. The dispersion of population and economic activities would likely make the problems worse rather than better.

From a demographic standpoint, not only do dense settlements have greater capacity than rural areas to absorb large populations sustainably, but urbanization itself is a powerful factor in fertility decline. Urbanization provides few incentives for large families, and numerous disincentives.

Nonetheless, urban areas are associated with significant environmental challenges. Rapid expansion of urban areas changes land cover and causes habitat loss. The combination of urban population growth, decreasing densities, and peri-urbanization could convert large chunks of valuable land to urban uses in coming decades.

The environmental challenges posed by the conversion of natural and agricultural ecosystems to urban use have important implications for the functioning of global systems. How serious they are depends on where and how urban localities will expand. They depend even more on the patterns of consumption that city populations impose.

"Urban footprints" spread well beyond the immediate vicinity of cities, particularly in developed countries. Rising incomes and consumption in urban areas lead to increasing pressure on natural resources, triggering land-use and land-cover changes in their zones of influence, sometimes over vast areas. This typically causes much greater losses of habitat and ecosystem services than urban expansion itself.

For example, tropical forests in Tabasco were razed to provide space for cattle, in response to rising demand for meat in Mexico City, 400 kilometers away. Rising demand for soybeans and meat in China's urban areas, added to the demand from Japan, the United States, and Europe, is accelerating deforestation in the Brazilian Amazon.[3]

The need for freshwater is among the greatest challenges for urban areas. The dependence of cities on a guaranteed supply of water makes significant demands on global freshwater supplies. Cities already compete with the much larger demands of agriculture for scarce water resources in some regions, such as the southwestern United States, the Middle East, southern Africa, parts of central Asia, and the Sahel. Ultimately, cities outbid rural and agricultural users for available water supplies.[4]

Urban areas can affect water resources and the hydrological cycle in other ways: first, through the expansion of roads, parking lots, and other impervious surfaces, which pollute runoff and reduce the absorption of rainwater and aquifer replenishment; and second, through large-scale hydroelectric installations that help supply urban energy needs.[5]

CITIES AND CLIMATE CHANGE

Climate change has many implications for urban areas. Climate-related natural disasters are increasing in frequency and magnitude, and their consequences will depend on a number of factors, including the resilience and vulnerability of people and places.

Climate conditions have always shaped the built environment. Since the 1950s, however, traditional patterns adapted to local climatic conditions have been increasingly abandoned. Globalization and rapid technological developments tend to promote homogenized architectural and urban design, regardless of natural conditions. With this cookie-cutter architecture comes increased energy consumption from the transportation of exogenous materials and from the utilization of a single building design in a variety of environments and climatic conditions without regard to its energy efficiency.

The use of new architectural and urban forms, new materials, and innovations such as air conditioning have driven up both energy costs and cities' contributions to greenhouse gas emissions. Technological advances have also permitted the rapid growth of cities in places previously considered uninhabitable. For instance, the American city of Phoenix has boomed, thanks to engineering projects that diverted water from the Colorado River; the Saudi Arabian city of Riyadh thrives on water from desalinization plants.

Urban form and function also help define the nature of the interactions between cities and local climate change. For example, the "urban heat island effect" is an increase in temperatures in the urban core, compared with surrounding areas. As villages grow into towns and then into cities, their average temperature increases 2°C to 6°C above that of the surrounding countryside.[6] Urban designs and forms that neglect local climatic conditions and lose the cooling effects of green areas tend to aggravate this effect—especially in poor countries in the tropics.

Rapid urban growth, combined with the potent impacts of climate variability and climate change, will probably have severe consequences for environmental health in the tropics (causing, for example, heat stress and the buildup of tropospheric ozone), which can affect the urban economy (e.g., yield of labor and economic activities) and social organization. In a vicious circle, climate change will increase energy demand for air conditioning in urban areas and contribute to the urban heat island effect through heat pollution. Heat pollution, smog, and ground-level ozone are not just urban phenomena; they also affect surrounding rural areas, reducing agricultural yields,[7] increasing health risks,[8] and spawning tornadoes and thunderstorms.

Human health in urban areas may suffer as a result of climate change, especially in poor urban areas where inhabitants already suffer from a variety of problems associated with poverty and inequity. Climate change will aggravate these. For example, many poor people live in crowded neighborhoods that lack adequate health services, water supply, and sanitation. These conditions are ideal for spreading respiratory and intestinal conditions, and for breeding mosquitoes and other vectors of tropical diseases such as malaria, dengue, and yellow fever. Changes in temperature and precipitation can spread disease in previously unaffected areas and encourage it in areas already affected. Changes in climate and the water cycle could affect water supply, water distribution, and water quality in urban areas, with important consequences for waterborne diseases.

The impacts of climate change on urban water supplies are likely to be dramatic. Many poor countries already face accumulated deficiencies in water supply, distribution, and quality, and climate change is likely to increase those difficulties. The 2007 report of the Intergovernmental Panel on Climate Change underlines that cities in drier regions, such as Karachi in Pakistan and New Delhi in India, will be particularly hard hit.[9]

POVERTY AND VULNERABILITY TO NATURAL DISASTERS

Cities are highly vulnerable to natural disasters: Sudden supply shortages, heavy environmental burdens, or major catastrophes can quickly lead to serious emergencies. The consequences of such crises are multiplied by poorly coordinated administration and planning. Such disasters have become more frequent and more severe during the last two decades, affecting a number of large cities. In 1999, there were over 700 major natural disasters, causing more than US$100 billion in economic losses, and thousands of victims. Over 90 percent of losses in human life from natural disasters around the world occurred in poor countries.

The impacts of global environmental change, particularly climate-related hazards, disproportionately affect poor and vulnerable people—those who live in slum and squatter settlements on steep hillsides, in poorly drained areas, or in low-lying coastal zones.[10] For example, informal settlements on hillsides surrounding Caracas, Venezuela, were devastated in the December 1999 flash floods and landslides, which reportedly killed 30,000 people and affected nearly half a million others.[11] Hurricane Katrina's impact on New Orleans shows that developed countries are not immune to such wide-scale disasters; there, too, the brunt was borne by the poor.

Drought, flooding, and other consequences of climate change can also modify migration patterns between rural and urban areas or within urban areas. For example, severe floods in the Yangtze basin in China in 1998 and 2002, caused by a combination of climate variability and human-induced land-cover changes, displaced millions of people, mainly subsistence farmers and villagers. Similar examples can be seen in India, Mexico, and other poor countries. Many such "environmental refugees" never return to the rural areas from which they were displaced.

SEA LEVEL RISE: NOT IF BUT WHEN, AND HOW MUCH?[12]

One of the alarming prospects of climate change is its impact on sea level rise and its potential consequences for coastal urban areas. Coastal zones have always concentrated people and economic activities because of their natural resources and trading opportunities. Many of the world's largest cities are on seacoasts and at the mouths of the great rivers. Both urban and rural areas of coastal ecosystems are the most densely populated of any in the world.

These populations, especially when concentrated in large urban areas within rich ecological zones, can be a burden on coastal ecosystems, many of which are already under stress. Sea level rise, especially if combined with extreme climatic events, would flood large parts of these areas. It would also introduce salt water into surface freshwater and aquifers, affecting cities' water supplies, and modify critical ecosystems supplying ecological services and natural resources to urban areas. It would inevitably provoke migration to other urban areas. Coastal settlements in lower-income countries would be more vulnerable, and lower-income groups living on flood plains would be most vulnerable of all.

The first systematic assessment of these issues shows that the low elevation coastal zone (LECZ) currently accounts for only 2 percent of the world's land area but 13 percent of its urban population.[13] Despite lower urbanization levels, Africa and Asia have much larger proportions of their urban populations in coastal zones than North America or Europe. Such differences reflect the colonial heritage of Africa and Asia, where major cities grew as ports and export centers of raw materials.[14] Asia stands out, as it contains about three-quarters of the global population in the LECZ, and two-thirds of its urban population.

Given the real and increasing threats of global environmental change in the LECZ, the continuation of present patterns of urban growth is of some concern. From an environmental perspective, uncontrolled coastal development is likely to damage sensitive and important ecosystems and other resources. At the same time, coastal settlement, particularly in the lowlands, is likely to expose residents to seaward hazards that are likely to become more serious with climate change.

MAKING THE MOST OF THE URBAN FUTURE

The process of globalization has drawn attention to the productive potential of cities and to the human cost. Yet the enormous scale and impact of future urbanization have not penetrated the public's mind. Until now, policymakers and civil society organizations have reacted to challenges as they arise. This is no longer enough. A preemptive approach is needed if urbanization in developing countries is to help solve social and environmental problems, rather than make them catastrophically worse.

Although megacities have received most of the attention, conditions in smaller urban areas call for even greater consideration. Contrary to general belief, the bulk of urban population growth is likely to be in smaller cities and towns, whose capabilities for planning and implementation can be exceedingly weak. Yet the worldwide process of decentralizing governmental powers is heaping greater responsibility on them. As the population of smaller cities increases, their thin managerial and planning capacities come under mounting stress. New ways will have to be found to equip them to plan ahead for expansion, to use their resources sustainably, and to deliver essential services.

Poor people will make up a large part of future urban growth. This simple fact has generally been overlooked, at great cost. Most urban growth now stems from natural increase (more births than deaths) rather than migration. But wherever it comes from, the growth of urban areas includes huge numbers of poor people. Ignoring this basic reality will make it impossible either to plan for inevitable and massive city growth or to use urban dynamics to help relieve poverty.

> The approaching urban millennium could make poverty, inequality, and environmental degradation more manageable, or it could make them exponentially worse.

First, preparing for an urban future requires, at a minimum, respecting the rights of the poor to the city. Many policymakers continue to try to prevent urban growth by discouraging rural-urban migration, with tactics such as evicting squatters and denying them services. These attempts to prevent migration are futile, counterproductive, and, above all, wrong—a violation of people's rights. If policymakers find urban growth rates too high, they have effective options that also respect human rights. Advances in social development, such as promoting gender equity and equality, making education universally available, and meeting reproductive health needs, are important for their own sakes. They will also enable women to avoid unwanted fertility and reduce the main factor in the growth of urban populations—natural increase.

Second, cities need a longer-term and broader vision of the use of urban space to reduce poverty and promote sustainability. This includes an explicit concern with the land needs of the poor. For poor families, having an adequate piece of land—with access to water, sewage, power, and transport—on which they can construct their homes and improve their lives is essential; providing it requires a new and proactive approach. Planning for such spatial and infrastructure requirements, keeping in mind poor women's multiple roles and needs, will greatly improve the welfare of poor families. This kind of people-centered development knits together the social fabric and encourages economic growth that includes the poor.

Similarly, protecting the environment and managing ecosystem services in future urban expansion requires purposeful management of space in advance of needs. The urban footprint stretches far beyond city boundaries. Cities influence, and are affected by, broader environmental considerations. Proactive policies for sustainability will also be important in view of climate change and the considerable proportion of urban concentrations at or near sea level.

Decisions taken today in cities across the developing world will shape not only their own destinies but also the social and environmental future of humankind. The approaching urban millennium could make poverty, inequality, and environmental degradation more manageable, or it could make them exponentially worse. In this light, a sense of urgency must permeate efforts to address the challenges and opportunities presented by the urban transition.

REFERENCES

1. Satterthwaite, D., 2006, Outside the Large Cities: The Demographic Importance of Small Urban Centres and Large Villages in Africa, Asia and Latin America, p. 1, Human Settlements Discussion Paper Number Urban03, International Institute for Environment and Development, London.

2. Cohen, B., 2006, Urbanization in developing countries: Current trends, future projections, and key challenges for sustainability, *Technology in Society* 28(1–2): 63–80.

3. Wallace, S., 2007, Amazon: Forest to farms, *National Geographic*, January.

4. Rosegrant, M. W., and C. Ringler, 1998, Impact on food security and rural development of transferring water out of agriculture, *Water Policy* 1(6): 567–586.

5. Vörösmarty, C., 2006, Box D.2: Water impoundment and flow fragmentation, p. 259–260, *Pilot 2006 Environmental Performance Index* by the Yale Center for Environmental Law and Policy and the Center for International Earth Science Information Network, Columbia University, New Haven, CT, and Palisades, NY.

6. According to the U.S. Environmental Protection Agency; see U.S. Environmental Protection Agency, Heat Island Effect, http://yosemite.epa.gov/oar/globalwarming.nsf/content/ActionsLocalHeatIslandEffect.html, accessed January 29, 2007.

7. Ashmore, M. R., 2005, Assessing the future global impacts of ozone on vegetation, *Plant, Cell and Environment* 28(8): 949–964.

8. Lo, C. P., and D. A. Quattrochi, 2003, Land-use and land-cover change, urban heat island phenomenon, and health implications: A remote sensing approach, *Photogrammetric Engineering and Remote Sensing* 69(9): 1053–1063.

9. Intergovernmental Panel on Climate Change, 2007, *Climate Change 2007: The Physical Science Basis: Summary for Policy Makers*, IPCC, Geneva, www.ipcc.ch/SPM2feb07.pdf, accessed February 6, 2007.

10. Perlman, J., and M. O. Sheehan, 2007, Fighting poverty and injustice in cities, Chapter 9, *State of the World 2007: Our Urban Future,* Worldwatch Institute, 2007; and de Sherbinin, A., A. Schiller, and A. Pulsipher, Forthcoming, The vulnerability of global cities to climate hazards, *Environment and Urbanization.*

11. Center for Research on the Epidemiology of Disasters, 2006, EM-DAT: The OFDA/CRED International Disaster Database, Center for Research on the Epidemiology of Disasters, Brussels, Belgium.

12. This section is based on McGranahan, G., D. Balk, and B. Anderson, Forthcoming, The rising risks of climate change: Urban population distribution and characteristics in low elevation coastal zones, *Environment and Urbanization*; and McGranahan, Gordon, Peter J. Marcotullio, Xuemei Bai, Deborah Balk, Tania Braga, Ian Douglas, Thomas Elmqvist, William Rees, David Satterthwaite, Jacob Songsore, and Hania Zlotnik, 2005, Urban Systems, p. 795–825, *Ecosystems and Human Well-Being: Current Status and Trends*, Rashid Hassan, Robert Scholes, and Neville Ash, eds., Island Press, Washington, DC.

13. McGranahan, G., D. Balk, and B. Anderson, Forthcoming, The rising risks of climate change: Urban population distribution and characteristics in low elevation coastal zones, *Environment and Urbanization*.

14. Gugler, J., 1996, Urbanization in Africa south of the Sahara: New identities in conflict, Chapter 7, *The Urban Transformation of the Developing World*, J. Gugler, ed., Oxford University Press, Oxford.

PART II

....

The Impact

CHAPTER 5

....

Climate Change and Population Growth

BRIAN C. O'NEILL

Climate change is moving to the top of the international agenda. Driven by a combination of high-profile reports, damaging extreme weather events, and observations of melting ice and other impacts occurring faster than expected, climate change is now viewed by some as "the single most destructive force confronting humanity," in the words of former UN Secretary General Kofi Annan. As awareness of that force has grown, environmentalists, demographers, and others are taking a closer look at the role of population growth in human-induced climate change.

Over the last century, greenhouse gas emissions have risen in tandem with an ever-larger human population; the curve of soaring carbon dioxide emissions is neatly matched by the curve of world population growth. But many questions remain about the relationship between these two phenomena. For example, to what extent is population growth a "driver" of climate change? Can slower population growth mitigate climate change, or help societies adapt to its ravages?

In 2001, my colleagues and I set out to answer those questions. We

Brian C. O'Neill is a Scientist III at the National Center for Atmospheric Research in Boulder, Colorado, and leads the Population and Climate Change Program at the International Institute for Applied Systems Analysis (IIASA) in Laxenburg, Austria.

surveyed the existing population-climate literature,[19] assessing what was known about demography as a driver of greenhouse gas emissions and the climate changes they cause, and whether demographic factors magnify or minimize the societal impact of changes in climate. We also examined the case that population-related policies could contribute to addressing climate change.

> Slower population growth would indeed lead to lower emissions, making the climate-change problem easier to solve. And slower growth would likely make societies more resilient to the impacts of climate change.

Here is what we found: Briefly, we concluded that slower population growth would indeed lead to lower emissions, making the climate-change problem easier to solve. We also found that slower population growth would likely make societies more resilient to the impacts of climate change. There were caveats in both cases: Population-related reductions in emissions would not be significant in the short term, and even in the long term must be seen in the context of much larger effects on emissions from economic growth and technological change. Similarly, population growth was not the most important determinant of resilience to any particular climate-change impact, such as effects on health, food and nutrition, or water scarcity. Nonetheless, we concluded that on balance, population-related policies that were desirable in their own right would also help mitigate climate change.

Since that time, there has been a new wave of scholarship on the role of demographic change in greenhouse gas emissions. The new work generally takes one of two approaches: statistical analyses of past trends, or modeling studies of future scenarios. In general, these analyses have employed more data, used better methodologies, and examined a wider range of demographic changes, including aging and urbanization. Here I assess that work, considering results from each approach in turn, and conclude by discussing these new findings in the context of population and climate-change policy.

STATISTICAL ANALYSES

Since the early 1970s, scientists have used the "IPAT" equation to decompose historical data on environmental impacts. IPAT posits that environmental impact (I) results from population growth (P), growth in per capita income or consumption as measures of affluence (A), and changes in technology (T): $I = P \times A \times T$. The IPAT equation was developed as part of a debate on the relative roles of population growth and technological change in environmental degradation in the United States.[19] Subsequently, this framework has been applied to the analysis of many environmental issues, including energy use and the emission of greenhouse gases.

Although IPAT can be a useful starting point for thinking about determinants of environmental impact, decomposition exercises tend to generate more heat than light.[20] One of the most common criticisms has been that these analyses were oversimplified. In particular, using the IPAT equation to apportion "blame" for a particular environmental outcome (such as greenhouse gas emissions) among its three determinants implicitly assumes that population size has a proportional effect on impact; that is, if population size increases 10 percent, impact increases 10 percent.

Recently, statistical analyses have become more sophisticated. Over the past several years, researchers have published studies that do not assume, but rather *test*, whether the effect of population growth on CO_2 emissions is proportional, using large data sets covering many countries over the past several decades. Overall these studies have concluded that population does indeed appear to have a proportional effect on emissions. The first study of this kind was by Dietz and Rosa,[10] which found that the "elasticity" of emissions with respect to population growth was 1.15. *Elasticity* is defined as the percent change in emissions associated with a 1 percent increase in population. An elasticity of 1.0 indicates a proportional effect; the Dietz and Rosa results indicated that population growth had an effect on emissions that was actually somewhat more than proportional—meaning that a 10 percent increase in population size would mean an increase in emissions of about 11.5 percent.

The Dietz and Rosa paper stimulated subsequent studies that employed better estimation techniques, examined a wider range of pollutants, and included more demographic variables, such as household size, urbanization, and age structure. These later studies also corrected some of the shortcomings of the Dietz and Rosa analysis, which relied on cross-sectional data for a single year. Results based on such data can be misleading if population size is correlated with some other variable that is actually at work—a so-called unobserved or latent variable. In contrast, the later studies used more reliable "panel" data: repeated cross-sectional data over time. Shi,[23] for example, employed panel data and found a population size elasticity substantially greater than 1.0.

Cole and Neumayer[5] used the same type of data but found a somewhat smaller elasticity of close to 1.0.

The later studies also suggest that other demographic variables—in addition to population size—affect emissions. Several find that urbanization is associated with higher emissions.[22] Cole and Neumayer find that a decrease in household size is associated with higher emissions.[5] The effect of age structure is less clear, with some studies finding a significant effect and others not.

Although these later studies affirmed the earlier finding that population growth and emissions are roughly proportional, there are several caveats to keep in mind. First, while population size is a driver of greenhouse gas emissions, it is not necessarily the most important driver. Increases in GDP also were found to have a roughly proportional effect on emissions, and technology effects were equally important. Second, the results include only direct effects, leaving out indirect effects of demographic change that may operate through other variables. For example, if population growth (or aging, or urbanization) affects per capita income growth, which in turn affects emissions, this effect is not captured in the elasticities assigned to the demographic variables.

> The impact of population growth on emissions is highest in low-income countries with less efficient energy systems.

Third, analyses that subdivided the sample of countries found that elasticities differed by subpopulation. For example, EU countries were found to have a much smaller population size elasticity (0.55) than estimates for all countries taken together, meaning that population size increases lead to a smaller increase in emissions. Even within the EU, differences between countries were substantial.[16] Similarly, countries with different incomes have substantially different elasticity estimates.[11, 23] These cases illustrate the importance of the context in which population growth occurs. The impact tends to be highest in countries with lower income levels, and with less efficient energy systems. Thus the single estimates of population elasticities found by studies that group all countries together should not be assumed to apply to all countries equally.

POPULATION IN EMISSIONS SCENARIOS

While statistical analyses look back at historical data, a complementary approach looks forward by simulating the effect of population variables on emissions within a structured modeling framework. In principle, this approach has the advantage of being able to incorporate a number of direct and indirect effects of demographic change. There is now a large literature on scenarios of future greenhouse emissions.[12] Population growth is widely recognized as an important driving force of future emissions, and virtually all quantitative emissions scenarios include population size changes as one driver. Here, we will focus on the few scenarios that test the independent effect of demographic change on emissions outcomes.

Early analyses were limited to relatively simple, back-of-the-envelope-style exercises[1, 2, 4, 15, 20] or to global models that did not distinguish among different regions.[13] More-sophisticated modeling frameworks, such as those currently used in the greenhouse gas emissions scenario community, have not explicitly tested the role of demographic change, with the exception of work that is now under way (discussed below).

Nonetheless, it is worth examining the relationship of population size to emissions in the existing scenarios literature. Perhaps the highest-profile set of scenarios was developed in the Intergovernmental Panel on Climate Change (IPCC) Special Report on Emissions Scenarios (SRES),[17] an effort to develop a common basis for climate-change projections, analyses of impacts and potential adaptation, and emissions mitigation assessments. The methodology involved developing four narrative "storylines" describing alternative trends in future development paths, then using six different models to generate quantitative interpretations of each storyline. Each storyline made different assumptions about emissions drivers such as population growth, economic growth, and rates of technological change. The models then calculated what emissions pathways might result, producing a set of forty individual scenarios, with four singled out as representative, or "marker," scenarios.

To quantify the effect of demographic drivers, it would have been ideal if some of the SRES scenarios had held all other assumptions constant and varied the assumptions about population growth. Such an exercise was in fact carried out to explore the effect of technological change, but not for any other factor. Nonetheless, we can examine the general relationship between population and emissions outcomes, keeping this limitation in mind.

Most of the SRES scenarios affirm the statistical analyses' finding that increased emissions are roughly proportional to population growth. In three-quarters of the scenarios—the B1, B2, and A2 series— higher population growth is associated with greater emissions. In fact, the scenarios indicate a relationship between population and emissions that is close to, but somewhat less than, proportional. That is, they suggest that when population is twice as large, emissions are nearly twice as large as well. This outcome is what one might intuitively expect, given the historical evidence reviewed above.

Other scenarios show that population and emissions need not grow in lockstep. The fourteen scenarios developed within the A1 storyline display an emissions range that is larger than that of all the scenarios from the other three storylines combined. Why? The A1 storyline is the one mentioned above, in which assumptions about technology varied while all others were held constant. Assuming low population growth rates, the A1 storyline spins out different scenarios for energy use: greater use of renewables, greater use of fossil fuels, or a balanced mix. The highest emissions scenarios of all were produced by robust economic growth and greater use of fossil fuels—even with slow population growth.

> Slower population growth alone is no guarantee of lower emissions; other factors, notably energy use, can easily outweigh the positive impact of slower growth.

These findings suggest that, on balance, slower population growth is associated with lower emissions. But slower population growth alone is no guarantee of lower emissions; other factors, notably energy use, can easily outweigh the positive impact of slower growth.

Since the publication of the SRES, many other scenarios have been developed and were reviewed most recently in the IPCC Fourth Assessment Report.[12] No assessments of demographic effects on global emissions appear in that literature either. Moreover, these newer scenarios have based their population assumptions on those used in the SRES, which are by now about ten years old. Newer demographic outlooks from the UN and the International Institute for Applied Systems Analysis (IIASA) anticipate less population growth than was predicted in the 1990s. The scenario community has not yet caught up with this new outlook; in particular, the low end of the current uncertainty range for future population has not been much explored.

CURRENT RESEARCH

To fill the data gap on demographic change and CO_2 emissions, my colleagues and I have launched a new research initiative. First, we are conducting country case studies that focus on emissions from the combustion of fossil fuels for energy production. Although emissions of non-CO_2 gases and aerosols, and of CO_2 from land-use change, are important, CO_2 from energy use is currently the main contributor to anthropogenic climate change and is expected to remain so over the course of the century. We have begun with three countries that are—or will soon be—major emitters: the United States, China, and India.

Our approach employs a neoclassical economic growth model, the PET (population-environment-technology) model, of the type that is commonly used in the energy and emissions scenario field. The details of the model are described elsewhere.[8, 9] In simplified terms, the model incorporates the activities of the household sector, final goods producers, and intermediate goods producers.

We adapted the basic model to address demographic issues by replacing the usual assumption in such models that all households are identical with a classification of households by age, size, and urban/rural status. This disaggregation allows different types of households to behave differently based on differences in preferences for consumption goods (e.g., residential energy, transportation, food), propensity to save, initial stock of assets, and labor supply. We developed new projections of households for each country that indicate the number of people living in households of different ages (where the age of a household is defined as the age of the household head), sizes (number of members), and urban/rural statuses.[14] These projections, anticipating shifts in the composition of the population according to these different categories, are then used as an input to the economic model and influence the emissions scenario outcomes.

Figure 5.1 shows results for the United States, in which we limited our analysis to aging.[9] In general, aging in the United States led to a reduction in projected emissions. For example, in a low–population growth scenario in which aging is more pronounced, emissions are reduced in the long run by more than a third. The effect is driven primarily by the implications of aging for the labor force: As a larger percentage of the population ages, the size of the labor force as a fraction of total population falls, leading to slower economic growth. The smaller overall scale of the economy leads to lower emissions.

In India and China, we included the effects of aging but focused mainly on urbanization.[8] We found relatively small aging effects (5 to 20 percent, acting in the same direction, and for the same reason, as in the U.S. analysis) but very large effects from urbanization (Figure 5.2).

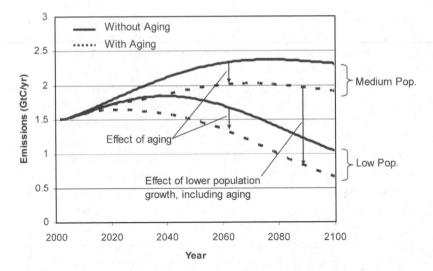

FIGURE 5.1 Effects of demographic change on U.S. CO_2 emissions, 2000–2100. Dashed lines include the effect of aging, solid lines do not. GtC/yr = gigatons of carbon per year. *Source*: Dalton et al., 2008, Population aging.

Adding urbanization to the model, driven by relatively modest urbanization projections, led to an increase in emissions of 30 to 60 percent by 2050, and 50 to 70 percent by 2100. Here again, the effect acted primarily through the labor force. Based on current data, urban workers are much more productive than rural workers in terms of the contribution of their labor to GDP growth. Thus, a shift in composition toward an increasingly urban workforce implies a substantial boost to overall productivity, faster economic growth, and more emissions.

> Urbanization implies greater productivity, faster economic growth, and more emissions.

In the U.S. case study, we also tested the effect of alternative population sizes. Figure 5.1 shows results for a medium path and a slow growth path producing relatively older and smaller households. The low–population growth path had a 21 percent smaller population in 2050 than the medium path and was 56 percent smaller in 2100. When accounting for the effects of aging in both scenarios, emissions turned out to be lower in the low-growth scenario by 21 percent in 2050 and 61 percent 2100. In 2050, the effect was proportional to the population reduction, reflecting the fact that the amounts of aging by 2050 in both sce-

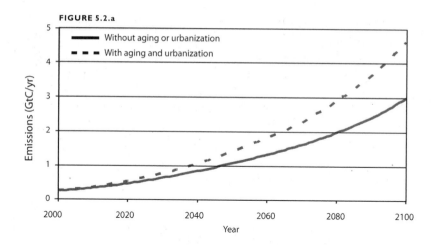

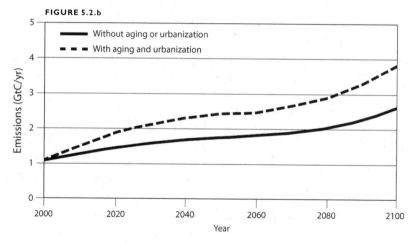

FIGURE 5.2 Effects of demographic change on CO_2 emissions in India (a) and China (b), 2000–2100. All projections are based on a medium population growth path. Curves show results when compositional changes are not considered (solid) and when both aging and urbanization are considered (dashed). GtC/yr = gigatons of carbon per year.

Source: Dalton et al., 2008, Demographic Change.

narios were very similar. In 2100, the emissions reduction is somewhat more than proportional to the population reduction, since aging in the low scenario outpaces aging in the medium scenario by that time. In that case, emissions are lower, not only because of a smaller population size, but also because the economy grows more slowly.

This work is in its initial stages, so there are still many caveats. First, we have not yet explored the effect of changes in retirement ages. Labor force participation at older ages may well increase as life expectancies lengthen, and this would dampen the effect of aging on economic growth. In addition, we have so far modeled each country in isolation, without international trade or capital flows. Including these might also dampen the effect of aging, if declining investment from an aging population is offset by investment from countries with younger populations. Despite these caveats, our initial results indicate that demographic change—including population size, aging, and urbanization—can have substantial effects on emissions.

DISCUSSION AND CONCLUSIONS

Research on demographic change and greenhouse gas emissions can be categorized, broadly speaking, into analyses of historical data and modeling of future emissions. Statistical analyses of past trends suggest that population has had a roughly proportional effect on emissions over the past several decades. These analyses are incomplete in some respects. For example, they leave out potential indirect effects such as the effect of population growth (or decline) on emissions through its effect on economic growth. They also tend to average over potentially substantial differences across country groups or subpopulations. Nonetheless, these studies provide important insight into the issue. So far, there is no statistical evidence to refute the thesis that larger population size is generally associated with higher emissions.

To date, modeling of future emissions scenarios has not measured the independent effect of demographic changes on emissions. Early analyses concluded that slower population growth could contribute to emissions reductions, at least in the long run, but these studies did not investigate interactions between population change and economic growth, nor did they account for demographic effects such as aging or urbanization. More-sophisticated modeling studies suggest that in general, lower population growth is associated with—but cannot guarantee—lower emissions.

Recently, we have begun to investigate the effect of demographic change in a model that is capable of incorporating the effects of population growth, aging, and urbanization on economic growth, consumption patterns, and emissions from fossil fuel–based energy use. We have found that, in the United States at least, the effect of smaller population size on emissions is somewhat more than proportional (in the long run) when the aging associated with slower population growth is taken into account. In India and China we have found that aging lowers emissions somewhat, but this effect is more than offset by the effect of urbanization, which can have a very large positive effect on future emissions growth. We have not yet assessed the net effect of lower population growth in India and China, but that work, as well as a global analysis, is under way.

What are the policy implications of this work? The new research confirms our earlier finding: Policy measures that slow population growth would likely reduce emissions, making the climate problem easier to solve. But again, there is a caveat: All of the existing literature on population growth and emissions takes changes in population size as a starting point. It does not consider *how* population growth might be slowed. Some policy measures or development trends that lower fertility may themselves have independent effects on emissions. Take, for example, girls' and women's education, which not only reduces fertility but also improves well-being and equity. By increasing labor productivity, improved education could spur economic growth even as it slowed population growth. This effect might offset, to some extent, the emissions reductions found in analyses that take population changes as a starting point. However, the magnitude of the effect has not been explicitly assessed.

Within the climate-change field, population policy has received very little attention, in part because of the political sensitivity of the issue and in part because many believe the demographic contribution to reduced emissions is relatively small.[3, 18] There is some truth to this: Slowing population growth is not the most important means of addressing the climate-change problem, according to current research. And the research is incomplete; an updated assessment of the potential contribution of population-related policies to emissions reductions at the global level is currently lacking.

Nonetheless, research shows that slowing population growth could make a contribution to solving the climate-change problem. To put that contribution in perspective, it is helpful to consider the "wedge" approach to mitigating climate change, proposed by Pacala and Socolow (Figure 5.3).[21] This approach compares a "business-as-usual" trajectory to one that would prevent greenhouse gas levels in the atmosphere from doubling. It then divides the gap between the two into a series of wedges, each representing an action begun today that would

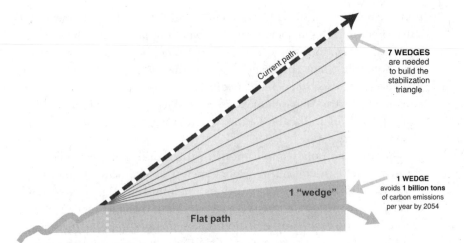

FIGURE 5.3 Stabilization wedges.
Source: Pacala and Socolow, 2004, *Stabilization wedges*.

lead to emissions reductions of a billion tons of carbon per year by 2054—for example, energy efficiency measures and reforestation. We estimate that slowing growth so that world population is below 8 billion by 2050 rather than above 9 billion, as assumed in the most recent UN medium scenario, would provide one—or even two—"wedges" of emissions reductions.

The effect of slowed population growth would be felt most dramatically during the second half of the century, when—according to mitigation scenarios—the deepest reductions in emissions must be made. These deep reductions are the ones that will likely be the most expensive to make, because the cheapest reductions are typically made first and each additional ton reduced is more expensive than the last. Slower population growth would therefore have an even larger impact when viewed from the perspective of costs, by eliminating the need for the most expensive reductions.

> We estimate that slowing growth so that world population is below 8 billion by 2050 . . . would provide one—or even two—"wedges" of emissions reductions.

Moreover, policies that lead to lower fertility have many benefits that are unrelated to climate change. Improved education of women and girls and better access to family planning and reproductive health are vitally important in their own right. The fact that these policies will also reduce fertility and lower greenhouse gas emissions only magnifies their benefits to human well-being. From a climate-change perspective, these are "win-win," "no-regrets" policies.

REFERENCES

1. Birdsall, N., 1994, Another look at population and global warming, p. 39–54, *Population, Environment, and Development*, UN, New York.

2. Bongaarts, J., 1992, Population growth and global warming, *Population and Development Review* 18(2): 299–319.

3. Bongaarts, J., B. C. O'Neill, and S. R. Gaffin, 1997, Global warming policy: Population left out in the cold, *Environment* 39(9): 40–41.

4. Cline, W. R., 1992, *The Economics of Global Warming*, Institute for International Economics, Washington, DC.

5. Cole, M. A., and E. Neumayer, 2004, Examining the impact of demographic factors on air pollution, *Population and Environment* 26(1): 5–21.

6. Cramer, J. C., 1998, Population growth and air quality in California, *Demography* 35(1): 45–56.

7. Cramer, J. C., and R. P. Cheney, 2000, Lost in the ozone: Population growth and ozone in California, *Population and Environment* 21(3): 315–338.

8. Dalton, M., L. Jiang, S. Pachauri, and B. C. O'Neill, 2008, Demographic Change and Future Carbon Emissions in China and India, paper presented at the Annual Meeting of the Population Association of America, March 28–31, 2007, New York, revised May 2008.

9. Dalton, M., B. C. O'Neill, A. Prskawetz, L. Jiang, and J. Pitkin, 2008, Population aging and future carbon emissions in the United States, *Energy Economics* 30: 642–675.

10. Dietz, T., and E. A. Rosa, 1997, Effects of population and affluence on CO_2 emissions, *Proceedings of the National Academy of Sciences* 94: 175–179.

11. Fan, Y., L.-C. Liu, G. Wu, and Y.-M. Wei, 2006, Analyzing impact factors of CO_2 emissions using the STIRPAT model, *Environmental Impact Assessment Review* 26: 377–395.

12. Fisher, B. S., N. Nakicenovic, K. Alfsen, J. Corfee Morlot, F. de la Chesnaye, J.-Ch. Hourcade, K. Jiang, M. Kainuma, E. La Rovere, A. Matysek, A. Rana, K. Riahi, R. Richels, S. Rose, D. van Vuuren, and R. Warren, 2007, Issues related to mitigation in the long-term context, *Climate Change 2007: Mitigation, Contribution of Working Group III to the Fourth Assessment Report of the Intergovernmental Panel on Climate Change*, B. Metz, O. R. Davidson, P. R. Bosch, R. Dave, L. A. Meyer, eds., Cambridge University Press, Cambridge, UK.

13. Gaffin, S. R., and B. C. O'Neill, 1997, Population and global warming with and without CO_2 targets, *Population and Environment* 18(4): 389–413.

14. Jiang, L., and B. C. O'Neill, 2007, Impacts of demographic trends on

U.S. household size and structure, *Population and Development Review* 33(3): 567–591.

15. MacKeller, F. L., W. Lutz, C. Prinz, and A. Goujon, 1995, Population, households, and CO_2 emissions, *Population and Development Review* 21(4): 849–865.

16. Matinez-Zarzoso, I., A. Bengochea-Morancho, and R. Morales-Lage, 2007, The impact of population on CO_2 emissions: Evidence from European countries, *Environmental and Resource Economics* 38: 497–512.

17. Nakicenovic, et al., 2000, *Special Report on Emissions Scenarios*, IPCC, Geneva, and Cambridge University Press, Cambridge, UK.

18. O'Neill, B. C., 2000, Cairo and climate change: A win-win opportunity, *Global Environmental Change* 10: 93–96.

19. O'Neill, B. C., 2001, I = PAT, p. 702–706, *Encyclopedia of Global Change*, volume 1, A. Goudie, ed., Oxford University Press, Oxford.

20. O'Neill, B. C., F. L. MacKellar, and W. Lutz, 2001, *Population and Climate Change*, Cambridge University Press, Cambridge, UK.

21. Pacala, S., and R. Socolow, 2004, Stabilization wedges: Solving the climate problem for the next 50 years with current technologies, *Science* 305: 968–972.

22. Parikh, J., and V. Shukla, 1995, Urbanization, energy use and greenhouse effects in economic development, *Global Environmental Change* 5(2): 87–103.

23. Shi, A., 2003, The impact of population pressure on global carbon dioxide emissions, 1975–1996: Evidence from pooled cross-country data, *Ecological Economics* 44: 29–42.

24. York, R., E. A. Rosa, and T. Dietz, 2003, Bridging environmental science with environmental policy: Plasticity of population, affluence, and technology, *Social Science Quarterly* 83: 18–33.

25. York, R., E. A. Rosa, and T. Dietz, 2003, STIRPAT, IPAT and ImPACT: Analytic tools for unpacking the driving forces of environmental impacts, *Ecological Economics* 46: 351–365.

CHAPTER 6

············

Fair Weather, Lasting World

ROBERT ENGELMAN

Population usually plays at best a cameo role in the public debate on global warming. That changed in the summer of 2007 when government officials of two major powers suggested provocative connections between climate change and human numbers. One comment was interesting; the other was radical—and hints not just at how to slow both warming and population growth, but at how to move toward a fairer, more equitable world.

What was interesting was that a spokesman for the government of China, which recently passed the United States as the world's top emitter of carbon dioxide (CO_2),[1] noted at a United Nations climate meeting that his country's one-child population policy has saved the planet's atmosphere 1.3 billion tons per year of the globe-warming gas. How? The policy has averted an estimated 300 million births—equivalent to the entire U.S. population—since its inception, said Su Wei, the senior foreign ministry official who headed China's delegation to the Vienna climate conference in late August of that year. He merely did the math using per capita emission rates.

You don't need to approve of China's one-child population policy (less draconian policies have had equally impressive impacts on fertil-

···········

Robert Engelman is Vice President for Programs at the Worldwatch Institute.

95

ity) to acknowledge Su Wei's larger point. All else equal, a smaller population will emit smaller amounts of greenhouse gases than a larger one will. Harlan Watson, the U.S. delegate to the climate conference, called this truth "simple arithmetic. If you look at mid-century, Europe will be at 1990 levels of population while ours will be nearing 60 percent above 1990 levels. So population does matter."[2]

Su Wei and Harlan Watson acknowledge the simple reality that population growth contributes powerfully to greenhouse gas emissions, but neither offers any useful prescriptions for reducing those emissions. Acknowledging the population-climate connection is better than denying it. But the world has a long way to go before that acknowledgment actually contributes to slowing human-induced climate change or population growth, let alone to a world that is sustainable and just.

> A climate agreement based on every human being's equal right to emit greenhouse gases may seem like a political fantasy today. . . . [But in] a rapidly warming world, anything is possible.

What was radical in the summer of 2007 was a suggestion by German Chancellor Angela Merkel, the day after the exchange in Vienna, that any future climate agreement should take account of the populations of all countries and their per capita emissions of greenhouse gases. Merkel proposed that developing countries be allowed to increase their national emissions until the average citizen's emissions reaches the average of industrialized nations. Per capita emissions in the latter group might fall in the meantime, but Merkel implied that governments of industrialized countries had no right to expect developing ones to restrain their emissions until per capita levels on both sides of the development divide were the same.

"At what point can we involve developing countries" in a global agreement to reduce greenhouse gas emissions, Merkel asked, "and what kind of measure do we use to create a just world?"[3] Those two questions emerge in one breath more rarely even than population and climate do. Merkel was not the first to join them; a comparable suggestion had been made earlier that year at a summit of the G8 nations by Indian Prime Minister Manmohan Singh. Merkel's endorsement was remarkable all the same, in that she heads the government of one of the world's wealthiest and most industrialized nations. A climate agreement based on every human being's equal right to emit greenhouse gases may seem like a political fantasy today, as the world's governments struggle to craft a climate-saving deal in Copenhagen in late 2009 that can actually be ratified by all or almost all nations. Merkel's agreement with Singh that such an agreement would be just, however difficult to reach politically, augurs an emerging reality: In a rapidly warming world, anything is possible.

A RIGHT TO THE ATMOSPHERE

One of the biggest challenges of addressing human-induced climate change is that it is a global problem in a world of nations. Nations and the numbers of people in them are accidents of history, geography, and demography. Some countries have many people; some have few. Whatever their populations, nations don't emit greenhouse gases; people do. We all produce CO_2 through the daily activities of living—traveling, cooking, heating and cooling homes and offices, making and using the products that transform drudgery into convenience. Climate change, then, is a global problem that results from the choices and actions of billions of individuals, each with needs and aspirations that are basic to human nature.

The world's nations encountered similar challenges in addressing population growth—another global concern that results from individual choices—and the lessons learned there may be useful in the effort to mitigate climate change. The 1994 International Conference on Population and Development offered an innovative and historically unprecedented approach: basing reductions in population growth on the right of women and couples to make their own choices about childbearing. At the conference, held in Cairo, representatives of almost all the world's governments agreed that population policies should be based on a recognition of human rights to development, sexual and reproductive health, and the ability of individual women and men to choose whether and when to have children. Few governments since have taken the agreement seriously (it lacks the status of a convention or other treaty), but the principles invoked nonetheless offer a model. Could such a rights-based approach become the basis of a global pact on climate change?[4]

The question Angela Merkel was asking in Japan was the right one: How can nations at all levels of development agree on how to share the atmosphere in ways that fairly allocate individual rights to use it? For unless we find a way to eliminate greenhouse gas emissions entirely (not likely, as any molecule bonding more than two atoms traps heat) or learn to suck the gases out of the atmosphere to the extent we emit them (possible, but a significant technological feat), people will always need to vent some heat-trapping gases into the air.

Though generally unrecognized in the United States and other industrialized countries, the long-term need for atmospheric equity is perhaps the biggest stumbling block to any compact aimed at the massive global reductions in greenhouse gas emissions that will ultimately be needed to stabilize the climate. The United Nations General Assembly in 1970 and the UN's Convention on the Law of the Sea in 1982 recognized that the deep oceans were the "common heritage of all mankind."[5] A similar recognition of humanity's common right to the atmosphere seems inevitable, despite the political challenges of achieving agreement to the principle, simply because there is no other defensible basis for insisting that low-emission countries restrain their emissions. With fossil fuel emissions so closely correlated with per capita wealth, what sane government would agree to cap its citizens' per capita emissions in perpetuity below those of other countries?

Some human beings add dozens of times their own body weight in carbon dioxide (a loose proxy for a host of greenhouse gases) to the atmosphere, and they feel entitled to keep doing so. Other human beings send skyward less than their own weight in CO_2—and they tend also to be the ones most vulnerable to the impacts of climate change. As climate change becomes a more urgent public issue and as national governments face the need to decrease the global emission total, more attention is likely to focus on the vast disparities in per capita emissions of carbon dioxide and other greenhouse gases.

As Figure 6.1 shows, the disparities are stark, ranging from more than 18 metric tons of CO_2 emissions per capita in the United States and Canada to little more than a ton in sub-Saharan African countries.[6] (This range excludes countries with very low populations, such as certain high-emission Middle Eastern oil states, and includes only "industrial" CO_2 emissions—those from fossil fuel combustion and cement production—ignoring those from deforestation and other land-use changes.) There is little practical political scope for addressing population's role in future climate change while the industrialized countries, whose populations are growing relatively slowly and represent only a fifth of the world's total, contribute two-thirds of all greenhouse gas emissions.[7]

Indeed, the critical link between population and climate change can best be addressed from a perspective of equal access to the

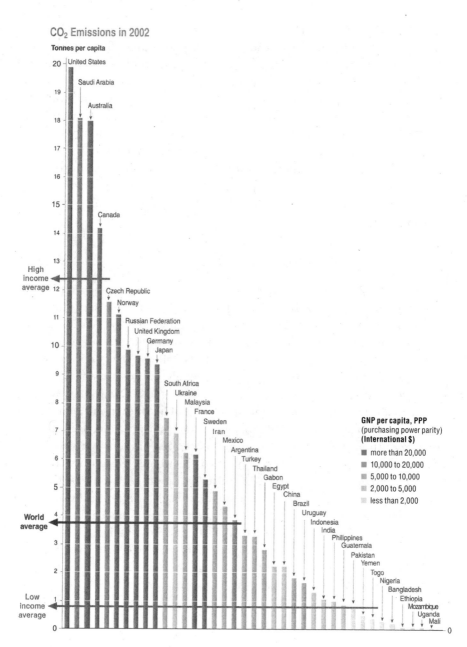

FIGURE 6.1 Per capita CO_2 emissions in 2002.

Source: World Bank, online database, 2004.

carbon-cycling properties of the atmosphere. From this perspective, population trends emerge as among the greatest sources of hope that humanity may actually succeed in resolving the problem of climate change before catastrophic ecological change has occurred. Only technological innovation seems as promising. The possibility that human population could peak before the middle of the century and then begin a gradual decline extending at least until the twenty-second century—as presented in one low projection by the United Nations Population Division[8]—offers a demographic scenario well suited to the need to bring greenhouse gas emissions to a peak in the next decade or less, followed by a fairly precipitous decline and a more gradual one extending at least until 2100.

SHARING THE AIR

Imagine, then, a global climate regime based in equal measure on two goals: avoiding catastrophic climate change and equitably allocating the right to emit greenhouse gases to all human beings. Assume a global emissions target that allows a small and short-lived increase in greenhouse gas emissions from today's levels, followed by steep yearly reductions aimed at restraining carbon dioxide concentrations at or below levels that would result in a global warming of 2°C (3.6°F) above the preindustrial global temperature average. The Intergovernmental Panel on Climate Change (IPCC) judges an atmospheric concentration of carbon dioxide of around 550 parts per million by volume (ppmv)—or, actually, concentrations of all greenhouse gases that have a warming equivalent to such a CO_2 concentration—to be about the most ambitious target that global climate policy can realistically aim for, though not necessarily a "safe" concentration for climate impacts.[9] The panel's findings stem from the work of hundreds of scientists from around the world tasked by the United Nations to develop agreed-on scientific assessments of climate change and its impacts. Some respected climatologists, such as NASA climate scientist James Hansen, nonetheless argue that much lower concentrations—possibly well below today's 385 ppmv, at least for CO_2 itself—would be needed for real confidence that catastrophic change can be avoided.[10]

Australian climate scientist W. L. Hare, working at the Potsdam Institute for Climate Impact Research, proposes a "safe landing path" for such low CO_2 concentrations that would require emissions to peak before 2020 and veer sharply down thereafter, reaching zero by the middle of the century and then "going negative"—with CO_2 actually being sucked out of the atmosphere at greater rates than it enters—for the rest of the century and beyond. Today's global average CO_2 industrial emissions of 31 billion metric tons of carbon[11] would be allowed to increase only slightly, to little more than 33 billion, before a concerted effort began to reduce global emissions effectively to zero by 2050.[12]

There wouldn't be much room for political compromise or missed annual targets along such a path; climate follows physical laws, not human ones. We can hope that within a few years, or at least a couple of decades, low-emission technologies will be so economically attractive, convenient, and ubiquitous that there will be no need to enforce global emissions controls. But what if that's no more the case in the near future than it is in the present? To avoid ecological disaster—which would translate into human disaster as well—something like the annual emissions limits Hare is suggesting may need to be enforced globally. But how?

The fair allocation to each member of the world's population is, in Harlan Watson's words, "simple arithmetic." Divide each year's needed global emission level by the number of emitters—every human being on the planet. Industrial carbon dioxide emissions—those from fossil fuel combustion and cement manufacture—are well quantified, as are national populations. The fossil-fuel CO_2 emissions could serve as a reasonable, though imperfect, proxy for all greenhouse gas emissions. Eventually, as data improve on carbon dioxide emissions from deforestation and other land-use changes and on emissions from other greenhouse gases, these, too, could be integrated into a global emissions trading system.

With a tally of each country's per capita emissions in relation to the targeted atmosphere-stabilizing level, country rankings would be publicized and over-emitters would be required to purchase the right to make their excessive emissions from countries whose citizens were under-emitting (and hence actually helping to solve the climate problem, albeit through their poverty). This could be accomplished through international agreement, with the price of tradable emissions permits based on public perceptions of the urgency of human-induced climate change and political will to slow it down. A global intergovernmental fund or bank could administer the financial transfers among countries, and public and political pressure would be applied to assure the transparency of the funds' allocation among and within countries. Countries that refused to participate fully might find themselves isolated in a globalized economy, denied credit, trading, travel, and migration privileges by participating countries, as well as ostracized by public opinion.

The revenue resulting from such a system could be used for any number of purposes in under-emitting countries. Health and education, for example, foster economic development and contribute to long-term reductions in fertility through women's intentions to raise fewer, healthier children, later in their lives. Development of non-carbon energy sources would absorb much of the new spending, so each country could continue its economic development without jeopardizing the revenue it would be gaining from other countries through the growing value of its per capita under-emissions. As the value of emissions trading grew, both over- and under-emitters would make every effort to reduce emissions without inducing poverty among their citizens—exactly the kind of incentives needed for a genuinely global effort to bring greenhouse gas emissions to sustainable levels as quickly as possible.

Both under- and over-emitters could raise objections about such a plan. (These are more accurate terms for dividing the world's countries for climate-change purposes than "developed" and "developing," especially since a number of the latter countries have surprisingly high per capita emissions, and more may in the future.) Under-emitters can accurately note that an emission-rights trading regime based on present emissions ignores the impact of past ones—overwhelmingly the legacy of over-emitters' past behavior and already disproportionately affecting under-emitters in the climate change already occurring. Over-emitters will note, accurately as well, that they will be taxing their own citizens and economies to transfer wealth abroad. The under-emitters' complaint can perhaps be addressed by adjusting the wealth transfers in their favor, or through separate agreements to finance climate-change adaptation in developing countries through generous development assistance. The over-emitters' complaint can only be addressed through the argument that without such wealth transfers, what chance is there of under-emitters' engaging in a global climate pact and restraining the growth of their emissions? And if global warming is real, and a real risk to humanity's future, what chance is there of stopping it without that participation? Indeed, most of the current growth in greenhouse gas emissions is coming from developing countries and is expected to do so for the foreseeable future—partly because of their more rapidly growing populations, partly because they are economically *developing*.[13]

That some kind of climate bargain along these lines is of interest to the under-emitters is already evident, as evidenced by the Indian prime minister's comments. An intriguing hint that financial assistance could help reduce emissions in developing countries came earlier from President Rafael Correa of Ecuador. In a radio address, Correa volunteered that his government would be willing to ban oil exploration in the Ecuadorian portion of the Amazon basin if the world's governments would compensate the country at roughly half the value of the extracted oil on the world market. He estimated the acceptable compensation at $350 million a year.

"If the world truly is interested in saving the planet," commented Lourdes Tiban, Ecuador's representative at a meeting of the UN's Permanent Forum on Indigenous Issues held in May 2007, "the government has decided to sell the oil, but keep it in the ground."[14]

Keeping oil—and coal and natural gas—in the ground may soon become a goal of international policy. Technological innovation and national voluntarism will help, but ultimately, as Tiban and Correa's remarks make clear, it will take money to spur economic and behavior change of the required magnitude. There will need to be a recognition, of course, that many poor people live in generally wealthy countries, just as some quite wealthy people live in generally poor countries. Indeed, one novel approach to addressing climate change bases countries' responsibility to contribute to emissions reductions and climate change adaptation in part on the cumulative per capita income in each country above a uniform global per capita threshold set significantly above poverty levels. Called *greenhouse development rights*, this concept specifically recognizes the need for economic development above poverty levels in all countries—and clarifies that individual responsibility to deal with climate change begins once this development occurs in the lives of all people.[15]

POPULATION SCENARIOS

Growing world attention to the idea of "atmospheric space" will also draw attention to numbers: How much space is there for future greenhouse gas emissions without their adding to warming, and into how many parts must this space be divided? A reasonable estimate of current global emissions is 31 gigatons, or billion metric tons, of carbon dioxide.[16] Those emissions are made by an estimated 6.8 billion human beings, for a per capita emission of just under 1.2 metric tons carbon per person. For comparison, current U.S. per capita emissions are 5.6 tons per person, while those of Ethiopia are a mere 30 kilograms and those of Chad are effectively zero.

> By imagining a global climate framework based on both equity and sustainability, we can see that slowing population growth is crucial to the goal of climate stability.

Let's imagine a global carbon-trading scheme in which countries with per capita emissions above the "sustainable path" would compensate countries with per capita emissions below that level, in effect purchasing these countries' emission rights at rates set to discourage such transactions and thereby driving global emissions as close as possible to the targeted level. The emissions path that Hare proposes for stabilizing the atmosphere, for example, implies a global emission of about 16 gigatons of CO_2 in 2035, roughly a 50 percent reduction from current levels.

Under this scenario, the amount of carbon each person can emit on average depends on the size of the world population. In other words, if the size of the "pie" is set at 16 gigatons, the size of each slice is determined by how many people are at the table. For 2035, the United Nations projects in its medium-variant scenario a world population of 8.6 billion people.[17] Divide Hare's 16 gigatons by this figure, and the "sustainable-path" per capita emission in that year will be 1,815 kilograms of carbon. That's less than one-twentieth of the U.S. per capita emissions in 2007 and is somewhat less than Peru's.[18]

The range of possible future population sizes makes a difference, and the difference grows with time. By 2050 the global emission must be effectively zero, and whether 7.8 billion human beings (UN low projection) or 10.8 billion (high projection)[19] are aiming to effectively zero out such emissions could prove decisive in making the gargantuan task manageable or a spur to widespread poverty. Intriguingly, the same mid-century point at which Hare argues net fossil-fuel CO_2 emissions should end is about the time that world population would peak and begin to decline under the UN's low-variant population projection. By contrast, under the high-variant scenario, world population rises for as long as demographers care to project, making more challenging year by year the goal of sustaining a world of zero net emissions.

Differences in future population growth can't by themselves determine whether we stabilize the climate or improve human well-being. But by imagining a global climate framework based on both equity and sustainability, we can see that slowing population growth is crucial to the goal of climate stability. So clear does the relationship become that special vigilance in defense of reproductive rights may be needed in a world of still-growing populations ever more worried about the future of the climate.

Indeed, some climate policy specialists have pointed to a demographic hazard of global carbon-trading regimes based on per capita emissions: Low-emitting countries may be tempted to try to boost their populations in an effort to gain more "low emitters" as sources of trading revenue. In theory, that could lead governments of such countries to discourage access to contraception and abortion in efforts to raise birthrates. One way around this danger would be to base per capita trading on population sizes frozen in some year specified by mutual agreement. (That might be 2015 in the unlikely event such a system became part of the climate agreement crafted in Copenhagen.) But such an arrangement would risk an equal and opposite pressure on reproductive freedom: restrictions on the right to reproduce in low-emitting countries, and possibly spurs to reproduction in higher-emitting ones worried about making payments into a climate fund when their populations are declining.

Best from the standpoint of reproductive rights would be to embrace the right of all human beings to utilize the atmosphere equally, as well as to choose for themselves if, when, and how often to have children. This requires some trust that most individuals will choose to be strategic and selective about childbearing in a time of rapid climate change, and that all nations will come to understand their self-interest in a smaller world population given the finite atmosphere.

Here, those working to stop climate change have much to learn from the experience of the population and family planning movement: Control and coercion may seem attractive options to some, but they have no hope of succeeding, whether the objective is slower growth of human population or of greenhouse gas emissions. Finding ways for people to improve their own well-being by contributing to the common good, by contrast, leads to a world that is both more sustainable and more equitable.

Whatever challenges may arise out of population stability or decline—and there may be some—they are unlikely to outweigh those of an atmosphere with progressively less room for each human being to exercise her or his right to occasionally allow a few heat-trapping molecules to escape into the air. As the problem of human-induced climate change presses more urgently on human well-being in the decades ahead, the link to population will become obvious. The more equitably the world is addressing the task of reducing global emissions, the more appealing a stable or declining population will become. With enough forethought by governments and the public, beginning as soon as possible, declines in both greenhouse gas emissions and human fertility will support each other. Both objectives can be achieved through frameworks based on human rights, health, and development.

REFERENCES

1. China's emissions surpass those of United States, Gregg, Jay S., Robert J. Andress, and Gregg Marland, 2008, China: Emissions pattern of the world leader in CO₂ emissions from fossil fuel consumption and cement production, *Geophysical Research Letters* 35, L08806, doi:10.1029/2007GL032887, 2008.

2. Reuters, 2007, China says one-child policy helps climate, August 30.

3. Kade, Claudia, 2007, Merkel backs climate deal based on population, Reuters, August 31.

4. Issues and concepts raised in this chapter are explored as well in Engelman, Robert, 1994, *Stabilizing the Atmosphere: Population Consumption and Greenhouse Gases*, Population Action International, Washington, DC; Engelman, Robert, 1998, *Profiles in Carbon: An Update on Population, Consumption and Carbon Dioxide Emissions*, Population Action International, Washington, DC; and Engelman, Robert, 1998, Population, consumption and equity, *Tiempo: Global Warming and the Third World* 30: 3–10, December.

5. UN Convention on the Law of the Sea, 1982, available on-line at http://www.un.org/Depts/los/convention_agreements/convention_overview_convention.htm, last accessed on October 31, 2008.

6. Bolin, Bert, and Haroon S. Kheshgi, 2001, On strategies for reducing greenhouse gas emissions, *Proceedings of the National Academy of Sciences*, PNAS early edition, www.pnas.org/cgi/doi/10.1073/pnas.081078998.

7. Population Action International, 2006 update, People in the Balance, Web database on population and natural resources, http://216.146.209.72/Publications/Reports/People_in_the_Balance/Interactive/peopleinthebalance/pages/?s=5, last accessed on October 31, 2008.

8. UN Department of Economic and Social Affairs, 2004, *World Population in 2300: Proceedings of the United Nations Expert Meeting on World Population in 2300*, UN, New York, ESA/P/WP.187/Rev. 1, available on-line at http://www.un.org/esa/population/publications/longrange2/2004worldpop2300report finalc.pdf, last accessed October 31, 2008.

9. Barker, Terry, et al., 2007, *Contribution of Working Group III to the Fourth Assessment Report of the Intergovernmental Panel on Climate Change: Technical Summary*, IPCC, Geneva, http://www.ipcc.ch/pdf/assessment-report/ar4/wg3/ ar4-wg3-ts.pdf, last accessed October 31, 2008.

10. Hansen, James, et al., 2008, Target atmospheric CO_2: Where should humanity aim? *Open Atmospheric Science Journal*, http://arxiv.org/abs/0804 .1126, last accessed October 31, 2008.

11. Global carbon dioxide emissions data (including preliminary estimates for 2007, converted from weight measured by carbon content to weight in carbon dioxide) from Carbon Dioxide Information Analysis Center, Oak Ridge National Laboratory, U.S. Department of Energy, http://cdiac.ornl. gov/ftp/trends/emissions/Preliminary_CO2_Emissions_2006_2007.xls, last accessed October 31, 2008.

12. Hare, W. L., 2009, A safe landing for the climate, *State of the World 2009: Into a Warming World*, Worldwatch Institute, Washington, DC.

13. Wheeler, David, and Kevin Ummel, 2007, Another Inconvenient Truth: A Carbon-Intensive South Faces Environmental Disaster, No Matter What the North Does, Working Paper Number 134, Center for Global Development, Washington, DC.

14. Rizvi, Haider, 2007, Environment: Ecuador seeks aid not to exploit Amazon oil, Inter Press Service, May 18, http://ipsnews.net/news.asp?idnews =37794, last accessed October 31, 2008.

15. Baer, Paul, and Tom Athanasiou, 2008, The Greenhouse Development Rights Framework: The Right to Development in a Carbon-Constrained World, second edition, Executive Summary, EcoEquity, Oakland, CA, http:// www.ecoequity.org/GDRs/GDRs_ExecSummary.html, last accessed October 31, 2008.

16. A metric ton amounts to 1,000 kilograms or 2,205 English pounds. To derive the equivalent weight in carbon dioxide, multiply the carbon weight by three and two-thirds. Carbon Dioxide International Analysis Center, Oak Ridge National Laboratory, U.S. Department of Energy, http://cdiac.ornl .gov/ftp/trends/emissions/Preliminary_CO2_Emissions_2006_2007.xls, last accessed October 31, 2008.

17. UN Population Division, 2007, World Population Prospects: The 2006 Revision, Web-based population database available at http://esa.un.org/ unpp/, last accessed October 31, 2008.

18. Calculation by the author based on UN population data and carbon dioxide emissions data from Carbon Dioxide International Analysis Center, Oak Ridge National Laboratory, U.S. Department of Energy, http://cdiac .ornl.gov/ftp/trends/emissions/Preliminary_CO2_Emissions_2006_2007 .xls, last accessed October 31, 2008.

19. UN Department of Economic and Social Affairs, 2007, World Population Prospects: The 2006 Revision Population Database, http://esa.un.org/ unpp/p2k0data.asp, last accessed October 31, 2008.

CHAPTER 7

....

Adapting to Climate Change

The Role of Reproductive Health

MALEA HOEPF YOUNG, ELIZABETH MALONE,
ELIZABETH LEAHY MADSEN, and AMY COEN

In 2008, weather-related disasters left a trail of devastation around the globe. Cyclone Nargis hit Burma in May, killing at least 130,000 people and leveling homes, farms, and hospitals.[1] In August and September, four back-to-back storms tore through the Caribbean, plunging Haiti—already the poorest country in the Western Hemisphere—into new depths of misery.

The grim realities in Burma and Haiti may offer glimpses of the future. While no single weather event can be attributed to climate change, it is clear that storms are intensifying, and stronger storms are likely a result of human-induced warming. The Intergovernmental Panel on Climate Change (IPCC) predicts that cyclones, typhoons, and hurricanes will become even stronger in the years to come. Even more likely are changes in precipitation patterns, leading to increased flooding and drought.[2] While a growing global movement seeks to reduce

....

Malea Hoepf Young was a research associate at Population Action International (PAI) from 2007 to 2009 before joining the Peace Corps, where she is a volunteer in Rwanda. Elizabeth Malone is a senior research scientist at the Joint Global Change Research Institute. Elizabeth Leahy Madsen is a research associate at PAI. Amy Coen is PAI's President and CEO.

greenhouse gas emissions and head off the worst impacts of climate change, a certain amount of warming is now inevitable. How, then, can we help societies—especially the most vulnerable—adapt to climate change?

In this chapter, we will examine the demographic, social, and environmental factors that make societies resilient or vulnerable to climate change. We will show that universal access to reproductive

> Universal access to reproductive health services—including family planning—can bolster resilience [to climate change] while improving human health and well-being.

health services—including family planning—can help bolster resilience while improving human health and well-being.

UNEQUAL IMPACTS

There is ample evidence that climate change is already occurring and will intensify in coming decades. The most severe impacts will occur in some of the world's poorest countries, including those in sub-Saharan Africa and the Asian megadeltas. These countries contribute only a tiny fraction of greenhouse gas emissions, yet their people will bear a disproportionate burden of the effects of climate change. The average American emits 200 times more carbon each year than the average Ethiopian, yet Ethiopia is expected to experience more severe impacts, including major changes in temperature, water availability, and malaria zones in coming years, with few resources to adapt.[3, 4]

Just as climate-change impacts are not felt equally throughout the world, they are not experienced equally within countries. The poor generally feel the greatest impact, and women and girls—who make up a disproportionate share of the world's poor—are therefore among the most vulnerable.[5] The stresses of a changing climate are being added to the many risks already facing women in developing countries, undermining the critical role that women play in the health and well-being of their families, the social cohesion of their communities, and the preservation of their fields, forests, and waterways.

In some cases, the impact of climate change on women is direct, and lethal. Data show that women nearly always outnumber men in deaths during disaster situations, ranging from heat waves to flooding and storms—the number and intensity of which will increase as climate change continues.[6, 7]

Climate change also adds to women's already heavy burdens. In much of the developing world, women work more hours each day than men: A study in rural Cameroon, for example, found that women work over 64 hours a week to men's 31 hours. The UN's Food and Agriculture Organization (FAO) estimates that women produce 60 to 80 percent of food grown in the developing world, often small-scale crops critical to their families' sustenance. And women and girls are responsible for collecting and carrying water—a time-consuming and physically demanding task in places where wells are not easily accessible.[8, 9, 10, 11] As communities cope with the effects of changes in climate, demands on women's time and workloads are likely to increase. For example, women will travel longer distances to collect water and fuel, and they will spend more hours coaxing crops from dry, depleted soil. To compensate for increased demands on their time, poor families may pull girls out of school, if they were are able to attend at all.[12]

FROM MITIGATION TO ADAPTATION

Recently, the movement to address climate change has expanded from a focus solely on mitigation through emissions reduction to include exploration of how people can adapt to a changing and increasingly hostile climate. An effective adaptation strategy must recognize the unequal impact of climate change and reinforce the adaptive capacity of the vulnerable groups described above.

As Brian C. O'Neill noted in Chapter 5, population growth and other demographic factors play an important role in greenhouse gas emissions, although they have generally been overlooked as a global climate-change mitigation strategy. Likewise, reproductive health,* family planning, and population policies have potentially important roles to play in increasing resilience to climate-change impacts among the most vulnerable, but they have been neglected in the literature and in practice. Decades of research have documented that when women can choose the size and spacing of their families through a broad range of reproductive health services and an enabling environment, the benefits are wide reaching and accrue from individual women to their families, communities, and nations.

---

*The definition of reproductive health used in this chapter is that used in the 1994 Program of Action of the International Conference on Population and Development.

MEASURING RESILIENCE

Developing successful adaptation strategies requires researchers and implementers to assess a society's vulnerability to climate-change impacts on one hand, and to identify its sources of resilience on the other. This requires attention to societal factors, not just physical atmospheric changes and impacts.

There are several definitions of *vulnerability to climate change*. The IPCC definition takes an ecosystem approach:

> Vulnerability is the degree to which a system is susceptible to, and unable to cope with, adverse effects of *climate change*, including *climate variability* and extremes. Vulnerability is a function of the character, magnitude, and rate of climate change and variation to which a system is exposed, its *sensitivity*, and its *adaptive capacity*.[13]

In terms of human societies, vulnerability can be analyzed by examining (1) the degree to which a society depends on climate for its well-being, (2) the damage inflicted by climate change, and (3) the society's coping and adaptation resources. A society with high climate dependence (e.g., reliance on rain-fed agriculture), damaging climate change (e.g., persistent drought), and low adaptive capacity (e.g., poor governance and high rates of poverty) will be highly vulnerable.

The inverse of vulnerability is resilience, or, as defined by the IPCC, "the ability of a social or ecological system to absorb disturbances while retaining the same basic structure and ways of functioning, the capacity for self-organization, and the capacity to adapt to stress and change."[14] A critical element of resilience is adaptive capacity, or "the ability of a system to adjust to climate change (including climate variability and extremes) to moderate potential damages, to take advantage of opportunities, or to cope with the consequences."[15]

More difficult than defining these concepts, however, is measuring them. Much of the early resilience literature assumed that wealthy, industrialized countries would be able to adapt to changes in climate while poor, less-industrialized countries would not, and they would experience widespread declines in quality of life.[16] The IPCC Third Assessment Report identifies features that "seem to determine . . . adaptive capacity: economic wealth, technology, information and skills, infrastructure, institutions, and equity"[17] but goes on to point out that all features are highly contingent on social conditions and that adaptive capacity is inextricably linked to overall development and sustainability.

Such contingency can be seen in how differently 2004's Hurricane

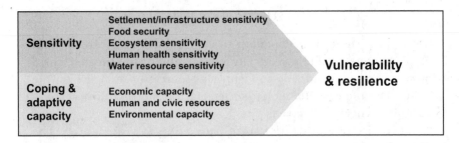

FIGURE 7.1 Conceptual model of vulnerability and resilience.

Jeanne affected the two countries that share the island of Hispaniola—the Dominican Republic and Haiti. Jeanne hit the Dominican Republic as a hurricane, causing significant damage but fewer than twenty deaths. However, even though the weakened storm did not hit Haiti directly, that country experienced much greater devastation. The storm's heavy rains caused flooding and mud slides in Haiti, where extreme poverty has led to near-total deforestation. As a result, 3,000 Haitians were killed, and many more were left far worse off than before the storm.[18, 19]

The Vulnerability-Resilience Indicators Model (VRIM) index developed at the Joint Global Change Research Institute[20, 21, 22] provides a framework to measure and rank country vulnerability and resilience, examining both sensitivity (how social and ecological systems could be negatively affected by climate change) and adaptive capacity (the capability of a society to maintain, minimize loss of, or maximize gains in welfare). The determinants of sensitivity and adaptive capacity are grouped into sectors. Adaptive capacity is determined by economic capacity, environmental capacity, and civic and human resources, while sensitivity is determined by the susceptibility of settlements and infrastructure, food security, ecosystems, human health, and water resources to climate-change impacts (see Figure 7.1). Each of these sectors is made up of one to three measurable proxies, for which national data are easily accessible.

In a global study, Malone and Brenkert sorted 160 countries according to the resilience indicator values, resilience in sensitivity, and adaptive capacity. Relative rankings generally align with expectations, with wealthy countries predicted to be able to protect themselves against climate-change impacts (see Figure 7.2).[23]

In the first quartile, Norway ranks first in adaptive capacity, because of robust economic capacity. Most of the Organization for Economic Cooperation and Development (OECD) countries are located in the

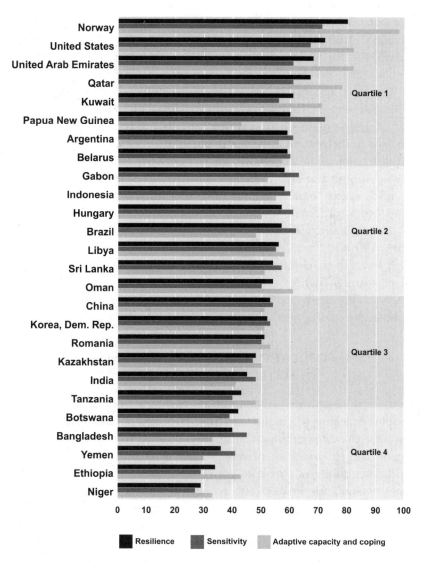

FIGURE 7.2 Selected countries ranked by resilience and vulnerability scores.

top quartile of countries (including the top ten spots), as are three oil-rich Middle Eastern countries. Some developing countries are in the top quartile as well: Papua New Guinea ranks high through its high resilience score in sensitivity (especially ecosystem resilience and water availability, due to its abundant natural resources and forest cover).

Countries in the second quartile score higher in their resilience in the sensitivity element than in their adaptive capacity, except for oil-rich Oman and Libya. This is a geographically diverse set, representing Africa, the Pacific islands, Latin America, eastern Europe, and Asia, with scores 15 percent lower on average than scores in the top quartile.

The third quartile is headed by China, followed by South Korea. Some former east European countries—Romania, Kazakhstan, Azerbaijan, and Georgia—have relatively high coping and adaptive capacity because of their highly educated populations. African countries in this quartile, such as the Republic of the Congo, Kenya, and Tanzania, have higher coping and adaptive capacities because of their high environmental capacity, including large areas of forest cover and unmanaged land. India, located in this quartile, scores 107th out of 160 countries.

African countries comprise most of the fourth quartile, along with Turkmenistan, Bangladesh, Haiti, Afghanistan, and Yemen. Yemen's high levels of economic inequality, water scarcity, and food insecurity place it in this highly vulnerable category. This is deeply troubling, as many fall into areas predicted to experience significant changes in climate in the coming years.

POPULATION AGE STRUCTURE AND CLIMATE-CHANGE RESILIENCE

Age structure provides a meaningful snapshot of a population's demographic trajectory at a moment in time. It shows a population's progress through the demographic transition—the shift from high birth and death rates to, first, longer lives and, later, smaller families. Countries at the beginning of the demographic transition have seen mortality rates decline, but fertility rates are still well above replacement level, meaning that younger age groups comprise a relatively large share of the population. Many countries in the developing world, particularly in sub-Saharan Africa and parts of southwest Asia, have such youthful age structures. Once declines in fertility rates have been sustained, however, working-age adults begin to comprise the largest share of the population, with lower dependency ratios relative to children and younger adults. Finally, at the end of the demographic transition, fertility rates reach replacement level or below, leading to a gradual aging of the population. Many countries in Europe, as well as some of east Asia and North America, have reached this point.

Demographic indicators of fertility and life expectancy are included in measurements of climate-change sensitivity, such as the VRIM model's analysis of human health. Another way to discern the influence of

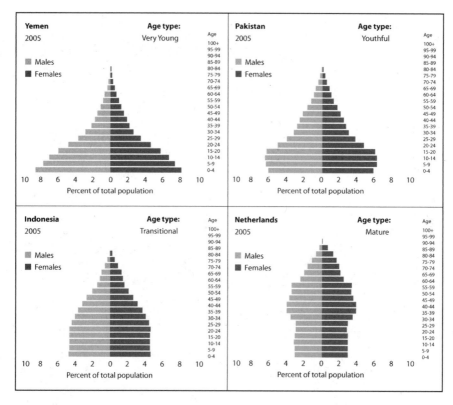

FIGURE 7.3 Population pyramids, or profiles, graphically display age structures. They demonstrate the comparative sizes of specific age groups relative to others for both the male and female populations.

demographic changes on resilience and vulnerability to climate change—and the potential of family planning and reproductive health programming as an adaptation strategy—is through an overlay of country resilience scores and population age structure.

Comparing Population Action International's (PAI's) classification system of age structures with the VRIM demonstrates a relationship between age structure and resilience. Resilience increases with each successively more mature age structure type; and as age structures mature, more are rated with high resilience (Figure 7.3).

Just as the VRIM groups countries into quartiles of resilience ratings, PAI's system places national populations into one of four major age structure types—very young, youthful, transitional, and mature (Figure 7.3). Eighty-eight percent of the countries ranked lowest in

a. Very young age structure

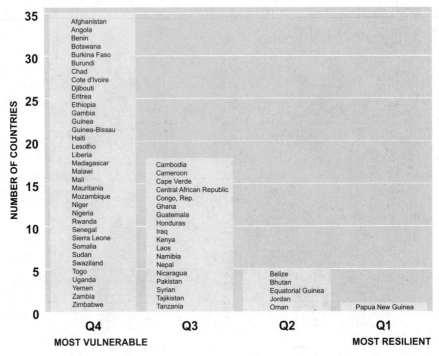

b. Youthful

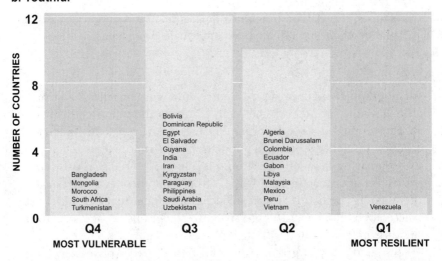

FIGURE 7.4 Resilience quartile by age structure category (quartile 4 indicates low resilience, quartile 1 indicates high resilience).

c. Transitional

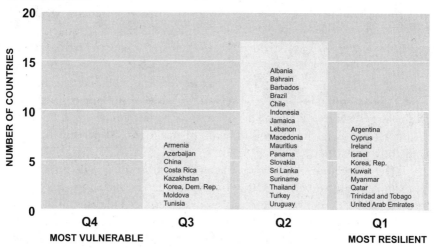

	Q4	Q3	Q2	Q1
	MOST VULNERABLE			MOST RESILIENT

Q3:
Armenia
Azerbaijan
China
Costa Rica
Kazakhstan
Korea, Dem. Rep.
Moldova
Tunisia

Q2:
Albania
Bahrain
Barbados
Brazil
Chile
Indonesia
Jamaica
Lebanon
Macedonia
Mauritius
Panama
Slovakia
Sri Lanka
Suriname
Thailand
Turkey
Uruguay

Q1:
Argentina
Cyprus
Ireland
Israel
Korea, Rep.
Kuwait
Myanmar
Qatar
Trinidad and Tobago
United Arab Emirates

d. Mature

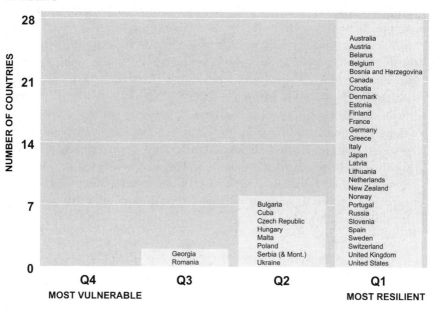

	Q4	Q3	Q2	Q1
	MOST VULNERABLE			MOST RESILIENT

Q3:
Georgia
Romania

Q2:
Bulgaria
Cuba
Czech Republic
Hungary
Malta
Poland
Serbia (& Mont.)
Ukraine

Q1:
Australia
Austria
Belarus
Belgium
Bosnia and Herzegovina
Canada
Croatia
Denmark
Estonia
Finland
France
Germany
Greece
Italy
Japan
Latvia
Lithuania
Netherlands
New Zealand
Norway
Portugal
Russia
Slovenia
Spain
Sweden
Switzerland
United Kingdom
United States

Eighty-eight percent of the countries ranked lowest in resilience to climate change have very young age structures.

resilience to climate change have very young age structures. Among countries in the third quartile of resilience, 75 percent have a very young or youthful age structure. However, the balance begins to shift once countries reach the second quartile of resilience, in which the largest share (42.5 percent) has a transitional age structure. Finally, among countries in the top quartile of resilience, 70 percent have a mature age structure.

The correlations between age structure and resilience do not imply a simple cause and effect. Other important factors, such as poverty, are certainly related to both, and more work is required to disentangle the effects of different variables. However, it is clear that countries with the most youthful populations and highest rates of population growth are also those facing the lowest levels of resilience to climate change, while those with more-balanced age structures demonstrate the highest levels of adaptive capacity. The relationship between population dynamics and resilience is worthy of further analysis and exploration.

For countries with large youthful populations, these findings underscore the importance of investing in young people. When young people are endowed with opportunities and rights, they—and the societies in which they live—will be more resilient.

Demography is not destiny, and age structure is not static. Progress along the demographic transition can be and has been promoted through comprehensive policies and programs. Governments and civil society can help lower fertility and improve the well-being of women, men, and their families, as well as promote development overall, by expanding access to voluntary sexual and reproductive health services, including family planning.

FAMILY PLANNING, REPRODUCTIVE HEALTH, AND CLIMATE-CHANGE RESILIENCE

The realization of women's reproductive rights—defined in the Cairo Program of Action as the right to decide freely and responsibly the number, spacing, and timing of their children and to have the information and means to do so—can bolster resilience to climate change. Achieving the goals set in Cairo would directly benefit women and also offer broader benefits to public health, economic resources, and environmental sustainability.

Health

A healthy population and workforce are essential to achieve the economic and agricultural productivity required to adapt to climate change. Today, sexual and reproductive ill health comprises one-third of the global burden of disease among women aged fifteen to forty-four, and nearly one-fifth of the global burden of disease, according to the World Health Organization.[24, 25] Women's morbidity and mortality is reduced when they can choose the number of children they have, limit childbearing to the safest reproductive years (preventing teenage pregnancies and pregnancies after age forty, which are more dangerous for both women and infants), and space births three or more years apart (infants born less than two to three years apart are significantly less likely to survive).[26] It is estimated that preventing unintended pregnancies would reduce maternal deaths by 25 percent, saving the lives of 150,000 women each year.[27] Ensuring prenatal care and safe delivery services would save more lives still. These programs also carry far-reaching benefits for future generations, as the death of a mother has profound impacts on her family: In developing countries, in families where a mother dies, her children are three to ten times less likely to survive, are more likely to be malnourished, and are less likely to enroll in school.[28, 29]

Economic Resources

Economic resources are a critical part of adaptive capacity—and thus resilience—to climate change, measured in the VRIM as per capita GDP and income equity. There are important economic advantages to reproductive health and family planning programs: Women who are empowered to make choices about their fertility are more likely to invest in their children's education and take advantage of economic opportunities outside the home, thus increasing the economic status of the household. Also, smaller families are better able to invest in each member, allowing more children—especially girls—to stay in school longer. This has important impacts later in life, as women who do not complete their education are more likely to live in poverty. Further, larger families are more likely than smaller ones to fall into poverty and are less likely to recover from economic and environmental shocks, such as those that will increase with global climate change.[30]

> When young people are endowed with opportunities and rights, they—and the societies in which they live—will be more resilient.

At the macro level, when nations make the transition to smaller families, they may reap the benefits of a "demographic dividend," which can translate to economic benefits for the country as a whole.[31] (See Chapter 11.) Investments in family planning and reproductive health also result in significant savings for national and local governments, as improved maternal and child health and slower population growth ease financial demands on the health sector, infrastructure development, educational sector, and water and sanitation services.[32]

Environment

The VRIM recognizes that environmental quality is critical in helping communities absorb the impact of changes in climate. A nation's population density can have a negative impact on environmental quality because more natural resources are required for a larger population, and pollution levels are higher in densely populated areas. Millions of people live in areas where natural resources are already depleted, and those resources may become even scarcer in a warming world. For example, more than 745 million people now face water stress or scarcity, and this number is estimated to grow to between 2.75 billion and 3.25 billion by 2025, depending on rates of population growth.[33] Food production is threatened, as the number of people living in countries where cultivated land is critically scarce is projected to increase from 448 million in 2005 to between 559 million and 706 million in 2025.[34] Limited arable land, often in tandem with social inequity, can push people to use marginal land, thus degrading fragile ecosystems.[35]

Biodiversity—the rich variety of species and systems that make up the web of life—is another indicator of environmental quality, and therefore of resilience to climate change. An estimated 1.3 billion people live in biodiversity "hot spots"—species-rich areas threatened by human activities. Comprising 12 percent of global land surface, they are home to nearly 20 percent of global population and are experiencing annual growth rates greater than the global average (1.8 percent vs. 1.3 percent).[36]

CONCLUSION

The effort to understand and build resilience to climate change is still in its early stages. Measures such as the VRIM provide us with only a snapshot of vulnerability to changes in climate, and they must be further honed and supplemented with country-level studies. More research exploring vulnerability to climate change is required, as well as operations research and program evaluation for a broad array of adaptation strategies.

As we have seen, poverty and gender inequity make women especially vulnerable to climate change. That is why adaptation research and interventions should pay close attention to the specific needs of women—best ensured by having women at the table.

Most importantly, while there are still many unknowns as to how climate change will unfold in each country in coming years, actions to build resilience must not wait for certainty in the climate science community, nor should they wait for all research to conclude. We know that the realization of women's rights and their well-being, including their reproductive health, is critical to improving many of the underlying factors affecting resilience and adaptability—including public health, economic security, and the environment. By implementing these strategies today, we not only save lives and improve the well-being of men, women, and children worldwide; we also help improve the ability of future generations to deal with the mounting challenges of climate change.

REFERENCES

1. BBC News, 2008, Burma aid effort "requires $1bn," BBC News, July 21.

2. IPCC, 2007, *Climate Change 2007: Synthesis Report. Contribution of Working Groups I, II and III to the Fourth Assessment Report of the Intergovernmental Panel on Climate Change,* IPCC, Geneva.

3. UN Statistics Division, Carbon Dioxide Emissions (CO_2), Metric Tons of CO_2 Per Capita, 209 countries from 1980–2004, available from http://unstats.un.org/unsd/cdb/cdb_topic_xrxx.asp?topic_code=29, last accessed April 29, 2008.

4. Boko, M., I. Niang, A. Nyong, C. Vogel, A. Githeko, M. Medany, B. Osman-Elasha, R. Tabo, and P. Yanda, 2007, Africa, p. 433–467, *Climate Change 2007: Impacts, Adaptation and Vulnerability. Contribution of Working Group II to the Fourth Assessment Report of the Intergovernmental Panel on Climate Change,* M. L. Parry, O. F. Canziani, J. P. Palutikof, P. J. van der Linden, and C. E. Hanson, eds., Cambridge University Press, Cambridge, UK.

5. UNDP, 2007, *Human Development Report 2007/2008: Fighting Climate Change: Human Solidarity in a Divided World,* UNDP, New York.

6. Confalonieri, U., B. Menne, R. Akhtar, K. L. Ebi, M. Hauengue, R. S. Kovats, B. Revich, and A. Woodward, 2007, Human health, *Climate Change 2007: Impacts, Adaptation and Vulnerability.*

7. UN Division for the Advancement of Women, 2004, Making risky environments safer, *Women 2000 and Beyond,* UN, New York, April.

8. FAO, FAO Focus: Women and Food Security. FAO, Rome, available

from http://www.fao.org/FOCUS/E/Women/Sustin-e.htm, last accessed May 14, 2008.

9. FAO, *Women, Agriculture, and Food Security*, FAO, Rome, available from http://www.fao.org/worldfoodsummit/english/fsheets/women.pdf, last accessed May 14, 2008.

10. WHO, 2001, *Water for Health: Taking Charge*, WHO, Geneva, available from http://www.who.int/water_sanitation_health/takingcharge/en/, last accessed May 14, 2008.

11. WHO, 2003, *The Right to Water*, WHO, Geneva, available from http://www.who.int/water_sanitation_health/rtwrev.pdf, last accessed May 14, 2008.

12. UNICEF UK, 2008, *Our Climate, Our Children, Our Responsibility: The Implications of Climate Change for the World's Children*, UNICEF UK, London.

13. IPCC, 2007, Summary for policymakers, 7–22, p. 21, *Climate Change 2007: Impacts, Adaptation and Vulnerability*.

14. IPCC, 2007, *Climate Change 2007: Impacts, Adaptation and Vulnerability. Contribution of Working Group II to the Fourth Assessment Report of the Intergovernmental Panel on Climate Change*, M. L. Parry, O. F. Canziani, J. P. Palutikof, P. J. van der Linden, and C. E. Hanson, eds., Cambridge University Press, Cambridge, UK, p. 880.

15. IPCC, 2007, *Fourth Assessment Report*, p. 869.

16. E.g., Magalhães, A. R., 1996, Adapting to climate variations in developing regions: A planning framework, *Adapting to Climate Change: An International Perspective*, J. Smith, N. Bhatti, G. Menzhulin, R. Benioff, M. I. Budyko, M. Campos, B. Jallow, and F. Rijsberman, eds., Springer-Verlag, New York.

17. IPCC, 2001, *Climate Change 2001: Impacts, Adaptation, and Vulnerability: Contribution of Working Group II to the Third Assessment Report of the Intergovernmental Panel on Climate Change*, James McCarthy, Osvaldo F. Canziani, Neil A. Leary, David J. Dokken Kasey, and S. White, eds., Cambridge University Press, Cambridge, p. 895.

18. Lawrence, Miles B., and Hugh D. Cobb, 2005, Tropical Cyclone Report: Hurricane Jeanne, 13–28 September 2004, National Hurricane Center, Miami, FL.

19. BBC News, 2004, Deadly Hurricane Season, September 28, online edition available at http://news.bbc.co.uk/2/hi/americas/3677022.stm, last accessed May 12, 2008.

20. Moss, Richard H., Antoinette L. Brenkert, and Elizabeth L. Malone, 2001, *Vulnerability to Climate Change: A Quantitative Approach*, PNNL-SA-33642, Pacific Northwest National Laboratory, Washington, DC.

21. Brenkert, Antoinette L., and Elizabeth L. Malone, 2005, Modeling vulnerability and resilience to climate change: A case study of India and Indian states, *Climatic Change* 72: 57–102.

22. Malone, Elizabeth L., and Antoinette Brenkert, In press, Vulnerability, sensitivity, and coping/adaptive capacity worldwide, *The Distributional Effects of Climate Change: Social and Economic Implications*, M. Ruth and M. Ibarraran, eds., Elsevier Science, Dordrecht.

23. Ibid.

24. Vlassof, M., S. Singh, J. E. Darroch, E. Carbone, and S. Bernstein, 2004, Assessing Costs and Benefits of Sexual and Reproductive Health Interventions, Occasional Report Number 11, Guttmacher Institute, New York.

25. WHO, Estimates of DALYs by sex, cause and WHO mortality subregion, available at http://www.who.int/whosis/en/, accessed August 29, 2007.

26. Tinker, Anny, and Elizabeth Ransom, 2002, Healthy Mothers and Healthy Newborns: The Vital Link, Population Reference Bureau and Save the Children, Washington, DC.

27. Cleland, J., S. Bernstein, et al., 2006, Family planning: The unfinished agenda, The Lancet 368(9549): 1810–1827.

28. UNFPA, 2005, *Reducing Poverty and Achieving the Millennium Development Goals: Arguments for Investing in Reproductive Health and Rights*, UNFPA, New York.

29. WHO, 2005, *The World Health Report 2005: Make Every Mother and Child Count*, WHO, Geneva, cited in Population Action International, 2007, *A Measure of Survival: Calculating Women's Sexual and Reproductive Risk: PAI Report Card 2007*, PAI, Washington, DC.

30. Cleland, J., S. Bernstein, et al., 2006, Family planning.

31. Bloom, David, David Canning, and Jaypee Sevilla, 2003, *The Demographic Dividend: A New Perspective on the Economic Consequences of Population Change*, Rand, Santa Monica.

32. Cleland, J., S. Bernstein, et al., 2006, Family planning.

33. Population Action International, 2006, *People in the Balance: 2006 Update*, PAI, Washington, DC.

34. Ibid.

35. Upadhyay, U., and B. Robey, 1999, Protecting the environment, *Population Reports* XXVII(2).

36. Population Action International, 2006, *People in the Balance: 2006 Update*.

CHAPTER 8

··········

Population Growth, Ecosystem Services, and Human Well-Being

LYNNE GAFFIKIN

Human beings are wholly dependent on nature; healthy ecosystems provide vital services such as food, freshwater, clean air, and protection from disease and disaster. But many of the planet's ecosystems are under great stress. In the last decade, it has become increasingly clear that degraded ecosystems are diminishing the quality of life for current and future generations.

Recognizing this challenge, in 2000, then UN Secretary-General Kofi Annan launched the Millennium Ecosystem Assessment (MA)—a comprehensive assessment of the condition of the world's ecosystems and their impact on human well-being. A four-year collaborative effort involving governments, nongovernmental organizations, the private sector, and scientists from a wide range of disciplines, the MA was conducted by 1,360 experts in ninety-five countries.[1] Those experts synthesized information from scientific literature and peer-reviewed data sets and models, and drew on the knowledge of local communities.

This chapter reviews important findings of the MA through a demographic lens, focusing on the relationships among population growth,

---

Lynne Gaffikin is a consultant to John Snow, Inc.; through EARTH, Inc. she focuses on links among biodiversity conservation, population dynamics, and human and ecosystem health.

ecosystem change, and the impact of that change on human well-being, now and in the future. The chapter also highlights how such change has particularly affected the most vulnerable people in society, who are often the least able to adapt and the most directly dependent on functioning ecosystem services.

STATE OF THE WORLD'S ECOSYSTEMS

In the last fifty years, human beings have modified ecosystems more rapidly and extensively than in any comparable period of human history. Those changes were made largely to meet soaring demands for food, freshwater, fiber, and fuel as the human population doubled between 1960 and 2000 and per capita consumption rose with increasing levels of urbanization and wealth in many parts of the world. As a result of our successful exploitation of ecosystems, human well-being has improved substantially over the past half century and continues to do so.

Those gains, however, have come at a substantial cost. The MA offers, in effect, an audit of the Earth's accounts, and there is much more red than black on the balance sheet. Nearly two-thirds of all ecosystem services—including fisheries and freshwater—are currently being degraded or used in ways that cannot be sustained.

> Nearly two-thirds of all ecosystem services—including fisheries and freshwater—are currently being degraded or used in ways that cannot be sustained.

Moreover, gains in well-being have not been equitably shared across nations nor among subgroups of people.[2] And negative consequences of ecosystem change have fallen, and will continue to fall, on the world's most vulnerable populations—often those least able to tolerate such consequences. At a national level, this translates into shortcomings in achieving Millennium Development Goals (MDGs), the world's operational commitment to sustainable development and reduced global poverty.

POPULATION: AN "INDIRECT DRIVER" OF ECOSYSTEM CHANGE

The MA defines "drivers" of ecosystem change as any factors, human-associated or naturally occurring, that contribute to change in any aspect of an ecosystem.[3] Drivers are characterized as either *direct* or *indirect*: the former representing unequivocal and measurable influences; the latter operating more diffusely, often through a direct driver.[4] Direct drivers are mostly physical, chemical, or biological—such as air and water pollution, harvesting of timber or fish stocks, or the conversion of forest to cropland. Demographic factors, including population growth, are among five key indirect drivers of ecosystem change. Other indirect drivers include economic, sociopolitical, scientific and technical, and cultural and religious factors.

Below, we examine the ways in which demographic processes—especially population growth—act as indirect drivers of change in six major ecosystems.

Cultivated Systems

Cultivated systems are ecosystems that have been transformed for the production of food, fiber, or fuel. These areas are home to a large majority of the world's population—almost 75 percent[5] lives within cultivated-system boundaries.*

Recent decades have seen remarkable growth in the extent and productivity of cultivated systems. As demand for agricultural products has grown, farmers have responded by expanding land under cultivation (extensification) and by improving yields (intensification). Between 1950 and 1980, more forests, savannas, and grasslands were converted to cropland than in the eighteenth and first half of the nineteenth centuries combined. Nearly a quarter of the planet's land surface is now cultivated.[6] More recently, the cultivation of new land has slowed in many areas, and, in the last 40 years, intensification has been the main source of increased output.

Demand for food and other cultivated products is driven by population growth as well as by increases in household income and changes in dietary preferences. In the last half century, population growth has been the single most important driver of demand for food, accounting for 60 to 70 percent of the increase in calories consumed in both developed and developing countries.[7] As growth rates decline in many areas, population growth accounts for a correspondingly smaller share of increased demand. However, growth rates remain high in many low-

---

*Because the boundaries of systems overlap (e.g., a cultivated system may also be classified as dryland), these percentages add up to more than 100 percent.

income developing countries, where increased demand puts ever-greater pressure on cultivated systems. Where poor farmers lack the means or the incentive (such as secure land ownership) to conserve soil and manage crops effectively, the result can be a downward spiral of poverty and land degradation.[8] Additionally, of critical importance, land cultivation for food production, as practiced in many parts of the world, has had substantial negative effects on *other* ecosystem services, including the provision of clean air and water.

Over the next half century, demand for food is estimated to grow by as much as 80 percent, mainly in developing countries. Meeting *local* demand will require both extensification and intensification of cultivated systems. Prospects for this are not good, especially in areas with limited arable land and/or degraded soil. Achieving food security *globally* will require sustained increased crop yields from highly productive cultivated land, which currently suffers from the cumulative negative effects of other ecosystem changes, including reductions in water flows and water pollution. Prospects for this are also questionable—see Chapter 12.

Coastal Systems and Marine Fishery Systems

The services provided by coastal and marine fishery systems provide more services *proportionately* to human well-being than any other. They are severely imperiled, however, suffering some of the most rapid ecological change.

Coastal areas—where sea and freshwater mix—are very densely populated. Nearly half of all major cities are located in these areas, and just under 40 percent of the world's population lives within 60 miles of a coast. Population density in coastal areas is almost triple that of inland areas.[9] High growth rates put increasing pressure on coastal zone ecosystems. For example, in just two decades, people have removed more than a third of the world's mangroves, which grow in the tidal mud of many tropical regions.[10] Mangrove forests provide an essential buffer against storms and sea level rise; their loss greatly magnified the human toll from the Indian Ocean tsunami and Cyclone Nargis.

Marine fishery systems are also under stress. Demand for fish and other seafood products is driven by population growth and increased demand for "luxury" seafood (with rising incomes). Increased migration to coastal areas also exerts pressure on the marine fishery system.[11]

Overfishing in marine areas is a critical problem: At least one-

In many areas, the total weight of fish available to be caught has fallen by 90 percent since before the onset of industrial fishing.

quarter of marine fish stocks are overharvested. The global fish catch increased until the 1980s but is now on the decline. In many areas, the total weight of fish available to be caught has fallen by 90 percent since before the onset of industrial fishing.[12] While overfishing is primarily driven by consumers in the affluent countries, declining fish supplies directly affect one-sixth of the world's population—mostly in poor communities in developing countries—people who depend upon fish as their main animal protein source.[13]

Inland Water Systems

Inland water, or aquatic, systems include lakes and rivers, marshes, swamps and floodplains, wetlands, small streams, and ponds. As with marine systems, inland waterways are also an important source of fish for a large number of the world's citizens. For some rural communities, especially in developing countries, inland fisheries may be their only source of animal protein.[14]

Many interrelated factors are driving change in these vital ecosystems, including population growth and increases in consumption. Habitat conversion and water extraction and diversion, principally for agriculture, have degraded many inland aquatic areas. Pollution from agricultural runoff has also negatively affected freshwater ecosystem functioning and fish yields—to the point that currently, fish take from inland freshwater systems exceeds sustainable yield.[15]

In addition to providing food, inland waterways also provide an important source of freshwater, a large percentage of which goes toward agriculture. The amount of water taken from rivers and lakes for agriculture, industry, and household use has doubled since 1960. As detailed in Chapter 14, water scarcity is a significant global problem.[16] As with other ecosystem changes, the world's poorest people suffer the most from water scarcity, as well as the heaviest burden of water-associated disease.[17]

Over the next half century, demand for freshwater is expected to increase up to 85 percent, concurrent with the increased demand for food.[18] This will occur mostly in developing nations. Given that freshwater areas are already the most degraded of all systems, meeting this increased demand will be a significant challenge.

Dryland Systems

Roughly one-third of the world's population lives in arid grasslands, woodlands, deserts, and desert margins, which cover some 40 percent of the planet's land area.[19] Desertification or other land degradation, including soil erosion, has affected 10 to 20 percent of the most vulnerable dryland areas.

Drylands are home to some of the poorest, most marginalized world citizens.[20] The lack of access to freshwater severely diminishes human well-being in these areas. While not generally densely populated, drylands have some of the fastest population growth *rates* in the world. In the last decade of the twentieth century, the number of people living in drylands grew by nearly 20 percent. Population increase is a particularly key driver of ecosystem change in the least arid drylands, where growth tends to extend and/or intensify the use of cultivated land.[21]

Where migrating populations contribute to population growth in arid areas, one often finds tension between nomad and farmer populations—deriving from conflicts over water and the use of land as pastoral rangelands versus for cultivation.[22] In the future, such conflicts are likely to increase as growing dryland populations put more land under the plow. On a positive note, the MA has observed that where indigenous, adaptive traditions have come into play and/or water conservation and soil enhancement technologies have been embraced, negative effects of population growth have been partially offset.

Forest and Mountain Systems

Forests cover approximately a quarter of the world's land area (not including Greenland and Antarctica).[23] In addition to providing food, wood for timber and fuel, and watershed functions, forests are a crucial "sink" for carbon, and they contain half the world's biodiversity.[24, 25] The loss of forests since preagricultural times has been dramatic—estimated at 20 to 50 percent.[26] Forest loss and degradation are driven by a multitude of factors, including population growth.[27]

Mountain systems are valued for their ecosystem services and biodiversity, but they are also exceptionally fragile. At least half the world's population, more than 3 billion people, depends on mountain system resources. Billions get their drinking water from the world's great mountain ranges—including a half billion people in the Himalaya–

Hindu Kush region and a quarter billion downstream who rely on gla-cial meltwaters.[28] But, because of climate change, those glaciers are melting rapidly.

One in five of the world's people, including some of the planet's poorest and most vulnerable citizens, live in mountainous systems. Many mountain dwellers lack access to public services and suffer from malnutrition and food insecurity. This is especially the case for those living at higher elevations. The rate of population increase in moun-tain areas varies globally but averaged 16 percent between 1990 and 2000.[29] Subsistence farmers are often driven into the mountains by soil degradation and poor crop yields in lowland areas, which are exacer-bated by population increase. Migration combines with natural popu-lation growth in mountainous areas to produce some of the world's highest rural population densities.[30]

Tropical *mountain forests*—where the two systems overlap—have recently experienced the fastest annual rate of deforestation and popu-lation growth.[31] A key direct driver of such deforestation is conversion of land for agriculture, especially where deforestation reflects underly-ing poverty.

Urban Systems

Urban systems generally consume, rather than generate, ecosystem ser-vices.[32] Although they cover less than 3 percent of the Earth's land area, urban areas are home to over half the world's population.[33] While the rate of urban population growth in the past few hundred years has been substantial, the increase in flow of ecosystem services *to* urban areas has increased at an even higher rate.[34] This is particularly the case in densely populated urban coastal zones.

As the trend toward urbanization increases globally, logically the effect that urban systems have on ecosystem health will correspond-ingly increase. Importantly, the MA noted that urbanization itself is not intrinsically bad for the functioning of ecosystems on which cities depend. Rather, it is the growth of urban populations and economies—and consumption patterns associated with rising incomes—that increasingly exerts negative pressures on local and global ecosystems.[35] This is compounded by the fact that in many developing countries, a substantial proportion (25 to 50 percent) of the total population lives in settlements around urban centers with few public services and no system for governing their effect on ecosystem degradation.

PROSPECTS FOR THE FUTURE

While the causes of environmental change are complex and interrelated, increased pressure on the world's ecosystems generally results from two factors: more people inhabiting the planet and ever-higher rates of consumption. The MA shows that population growth contributes—in ways both known and unknown—to the degradation of many of the world's ecosystems. The magnitude of ecosystem change is expected to grow along with the expanding human enterprise.

The fruits of that enterprise are not equitably shared among the world's people. And, as demand for ecosystem services degrades these vital natural systems, the poor suffer the most. Wealthier populations exert disproportionately more pressure on global ecosystems and have the ability to adapt more readily to ecosystem changes. Conversely, those most directly dependent on ecosystem services are the least able to adapt. The regions that are most vulnerable to ecosystem change are, not coincidentally, the same regions that are struggling to achieve the Millennium Development Goals.

What do these trends portend for the future? The MA offers four plausible scenarios—explorations of how events may unfold between now and 2050, considering different approaches to development and ecosystem management. Regardless of assumptions, the scenarios consistently identify three regions that are likely to see a rapid decline in per capita ecosystem services: sub-Saharan Africa, the Middle East, and south Asia.[36] The vast majority of population growth will also be concentrated in these areas, especially in their poorest urban communities.

For millions of people in those regions, the effects of degraded ecosystem services will be direct and severe: hunger, malnutrition, and disease. Other impacts will be spread more broadly. Continued watershed degradation will cause regional shortages of freshwater. The loss of mangroves and other wetlands will diminish the ability of coastal ecosystems to buffer natural disasters, including hurricanes and typhoons. Land degradation will cause deserts to grow and cropland to shrink. Globally, everyone on Earth will experience the effects of climate change.

The MA also shows that ecosystem changes can be "non-linear" and potentially irreversible.[37] Natural systems have a tendency to reach "tipping points" before we can understand or act on the warning signs.

Moreover, the thresholds for abrupt change cannot be reliably forecast by existing science. The difficulty of predicting those thresholds justifies the use of a "precautionary approach" to ecosystem management: Where there is a threat of irreversible harm, the precautionary approach urges action to address the threat even in the face of scientific uncertainty.[38] What is not difficult to predict is that ecosystem degradation will continue to have a disproportionately negative effect on the world's poorest and most vulnerable people, including women and children. In this there is no justice.

The outcomes described above are not, however, fixed and immutable. The MA recognizes that decisions made at all levels—local, national, and international—will shape the future of ecosystems and human well-being. This is true for direct drivers of ecosystem change, such as air pollution, and for indirect drivers, such as population growth. Decisions made today—by individual women and men, by health ministers and presidents, by aid officials and heads of multilateral organizations—will determine the future size, distribution, and consumption patterns of the human population. Indirectly, those decisions will affect the health of ecosystems on which human well-being depends.

The definition of *human well-being* in the MA includes freedom of choice and action. Certainly, that definition must include the freedom to choose the number and timing of one's children. At the same time, enabling people to achieve their own reproductive goals results in improved health, reduced poverty, and slower population growth. These outcomes, in turn, would reduce negative pressures on fragile ecosystems. For these reasons, universal access to family planning and other health services, girls' education, and other programs that enable people to control their reproductive destiny are crucial to human well-being. They also must be an integral part of the precautionary approach to ecosystem management.

REFERENCES

1. Millennium Ecosystem Assessment, 2005, *Ecosystems and Human Well-being: Biodiversity Synthesis*, World Resources Institute, Washington, DC, accessed online at http://www.millenniumassessment.org/documents/document.354.aspx.pdf.

2. Hassan, Rashid, Robert Scholes, Neville Ash, eds., 2005, *Ecosystems and Human Well-being: Current State and Trends,* Volume 1, *Findings of the Condition and Trends Working Group of the Millennium Ecosystem Assessment,* Island Press,

Washington, DC, accessed online at http://www.millenniumassessment.org/documents/document.766.aspx.pdf.

3. Nelson, Gerald, 2005, Drivers of ecosystem change: Summary chapter, Chapter 3, *Ecosystems and Human Well-being: Current State and Trends,* accessed online at http://www.millenniumassessment.org/documents/document.272.aspx.pdf.

4. Millennium Ecosystem Assessment, 2003, *Ecosystems and Human Well-Being: A Framework for Assessment,* Island Press, Washington, DC, accessed online at http://www.millenniumassessment.org/documents/document.48.aspx.pdf.

5. Cassman, Kenneth, and Stanley Wood, 2005, Cultivated systems, Chapter 26, *Ecosystems and Human Well-being: Current State and Trends,* accessed online at http://www.millenniumassessment.org/documents/document.295.aspx.pdf.

6. Ibid.

7. Ibid.

8. Ibid.

9. Agardy, Tundi, and Jacqueline Alder, 2005, Coastal systems, Chapter 19, *Ecosystems and Human Well-being: Current State and Trends,* accessed online at http://www.millenniumassessment.org/documents/document.288.aspx.pdf.

10. Millennium Ecosystem Assessment Board of Directors, *Living Beyond Our Means: Natural Assets and Human Well-being,* Island Press, Washington, D.C, 2005, accessed online at http://www.millenniumassessment.org/documents/document.429.aspx.pdf.

11. Pauly, Daniel, and Jacqueline Alder, 2005, Marine fisheries systems, Chapter 18, *Ecosystems and Human Well-being: Current State and Trends,* accessed online at http://www.millenniumassessment.org/documents/document.287.aspx.pdf.

12. Millennium Ecosystem Assessment Board of Directors, op. cit.

13. Pauly and Alder, 2005, Marine fisheries systems, accessed online at http://www.millenniumassessment.org/documents/document.287.aspx.pdf.

14. Ibid.

15. Ibid.

16. Millennium Ecosystem Assessment, 2005, *Ecosystems and Human Well-being: Biodiversity Synthesis,* accessed online at http://www.millenniumassessment.org/documents/document.354.aspx.pdf.

17. Millennium Ecosystem Assessment, 2005, *Ecosystems and Human Well-being: Current State and Trends,* accessed online at http://www.millenniumassessment.org/documents/document.289.aspx.pdf.

18. World Resources Institute, 2000, *A Guide to World Resources 2000–2001: People and Ecosystems: The Fraying Web of Life,* World Resources Institute, Washington, DC, accessed online at http://pdf.wri.org/world_resources_2000-2001_people_and_ecosystems.pdf.

19. Millennium Ecosystem Assessment, 2005, *Ecosystems and Human Well-being: Desertification Synthesis,* World Resources Institute, Washington, DC,

accessed online at http://www.millenniumassessment.org/documents/
document.355.aspx.pdf.

20. Harrison, Paul, and Fred Pearce, 2001, *American Association for the Advancement of Science AAAS Atlas of Population and Environment*, University of California Press, Berkeley, CA, accessed online at http://atlas.aaas.org/index.php?sub=intro.

21. Safriel, Uriel, and Adeel Zafar, 2005, Dryland systems, Chapter 22, *Ecosystems and Human Well-being: Current State and Trends*, accessed online at http://www.millenniumassessment.org/documents/document.291.aspx.pdf.

22. Millennium Ecosystem Assessment, 2005, *Ecosystems and Human Well-being: Desertification Synthesis*, accessed online at http://www.millennium assessment.org/documents/document.355.aspx.pdf.

23. Shvidenko, A., C. V. Barber, and R. Persson, 2005, Forest and woodland systems, Chapter 21, *Ecosystems and Human Well-being: Current State and Trends*, accessed online at http://www.millenniumassessment.org/documents/document.290.aspx.pdf.

24. World Resources Institute, 2000, *A Guide to World Resources 2000–2001*, accessed online at http://pdf.wri.org/world_resources_2000-2001_people_and_ecosystems.pdf.

25. Harrison, Paul, and Fred Pearce, 2001, *AAAS Atlas*, accessed online at http://atlas.aaas.org/index.php?sub=intro.

26. World Resources Institute, 2000, *Guide to World Resources 2000–2001*, accessed online at http://pdf.wri.org/world_resources_2000-2001_people_and_ecosystems.pdf.

27. Gardner-Outlaw, Tom, and Robert Engelman, 1999, *Forest Futures: Population, Consumption and Wood Resources*, Population Action International, Washington, DC.

28. UNEP, 2008, Meltdown in the mountains: Record glacier thinning means no time to waste on agreeing new international climate regime, March 16.

29. Körner, Christian, and Ohsawa Masahiko, 2005, Mountain systems, Chapter 24, *Ecosystems and Human Well-being: Current State and Trends*, accessed online at http://www.millenniumassessment.org/documents/document.293.aspx.pdf.

30. Ibid.

31. Ibid.

32. McGranahan, Gordon, and Peter Marcotullio, 2005, Urban systems, Chapter 27, *Ecosystems and Human Well-being: Current State and Trends*, accessed online at http://www.millenniumassessment.org/documents/document.296.aspx.pdf.

33. UN, 2002, *World Urbanization Prospects: The 2001 Revision*, E/ESA/WP.191, UN, New York.

34. Millenium Ecosystem Assessment, 2005, *Ecosystems and Human Well-being: Health Synthesis*, World Health Organization, accessed online at http://www.millenniumassessment.org/documents/document.357.aspx.pdf.

35. McGranahan, Gordon, and Peter Marcotullio, 2005, Urban systems, Chapter 27, *Ecosystems and Human Well-being: Current State and Trends*,

accessed online at http://www.millenniumassessment.org/documents/
document.296.aspx.pdf.

36. Millenium Ecosystem Assessment, 2005, *Ecosystems and Human Well-being: Health Synthesis*, accessed online at http://www.millenniumassessment
.org/documents/document.357.aspx.pdf.

37. Ibid.

38. UNEP, *The Rio Declaration on Environment and Development*, accessed
online at http://www.unep.org/Documents.multilingual/Default.asp?
DocumentID=78&ArticleID=1163.

CHAPTER 9

··········

Numbers Matter

Human Population as a Dynamic Factor in Environmental Degradation

JOHN HARTE

It is widely assumed that environmental degradation grows in proportion to population size, if per capita consumption and technology are held constant. But that assumption is hugely optimistic. The environmental consequences of increasing human population size are dynamic and nonlinear, not passive and linear. In this chapter, we will see how complex dynamics—in particular, feedbacks, thresholds, and synergies—can amplify the environmental impact of population growth.

> It is widely assumed that environmental degradation grows in proportion to population size, if per capita consumption and technology are held constant. But that assumption is hugely optimistic.

For the past several centuries, humanity has been polluting air and water, altering Earth's climate, eroding the soil, fragmenting and eliminating the habitat of plants and animals, and depleting the natural bank account of nonrenewable resources. Of especially great long-

John Harte holds a joint professorship in the Energy and Resources Group and the Ecosystem Sciences Division of the College of Natural Resources at the University of California, Berkeley.

term concern, we are degrading the capacity of natural ecosystems to regenerate or maintain renewable resources and "ecosystem services," such as the provision of clean air and water, control of flooding, the maintenance of a tolerable climate, the conservation and regeneration of fertile soil, and the preservation of genetic and other forms of biological diversity.[1, 5, 7, 8, 10, 12, 13, 14, 15, 16, 18, 20]

The linkages between human activity and environmental degradation are myriad, but at the risk of some oversimplification, one can usefully group the contributing factors in three categories: human population size, the per capita rate of consumption of energy and materials that contribute to our affluence, and the impacts stemming from the technologies used to provide that per capita rate of consumption. An equation is sometimes used to express this:

Environmental **I**mpact = (**P**opulation size) ×
(per capita **A**ffluence level) × (impact from the **T**echnologies
used to achieve that level of per capita affluence)

$$I = P \times A \times T$$

This "IPAT equation"[2] is a useful reminder that population, affluence, and technology all play a role in determining environmental impacts. But it is also misleading if taken too literally, for it conveys the notion that population is a linear multiplier. In particular, it suggests that if per capita affluence—and the technologies and other means used to achieve that level of per capita affluence—is held constant, then the impacts simply grow in proportion to population size; if the population doubles in size, then the impacts double in magnitude.

In reality, population size plays a much more dynamic and complex role in shaping environmental quality. The notion that impacts are roughly proportional to our numbers ignores a host of threshold effects, synergies, feedbacks, and other nonlinear phenomena that can amplify the environmental impact of human numbers.

THRESHOLDS, SYNERGIES, FEEDBACKS: AN OVERVIEW

Thresholds

There are many situations in which a small or intermediate-sized stress to a system generates little or no impact, but when the stress exceeds a certain level (the threshold), the impact increases dramatically. A prominent example is the response of surface waters to acid rain. Below the threshold, alkalinity prevents acidification of the lake or stream, but above a certain level of added acid, the pH falls rapidly.

Suppose we assume that the amount of acid rain falling to Earth each year is simply proportional to the amount of coal being burned and that the amount of coal being burned is proportional to population size (because affluence and technology are held constant). Then an increase in population size from a level below the threshold to one above the threshold can generate a disproportionately large decrease in the pH of surface waters. Quite suddenly, lakes and streams will acidify and lose their ability to sustain life.

But the situation is even more complex. Of the major fossil fuels used today (coal, petroleum, and natural gas), coal generally produces the most acidity per unit of energy. Thus, environmentalists urge policymakers to replace coal with the cleanest of these fuels, natural gas. But unfortunately, natural gas supplies are far more limited than are coal supplies. Hence, with every increase in the size of the human population, the day arrives sooner when more coal, per capita, must be used. In the IPAT formulation, one would express this by stating that T, the technology, changed at the same time that P increased. But the point is that this change in T is caused by a change in P, and thus T should be considered as a function of P. And this means that P is not simply a linear multiplier.

For another example, consider the effect of habitat loss on the survival of species. In places such as Amazonia, where roughly a quarter of the rain forest has been destroyed, numerous species have been driven to extinction. Species may become extinct because they depended on a certain amount of intact habitat: Past a threshold of habitat loss or fragmentation, the species can no longer be sustained. Or a species may be lost because it was "endemic"—unique—to a particular habitat, which was destroyed. Although we lack the knowledge needed to predict accurately how some species become extinct when a specified area of original habitat is lost, general ecological principles do provide a way of characterizing the qualitative features of the relationship between habitat area and species diversity. Virtually all ecological models predict that the fraction of extinct species will increase faster than linearly with habitat loss.

Thresholds abound in nature. For example, thresholdlike nonlinearities characterize the relationship between population density and the probability that an infectious disease will become epidemic in a region;[11] a similar nonlinearity is likely to characterize the relation between the area of a monoculture crop and the probability that a plaguelike crop-pest outbreak will occur there. In some instances, contingent and unpredictable factors blur the actual threshold level of a stress, rendering it difficult to predict the level of an activity that will cause a sudden increase in impacts. For example, in climate science, it is difficult to determine at what level of heating Greenlandic and west Antarctic ice will rapidly melt, resulting in a rapid and very dangerous rise in sea level. These difficulties point to a gap in predictive capability, however, and should not obscure the fact that such "tipping points" exist.

Synergies

Synergy occurs when the combined effect of two causes is greater than the sum of the effects of the two causes acting in isolation of each other. Clearly, if two environmental stresses act synergistically and each stress grows in proportion to population size, then the combined effect of both stresses may grow faster than linearly with increasing population size.

Consider the following example. Both climate warming and deforestation stress biodiversity in currently forested areas of the world. The former can result in forest dieback due to increased frequency of drought conditions, while the latter releases carbon dioxide, a greenhouse gas, to the atmosphere, exacerbating climate warming. Thus the sum of the effects of each of these two stresses—climate warming and deforestation—generates additional stresses that reinforce the total response.

Deforestation can contribute to the problem of acid rain, as has been observed in Africa where nitric acid is formed from burning vegetation. To make matters worse, plants and animals weakened by any one of these threats will generally be more vulnerable to the others. Fish weakened by radiation have been shown to be more easily damaged by thermal pollution than are healthy fish, and trees subjected to some air pollutants become more susceptible to insect damage.

The previous subsection, Thresholds, showed that even if habitat loss is proportional to the size of the human population, loss of habitat can cause more extinctions than expected, because species depend on intact habitat. But the situation is even worse than this because habitat loss is likely to increase disproportionately with population size.

To see this, consider Figure 9.1, which shows how a doubling of population can lead to a greater-than-double loss of habitat. Imagine that

Non-linear habitat implications of population growth

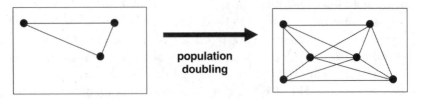

FIGURE 9.1 A doubling of population can lead to a greater-than-double loss of habitat.

initially a nation has three cities indicated by the three dots in the left panel. Between each of the cities are highways and transmission line rights-of-way, which are barriers to wildlife movement, death traps for some species, and opportunities for invasive species (that may displace native species) to spread. The right-hand panel shows the situation when the population has doubled and there are six cities instead of three. Now there are twelve connecting lines, implying a fourfold increase in these harmful stresses on plants and animals. While the illustration is a schematic and not to be taken too literally, it does indicate how the web of infrastructural interconnections in human society, which becomes increasingly dense as population grows, can result in synergistic land-use effects that are damaging to biodiversity.

Feedback

When the global climate system, an ecosystem, an organism, or any other complex entity is disturbed by some perturbation, the effect on the entity (expressed, for example, as a temperature change, an alteration of species diversity, or a change in life span) is the sum of both direct and indirect effects. For example, the buildup of greenhouse gases in the atmosphere is a direct cause of increased surface temperature due to the extra heat-absorbing capacity of these gases. As models consistently show, this direct effect is roughly 1°C for a doubling of atmospheric carbon dioxide.[8]

But this direct effect can also trigger secondary or indirect "feedback" effects.[9] In the case of climate change, the area of Earth that is covered by ice or snow decreases under warming. Ice and snow reflect sunlight back to space; without that reflective covering, more sunlight is absorbed by the Earth's surface. This is called a positive feedback effect.

In some ecological processes, feedback grows with the size of the human population. For example, in a warmer climate, people may rely more on air conditioning, thereby burning more fossil fuel and emitting more carbon dioxide, causing a positive feedback. A larger population would amplify that feedback effect. (A decreased use of heating fuels during warmer winters associated with global warming would have the opposite effect and result in a negative feedback.)

More significant feedback effects involve agriculture and climate change. For example, warming is likely to accelerate the decomposition of soil organic matter, particularly in tilled, fertilized, and irrigated soils. That decomposition, in turn, will accelerate the release of carbon dioxide into the atmosphere.[17] Because the total area of cultivated land increases with population size, population growth affects the intensity of this positive feedback effect.

Another example concerns wildfire, which will become more frequent and intense in the drier conditions created by global warming. The positive feedback arises because forest fires release significant quantities of carbon dioxide to the atmosphere and thus contribute to further warming. Here, too, the risk increases with the size of the human population because, even though they are exacerbated by drought, fires are often inadvertently—or sometimes intentionally—triggered by people.

OTHER EFFECTS OF POPULATION GROWTH

In addition to thresholds, synergies, and feedbacks, there is an assortment of other reasons why population growth can have outsize effects on the environment. One reason might be described as the "low-hanging fruit" phenomenon. Just as the low-hanging fruit gets picked first, human societies tend to use the most fertile soil, the cleanest water, the least polluting fuels first. As we draw down these desirable resources over time, our options are increasingly limited to second-rate resources.

Consider, for example, the exhaustion of the natural processes that result in "sinks" for our pollutants. Each year, the oceans and forests remove from the atmosphere a net quantity of carbon dioxide, thus alleviating somewhat the problem of global warming. Currently a quantity of carbon dioxide equal to about a third of what we emit is removed each year, with somewhat more than half of that going into the oceans and much of the remainder going into net growth of forests. But these natural sinks have only a limited capacity to take up carbon, and like kitchen sinks, they can partially clog if too much is fed into them.

Natural carbon sinks can be thought of as a nonrenewable resource. As with natural gas, a cleaner alternative to coal that we are rapidly exhausting, environmental problems increase when we use up the carbon sinks. With both natural gas and the carbon sinks, the faster our population grows, the quicker we use up these desirable resources, and thus the greater the impacts on the environment. Similarly, when natural supplies of clean water, either in aquifers or in surface streams, are exhausted, we must turn to new technologies to provide water, and many of these, such as dams and desalinization, carry environmental costs.

Or consider the effect of climate change on water supplies. Hotter temperatures and more frequent and intense drought increase the need for irrigation in agriculture. At the same time, population growth increases both the magnitude of climate warming and the amount of agricultural land and water needed to grow food. As a result, we will be forced into more energy-intensive means of obtaining water and thus will accelerate the warming in order to sustain a fixed level of per capita affluence. Hence we can think of the term T in IPAT, the impacts from the technologies needed to sustain a given per capita level of affluence, as inevitably a function of population, P. These two drivers cannot be disentangled.

It is not just the magnitude of environmental impacts, I, that increases disproportionately with P when A and T are fixed. In addition, our efforts to solve social problems, such as environmental injustice arising from inequitable distribution of impacts and of resources across income, cultural, and racial groups, are hindered by rapidly growing numbers. High rates of population growth make it more difficult to ensure adequate schooling, material resources, and civic order, thereby worsening social conditions. And again there is a pernicious feedback at work here: The existence of these social problems makes it more difficult to solve environmental problems. For example, in an equitable society with only small income disparities, a carbon tax to discourage fossil fuel consumption would make a great deal of sense. While it would be a sales tax, its burden would not fall on a particular group (unlike in an inequitable society, where the poor spend a larger fraction of their income on fuel than do the rich), and thus it could replace an income tax without exacerbating inequality. Thus the population trap catches us twice. Growing population size exacerbates social problems, and growing social problems frustrate efforts to confront environmental problems. Unsolved environmental problems further exacerbate injustice and inequity, again weakening the social order. The overall effect is an intensifying downward spiral.

CONCLUSION

To conclude, humanity is degrading environmental goods and services such as clean water, air, soil, and biodiversity and simultaneously reducing the capacity of natural processes to replenish these contributors to the quality of life. As abhorrent as is the current inequity in the distribution of resources between north and south, rich and poor, it pales in comparison with the impending inequity between us, living today, and those who will be born tomorrow and who, under current trends, will inherit a rapidly deteriorating planetary life-support system.

> As abhorrent as is the current inequity in the distribution of resources between north and south, rich and poor, it pales in comparison with the impending inequity between us, living today, and those who will be born tomorrow.

The future habitability of our planet will be shaped by highly nonlinear dynamical mechanisms that have the potential to generate unintended and undesirable consequences for future generations. As the stewards of our descendants, it is our moral obligation to better understand that landscape and to seek pathways to the future that avoid such rapidly escalating damages. Implementation of family planning and other policies throughout the world that give people greater control over reproduction, in developed and developing nations alike, is a critical step toward that end.

REFERENCES

1. Daily, G., ed., 1997, *Nature's Services*, Island Press, Washington, DC.

2. Ehrlich, P., and J. Holdren, 1971, The impact of population growth, *Science* 171: 1212–1217.

3. Harte, J., 1988, *Consider a Spherical Cow: A Course in Environmental Problem Solving*, University Science Books, Mill Valley, CA.

4. Harte, J., 1996, Feedbacks, thresholds, and synergies in global change: Population as dynamic factor, *Biodiversity and Conservation* 5: 1069–1083.

5. Harte, J., 1993, *The Green Fuse: An Ecological Odyssey*, University of California Press, Berkeley, CA.

6. Harte, J., C. Holdren, R. Schneider, and C. Shirley, 1991, *Toxics A to Z: A Guide to Everyday Pollution Hazards*, University of California Press, Berkeley.

7. Harte, J., M. Torn, and D. Jensen, 1992, The nature and consequences of indirect linkages between climate change and biological diversity, p. 325–343, *Global Warming and Biological Diversity*, R. Peters and T. Lovejoy, eds., Yale University Press, New Haven, CT.

8. IPCC, 2001, *Climate Change 2001: The Scientific Basis. Contribution of Working Group I to the Third Assessment Report of the Intergovernmental Panel on Climate Change*, Cambridge University Press, Cambridge, UK.

9. Lashof, D., B. DeAngelo, S. Saleska, and J. Harte, 1997, Terrestrial ecosystem feedbacks to global climate change, *Annual Review of Energy and the Environment* 22: 75–118.

10. L'Vovich, M., and G. White, 1990, Uses and transformations of terrestrial water systems, p. 235–252, *The Earth as Transformed by Human Action*, B. Turner, ed., Cambridge University Press, Cambridge, UK.

11. Murray, J. D., 1989, *Mathematical Biology*, Springer-Verlag, Berlin.

12. Myers, N., 1983, *A Wealth of Wild Species*, Westview, Boulder.

13. NRC/NAS, 1995, *Science and the Endangered Species Act*, Committee on Scientific Issues in the Endangered Species Act, Commission on Life Sciences, National Research Council, National Academies Press, Washington, DC.

14. Postel, S., 1993, Water and agriculture, p. 56–66, *Water in Crisis: A Guide to the World's Fresh Water Resources*, P. Gleick, ed., Oxford University Press, Oxford.

15. Postel, S., G. Daily, and P. Ehrlich, 1996, Human appropriation of renewable fresh water, *Science* 271: 785–788.

16. Rozanov, B., V. Targulian, and D. Orlov, 1990, Soils, p. 203–214, *The Earth as Transformed by Human Action*, B. Turner, ed., Cambridge University Press, Cambridge, UK.

17. Schlesinger, W., 1991, *Biogeochemistry: An Analysis of Global Change*, Academic Press, San Diego.

18. Shiklomanov, I. A., 1993, World fresh water resources, p. 13–24, *Water in Crisis: A Guide to the World's Fresh Water Resources*, P. Gleick, ed., Oxford University Press, Oxford.

19. Torn, M., and J. Harte, 2006, Missing feedbacks, asymmetric uncertainties, and the underestimation of future warming, *Geophysical Research Letters* 33, L10703. doi: 10.1029/2005GL025540M.

20. Westman, W., 1977, How much are nature's services worth? *Science* 197: 960–964.

CHAPTER 10

···········

Environmental Justice in an Urbanizing World

GORDON MCGRANAHAN

Urbanization—and the economic growth that accompanies it—contributes to environmental change at local, regional, and global levels. But not all urbanites are equal: While poor urban settlements create local impacts that affect the health of their residents, it is wealthier city dwellers whose environmental "footprints" are felt around the globe.

In the popular imagination, squatter settlements and slums are emblematic of urban environmental problems. It is easy to see why. Such settlements are often in environmentally inappropriate locations: on floodplains, on steep erosion-prone hillsides, or near railway lines, high-tension lines, or waste dumps.[4] Moreover, they often lack "environmental services": water, sanitation, and waste collection.[14] Such settlements are often unhealthy to live in and are a perceived environmental and social threat to those living nearby. But it is important to recognize that the immediate, visible, and health-threatening problems associated with poor informal settlement are just one form of urban environmental burden.

> While poor urban settlements create local impacts that affect the health of their residents, it is wealthier city dwellers whose environmental "footprints" are felt around the globe.

Gordon McGranahan is head of the Human Settlements Group at the International Institute for Environment and Development.

URBAN ENVIRONMENTAL BURDENS:
LOCAL, REGIONAL, AND GLOBAL

Economic growth is associated with a transition from predominantly rural populations and agricultural employment to urban populations and industrial or service employment. It is also associated with a transition from local environmental health burdens that impact health directly toward global environmental burdens that impact health through their effects on life-support systems.[5, 6, 8] (See Figure 10.1.) This urban environmental transition is summarized below.

Much empirical research on the relationship between environmental burdens and economic status has focused on what is called the "environmental Kuznets curve" (EKC),[1, 9, 12, 17] which takes the form of the inverted U-shaped curve in the center of Figure 10.1. The term harks back to the suggestion by Nobel Prize–winning economist Simon Kuznets that income inequality first increases and then decreases in the course of economic growth. Proponents of EKC theory believe that rising affluence is first accompanied by an increase in environmental burdens but that those burdens decline over time as, for example, wealthy citizens demand improved environmental quality. Most research has failed to confirm the existence of the EKC.[3, 12]

As shown in Figure 10.1, the relationship between affluence and urban environmental burdens is considerably more complex than the simple curve of the EKC. Generally, the main *household-level* burdens, including inadequate access to water and sanitation, and indoor air pollution all fall monotonically with income—the "local" curve in the figure. The main *global* burdens, including greenhouse gas emissions and ecological footprints, tend to rise with income, as shown in the "global" curve. On the other hand, many of the *regional* burdens, including urban air pollution, have been found to display the inverted U-shape of the EKC.

The figure can also be viewed as reflecting change over time. For countries that industrialized in the nineteenth century, the declining local curve can be taken to represent the sanitary revolution of the late nineteenth and early twentieth centuries; the declining part of the city-regional curve represents the pollution revolution of the mid/late-twentieth century. (A hope that the global curve will turn is implicit in the contemporary aspirations for a sustainability revolution.) Countries that have industrialized and motorized more recently have experienced somewhat different trajectories.

Historically, the urban transition has meant the physical displace-

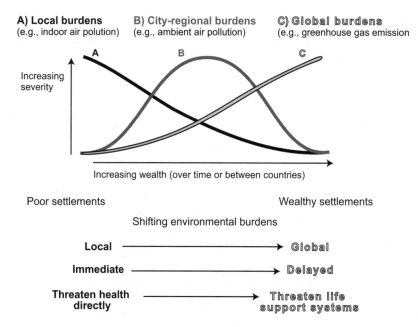

FIGURE 10.1 The shifting scale of urban environmental burdens associated with increasing affluence.

Source: McGranahan et al., 2001, *The Citizens at Risk.*

ment of environmental burdens.[7] Sewers are used to collect fecal material and release them beyond urban boundaries. Higher stacks are used to disperse air pollution. Water and other resources are brought longer distances to urban centers as local sources run out. More generally, increasing consumption puts more pressure on the environment, but growing affluence gives people the capacity to avoid the negative environmental consequences of their own consumption.

> Increasing consumption puts more pressure on the environment, but growing affluence gives people the capacity to avoid the negative environmental consequences of their own consumption.

One recent study estimated environmental health impacts by measuring disability adjusted life years (DALYs) lost for different income groups.[11] The results supported the claim above: Affluence is associated with lower local environmental health risks, higher global risks (i.e., greater contributions to climate change), and first rising and then declining "community environmental risks." The results implied, further, that while contributions to climate change increase with economic status, the burden *borne* per capita actually declines. For example, the climate-change-related health burden suffered per unit of carbon emitted by sub-Saharan Africans was over *7,000 times greater* than for North Americans.

It is important to recognize that neither the improvements in local burdens nor the turnaround of regional burdens, nor the hoped-for turnaround in global burdens, are driven by affluence alone. The sanitary revolution of the urbanizing world of the nineteenth century was driven by a social movement, particularly in its early stages, as was the pollution revolution of the twentieth century. We cannot assume that these problems will be solved by market forces, which tend to neglect environmental burdens (or, as economists call them, externalities).

SEIZING THE ENVIRONMENTAL OPPORTUNITIES OF URBANIZATION

While it is important not to blame urbanization for all the ills that arise in cities, in practical terms it is important to seize the opportunities that urbanization provides. Such opportunities arise at every scale, from the local environmental burdens associated with poverty to the global burdens of affluence.

Local Environmental Burdens

When Europe and North America began to urbanize, urban dwellers were far less healthy than their rural counterparts.[16] Economic growth did little to improve the situation until governments were compelled to improve living conditions in deprived areas.[13] The sanitary movement of the nineteenth and early twentieth centuries played an important role in documenting the dangers of poor sanitation and identifying the means for implementing improvements. Urbanization sped the sanitation revolution: Water and sanitation systems cost less (per capita) in dense urban settlements than in dispersed rural settlements. Innovations in the technologies and governance of networked water and sanitation provision eventually helped to turn the urban health penalty into an urban health advantage. But today, many urban dwellers still live in very unhealthy conditions. For the most part, however, the barriers to improvement are more social and political than technical and financial.[2]

Regional Environmental Burdens

Many of the activities that drive regional environmental problems are concentrated in and around urban areas, but just as urban slums and squatter settlements should not be taken to imply that poor people should stay in rural areas, so urban concentrations of polluting and resource-depleting enterprises should not be taken to imply that these enterprises should be located in rural areas. When, for example, China promoted township and village enterprises to stimulate the rural economy, they proved to be especially polluting and resource inefficient.[10]

Urban activities are often easier to monitor and regulate, and when economic enterprises are located in the vicinity of large human populations, the political pressure to reduce pollution is likely to be higher. The environmental movement of the mid-twentieth century, which focused heavily on pollution issues, was essentially an urban movement, like the sanitary movement before it. When it comes to resource-consuming enterprises, and the interruption of ecological cycles, urban location is more problematic. Again, however, it is also important to recognize and exploit the advantages that urban location can bring, and not to assume that the burdens imposed by urban activities are inherent to their urban location.

Global Environmental Burdens

The consumption that drives most global environmental burdens is also concentrated in urban areas, but urban density provides opportunities for reducing those burdens that rural dispersion does not. For example, urban settlement is generally more suitable for public transportation and reduced travel distances. This advantage is lost when urban development is characterized by sprawl, and density declines. However, with the right transport infrastructure, and fuel prices that reflect environmental costs, urban transport demands can be greatly reduced. Urban transport infrastructure is in many ways the equivalent for global sustainability to what urban water and sanitation infrastructure is for local environmental health conditions.

ENVIRONMENTAL JUSTICE AND URBAN TRANSITIONS

Urbanization is associated with a model of development that imposes serious environmental burdens, and it shifts the locus of these burdens from the local to the regional to the global scale. This should not be taken to imply that de-urbanization will reduce environmental burdens, however. At least in the short run, it is critical to seize the opportunities that urbanization provides at every scale, from the local up to the global. Politically, the challenge of organizing improvements in water and sanitation in the world's slums is quite different from that of organizing reductions in carbon emissions driven by urban affluence, but in both cases there are important issues of environmental justice to consider.

A concern for environmental justice emerged out of North American civil rights politics, but the concept is now used in a wide range of political settings, and it is being extended to international and intergenerational burdens.[15] Environmental justice provides a powerful lens for viewing the relations between environmental burdens and economic status.

The spatial displacement of environmental burdens amplifies inequalities in economic status. Locally, the poorest urban households pollute each other with bad sanitation and smoky cooking fuels, while also living downstream or downwind of pollution from sewers or power stations serving the more affluent. Internationally, the urban centers of low-income countries are more vulnerable to climate change but contribute far less to greenhouse gas emissions. Not only is the distribution of consequences of this spatial displacement unjust, but many of the underlying actions—from building a latrine too close to a neighbor's well to emitting large quantities of greenhouse gases—are widely considered immoral.

The temporal displacement inherent in global burdens, such as climate change, also raises issues of environmental justice. On one hand, it would be unjust to create poverty in the future in order to secure affluent lifestyles today. On the other, it would be equally unjust to force poor groups today to make sacrifices for better-off generations in the future. There are inherent uncertainties in the trade-offs being made, but there is clearly a moral case to be made for redressing the injustice evident in urban environmental burdens.

Population growth complicates issues of environmental justice but does not radically alter these conclusions. Population growth is higher among poorer groups—and the implications of this are different in the short and long runs.

In the short run, the impact of existing population growth patterns on global resource and climate issues is minimal—the multifold differences in the ecological footprint of affluent and poor populations swamps the differences in their population growth rates. The more obvious concern is whether population growth is undermining efforts to secure adequate home and work environments for the urban poor or adding to regional environmental burdens.

There is a danger that prescriptive approaches to reducing fertility among poor urban residents will simply reinforce prejudice and add to the burdens they face. On the other hand, better reproductive health, women's education, and support for voluntary family planning will improve living conditions in deprived neighborhoods. The challenge is to distinguish one from the other, and to support only the latter.

In the long run, the impact of population growth on global resource scarcity and climate change depends heavily on whether the growing populations go on to succeed economically and to adopt environmentally burdensome affluent lifestyles. This certainly does not provide a legitimate moral basis for introducing coercive population policies in the present. Indeed, many would argue that it would be highly immoral to prevent poor people from procreating, on the grounds that their progeny might become affluent. True, environmental considerations remove any residual credibility from the claim that the poor can just follow the path of the currently affluent. However, this does not negate the primary role that the affluent have played in undermining the ability of the planet to meet the needs of others, in either the present or the future.

Environmentalists sometimes argue that reducing global burdens must take precedence over distributional concerns, since otherwise humans won't survive. The extreme moralist could reply that if humans don't take distributional concerns just as seriously, they don't deserve to survive. Hopefully, we will not only survive but also demonstrate that we deserve to.

REFERENCES

1. Dinda, Soumyananda, 2004, Environmental Kuznets curve hypothesis: A survey, *Ecological Economics* 49(4): 431.

2. Dye, Christopher, 2008, Health and urban living, *Science* 319(5864): 766–769.

3. Ekins, Paul, 2000, *Economic Growth and Environmental Sustainability*, Routledge, London.

4. Hardoy, Jorge E., Diana Mitlin, and David Satterthwaite, 2001, *Environmental Problems in an Urbanizing World*, Earthscan, London.

5. McGranahan, Gordon, 2007, Urban Environments, Wealth and Health:

Shifting Burdens and Possible Responses in Low- and Middle-Income Nations, Urban Environment Discussion Paper, International Institute for Environment and Development, London.

6. McGranahan, Gordon, 2007, Urban transitions and the displacement of environmental burdens, p. 18–44, *Scaling Urban Environmental Challenges: From Local to Global and Back*, Peter J. Marcotullio and Gordon McGranahan, eds., Earthscan, London.

7. McGranahan, Gordon, Pedro Jacobi, Jacob Songsore, Charles Surjadi, and Marianne Kjellén, 2001, *The Citizens at Risk: From Urban Sanitation to Sustainable Cities*, Earthscan, London.

8. McGranahan, Gordon, Peter J. Marcotullio, Xuemei Bai, Deborah Balk, Tania Braga, Ian Douglas, Thomas Elmqvist, William Rees, David Satterthwaite, Jacob Songsore, and Hania Zlotnik, 2005, Urban systems, p. 795–825, *Ecosystems and Human Well-Being: Current Status and Trends*, Rashid Hassan, Robert Schole, and Neville Ash, eds., Island Press, Washington, DC.

9. Nahman, A., and G. Antrobus, 2005, The environmental Kuznets curve: A literature survey, *South African Journal of Economics* 73(1): 105–120.

10. Smil, Vaclav, 1997, China shoulders the cost of environmental change, *Environment* 39(6): 6–9.

11. Smith, K. R., M. Ezzati, 2005, How environmental health risks change with development: The epidemiologic and environmental risk transitions revisited, *Annual Review of Energy and Resources* 30:291–333.

12. Stern, David I., 2004, The rise and fall of the environmental Kuznets curve, *World Development* 32(8): 1419–1439.

13. Szreter, Simon, 2005, *Health and Wealth: Studies in History and Policy*, University of Rochester Press, Rochester, NY.

14. UN-Habitat, 2003, *The Challenge of Slums: Global Report on Human Settlements 2003*, UN Human Settlements Programme, Nairobi.

15. Walker, Gordon, and Harriet Bulkeley, 2006, Geographies of environmental justice, *Geoforum* 37(5): 655–659.

16. Woods, Robert, 2003, Urbanisation in Europe and China during the second millennium: A review of urbanism and demography, *International Journal of Population Geography* 9: 215–227.

17. Yandle, Bruce, Madhusudan Bhattarai, and Maya Vijayaraghavan, 2004, *Kuznets Curve for the Environment and Economic Growth*, Poverty and Environment Research Centre (PERC), Bozeman, MT.

CHAPTER 11

....

The New Economics of
Population Change

RACHEL NUGENT

Population change and economics are so closely intertwined that it is difficult to know which has greater influence on the other. Nonetheless, some generalizations can be made: Around the world and over the years, rapid population growth has consistently been associated with slow economic growth. This means that economic planning needs to take into account population change and its implications.

The links between population change and economics are not obvious at first glance, because population change comes from three different sources: fertility, migration, and mortality. Changes in each have different economic consequences. This chapter will focus on the fertility component of population change, because the fertility rate is a critical factor in development for poor countries, especially in the twenty-eight developing countries that are still experiencing very high fertility.

First, we will try to disentangle the links between fertility-related population change and economic outcomes in countries and in families. The goal is to describe what is known about that relationship and then to indicate where population policy can improve both microeconomic and macroeconomic outcomes.

Rachel Nugent is Deputy Director for Global Health at the Center for Global Development.

POPULATION POLICY: NOT JUST FAMILY PLANNING

Population policy and family planning are sometimes conflated, but effective population policy must encompass much more than family planning. For example, policies that improve public health—particularly of women and children—are also crucially important in affecting population growth rates and human well-being. Less obvious are the ways in which governmental policies affect the economic, social, and educational choices of young people, which in turn help determine their family planning and reproductive health behaviors.

> Persistently high fertility rates impede economic progress, especially for women.

It is clear that countries vary enormously in their needs and their ability to absorb additional people and provide them with a satisfying life. However, significant economic benefits can be derived from maintaining fertility rate at or near replacement level. Those benefits are not guaranteed—they depend on employment and educational opportunities being widely available—but it is virtually certain that persistently high fertility rates impede economic progress, especially for women.

Population trends and economics interact at both a macro level—that of the whole country—and the micro level of individuals and households. At the macro level, a large and growing economy can provide for a large and growing population, and a simple measure of whether they are in balance is to divide total economic output (GDP or gross domestic output) by the number of people it takes to produce per capita GDP. If that figure is enough to provide all members of a society with adequate resources, then there may be no economic justification for active governmental policies to influence population change. But if per capita GDP is low relative to needs, or GDP is distributed very unevenly across the population of a country, there may be reason for concern about population growth.

THE FERTILITY-POVERTY CONNECTION

Research does not show conclusively that rapid population growth causes poverty, but we see across all developing countries over time a strong inverse relationship between fertility and per capita income, and between fertility and life expectancy—two common indicators of well-

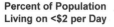

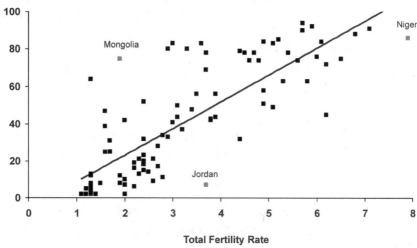

FIGURE 11.1 Association between fertility and poverty.

Source: Population Reference Bureau, *Population and Economic Development Linkages 2007 Data Sheet* © Population Reference Bureau.

being. Lower fertility is associated with higher per capita income and higher life expectancy. There is also a clear connection between high fertility and poverty, as Figure 11.1 demonstrates. These associations indicate that a *concomitant* of lower fertility is improved well-being—through higher incomes, higher life expectancy, and lower poverty rates. However, they do not indicate which comes first: low fertility or higher income and life expectancy.

The fertility-poverty link is even stronger in the subsample of twenty-eight countries with very high fertility—more than five children per woman. Those countries have lower per capita income and lower life expectancy than developing countries as a whole. In fact, despite some variation across those countries in their fertility rates, there is the appearance of a high-fertility poverty trap, in which low incomes may exacerbate high fertility rates and vice versa. Almost every country in this category shows life expectancy of less than

> Many countries are caught in a high-fertility poverty trap, in which low incomes may exacerbate high fertility rates and vice versa.

fifty-five years, implying a vicious cycle wherein low life expectancy leads to high fertility and vice versa.

These traps are plausibly caused by poor health (high infant and maternal mortality, high infectious diseases rates, inadequate public health services), and low standards of living (poor employment opportunities, malnutrition, low rates of educational achievement), all of which may be exacerbated by—or causes of—high fertility.

Policies in these countries must strive to break the cycles that create poverty and low–life expectancy traps. There are many routes to achieve this, including poverty alleviation programs, improved educational and employment opportunities, and improved access to family planning and other health services.

FERTILITY AND ECONOMICS AT THE HOUSEHOLD LEVEL

Personal choices about family size have significant impact on individual economic circumstances—and vice versa. Understanding how people make these choices is key to designing appropriate policies that can help people achieve the family size they want.

The family size decision is usually seen by economists as a choice between quantity (number of children) and quality (investment in each child). This view holds that a larger family size means that each child receives a smaller share of a fixed household pie (i.e., lower quality). Smaller numbers of children allow for greater investment of family resources in each (i.e., higher quality). These contributions can be financial investments (for schooling, health care, etc.), as well as parental time and emotional nurturing. The quality/quantity trade-off framework offers a useful first cut at understanding some of the factors that determine people's choices, but it has significant limitations.

For example, the quality/quantity model assumes that fertility decisions are made jointly by both partners in a couple. But in reality, men and women often have different preferences about family size. For women, the time commitment involved in having and raising children is almost inevitably greater than for men; hence, women face a higher opportunity cost. Less chance at schooling, less paid employment, and even greater health risks factor into women's childbearing decisions. Women often desire fewer children than their husbands, and in many countries, both spouses want fewer children than they actually have. The gap between desired and actual fertility hints at another problem with this model: Much conception and childbearing is unplanned and even unwanted, and it often does not reflect a rational decision-making process of any kind.

Moreover, the choice between quality and quantity is not always so stark. Recent research shows that larger caregiving networks and more gender-equitable conditions within the family affect child outcomes—perhaps even more than the number of children in the family.[1] For instance, in countries like Kenya, where the costs of unanticipated childbearing are spread across large family networks, the "quality" costs of large families are not as pronounced.[2]

Still, compared with children from small families, children in large families generally do worse in school and on tests, have poorer health, and are less likely to survive to adulthood. Further, larger families invest less in each child's education. While there is consistency in the negative correlations found in these studies, they do not firmly establish that large family size causes the negative individual outcomes.[3] Further research is underway to better understand the interactions.

A still-disputed question is whether poor couples choose large families because they rely on the work and income potential of children, especially as a form of old-age insurance. But research indicates that lower fertility requires access to education, health, and other services that make it feasible for people to invest in their children. Parents need a reasonable expectation that the returns on those investments might eventually reach them, especially in times of emergency or in their old age. Infant and child mortality also affect childbearing decisions; where many children die before reaching maturity, parents may choose greater fertility in order to ensure that a minimum number of children survive.

> Lower fertility requires access to education, health, and other services that make it feasible for people to invest in their children.

Rwanda provides an example of many factors that influence fertility. With many of its citizens living in poverty (57 percent), a high proportion of people engaged in agriculture (80 percent), and rapid population growth (3.5 percent per year), Rwanda's fertility rate remains high at 6.1 children per woman. This high fertility persists despite free education, high enrollment in the primary grades, and a fast-growing economy. However, infant and child mortality are extremely high (8.9 percent and 15.2 percent in 2005), and the memory of war lingers.[4] In these circumstances, many people still see children as a necessary way to increase household income, and they try to replace children lost through premature death. Recent data indicate that desired family size is much smaller (4.5) than the actual size and suggest that improved reproductive health care and child survival could alter choices about replacement fertility.

POPULATION POLICIES CAN
IMPROVE MICROECONOMIC OUTCOMES

The above discussion suggests many ways in which policies can affect household and individual choices about family size. First, policies that alter the costs or perceived benefits of having children will affect the quantity/quality trade-off made by couples. For instance, the presence of schools and health clinics better enables parents to materially improve well-being by investing in their children. Second, improved basic health care that reduces infant and child mortality will affect fertility rates by giving parents the confidence that their children will survive, thereby narrowing the gap between actual births and the desired number of children.

Finally, helping couples achieve their desired fertility through family planning is a sine qua non for development success. Research clearly shows that family planning is essential to girls' and women's involvement in education and economic activity—and therefore to sustainable economic development. Particularly, improved access to family planning among the poor is likely to have the greatest impact on poverty, as the poor tend to have greater unmet need and the lowest contraceptive prevalence rates (Figure 11. 2).

> Family planning is essential to girls' and women's involvement in education and economic activity—and therefore to sustainable economic development.

Bangladesh offers an example of how family planning programs and policies can vastly improve equity in access to contraception and thereby over the long run improve economic outcomes for individuals and households, even in the poorest countries.

A recent study looks at the development benefits associated with the Matlab family planning program in Bangladesh.[5] The long study period (twenty-five years) and experimental design give the findings special significance.

The authors found that fertility levels in the program area were 15

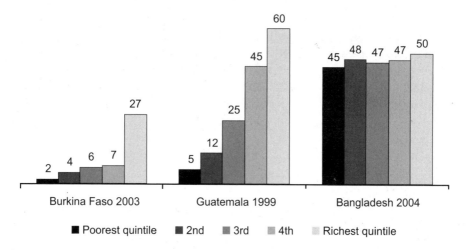

FIGURE 11.2 Use of modern contraception by wealth. Married women aged 15–49 using modern contraception (percent).

Source: Population Reference Bureau, *Population and Economic Development Linkages 2007 Data Sheet* © Population Reference Bureau.

percent lower than in the comparison area. Reducing the cost of contraceptives, particularly for poorer women who may not be able to buy contraceptives or for whom the costs of family planning are especially high, appears to make a significant difference for voluntary reduction of fertility (see Figure 11.2). The study's authors also found substantial non-fertility benefits from the Matlab program, including improvements to women's health and economic well-being—benefits not present in the comparison area.

In sum, good population policies support individuals and families in achieving their desired family size and structure, while also ensuring the economic resources to achieve a better quality of life. These conditions will be more likely if national and international development policies are providing the employment, educational, and financial opportunities that underpin a healthy economy.

MEASURING THE MACROECONOMIC OUTCOMES OF FERTILITY DECISIONS

The aggregation of all fertility decisions made by individuals will affect a country's overall economic conditions. There is no optimal population growth rate for economic well-being, but population growth that exceeds annual GDP growth leaves little room for increases in consumption, investment, or government services. Conversely, very slow or negative population growth will eventually leave a country bereft of a young, dynamic, innovative workforce that is the main building block of intellectual and economic progress. Thus, macroeconomic improvements and progress are determined directly by individual fertility choices.

A simple way to measure the impact of population growth—whether positive or negative—on a country's macroeconomy is the "dependency ratio," calculated as the number of dependent people per working-age adult. A country with a fast-growing population should worry more about its "youth dependency ratio," while a country with a very slow-growing population should be watching the "old-age dependency ratio." Again, there is no ideal number in either case, because the private and social costs of dependents, whether children or elderly, vary widely from place to place. But a higher dependency ratio implies that each worker is providing for a greater number of nonworkers in society.

African countries have high youth dependency ratios due to high fertility rates. At present, only six African countries have dependency ratios less than 0.5, meaning that in those countries each child has two workers or more supporting his or her needs; in the rest of the continent there are fewer than two workers supporting the needs of each child.* In Nigeria, for example, some 48 percent of the population is in the dependent categories, so a large share of the country's resources are being consumed rather than invested in long-term development.[6] It is fair to say that Nigeria's rapid population growth is slowing economic growth. However, the trend in Africa is toward a decline in dependency ratios, so by 2030 about half of the countries in Africa will have more than two workers supporting each child.[7]

In contrast, developing countries in Europe and the former Soviet Union have rapidly aging populations and very low youth dependency ratios. No country in this group has fewer than two working-age adults per child, and most have four to five adults per child. These countries should be able to support the needs of their young populations for edu-

---

*The youth dependency ratio overestimates the economic output available to each child, as it ignores the elderly and other dependent populations in society. However, it is a good measure of the demands placed by recent fertility trends on the macroeconomy.

cation and health care, as well as devote some economic resources to providing for infrastructure and other priorities. But these and other countries with aging populations face the risk that their populations will eventually be overburdened with nonworking elderly, and too few workers to support their needs.

THE DEMOGRAPHIC WINDOW OPENS

Fertility decline offers developing countries the opportunity to benefit from lower dependency ratios; that is, fewer dependents—either very old or very young—relying on the productive workforce for support. This is sometimes called a "demographic bonus" or "demographic window," and it occurs through two channels: a labor effect and a savings effect. When a country enters a period of lower dependency, a higher proportion of people are in their productive years, and GDP per capita rises. This is the labor part of the bonus. The savings component arises because workers have less need to support large numbers of children or (grand)parents and can therefore put more of their earnings into savings than into consumption. The macroeconomy benefits in the long run from greater savings per capita.

Countries may use the demographic bonus period to accelerate economic growth and create a mutually reinforcing "virtuous circle" of high economic growth and lower population growth. This is the opposite of the population/poverty trap that has mired many poor countries in economic stagnation or worse.

Figure 11.3 shows periods of past and future demographic bonus opportunity for selected developing countries. For example, about one-third of economic growth in east Asia in the 1980s–1990s has been attributed to the demographic bonus.[8] East Asian countries were able to capture that bonus by instituting beneficial policies—for example, substantial investments in education and public health. Latin America, with a similar demographic window, failed to benefit economically from a low dependency ratio, growing at an annual rate of only 2 percent compared with the 6 percent growth in east Asia. Many African countries have not yet entered the demographic window but have little time left to develop the policies and institutions needed to exploit it.

Countries can capture the demographic bonus by ensuring that adequate numbers of good jobs are available for their expanded labor forces and by investing the returns in social capital with long-term payoffs. As an example, the Republic of Korea met this challenge in the

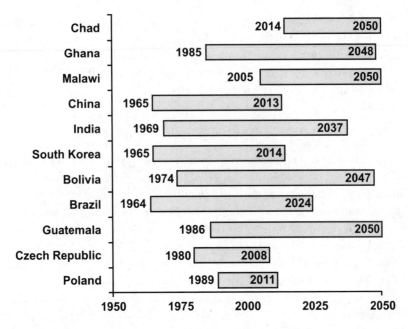

FIGURE 11.3 Periods of past and future demographic bonus opportunity for selected developing countries.

later part of the twentieth century by increasing the availability of secondary and tertiary education, after slowing birth rates lowered the need for primary education during the 1970s. Counterexamples from the Middle East are described in Box 11.1.

WHAT ABOUT COUNTRIES EXPERIENCING FERTILITY DECLINE?

The economic impacts of persistent fertility decline are less well known than the impacts of population growth, as developing countries are just beginning to reach this stage of the demographic transition. A large number of countries in the former Soviet Union, as well as some Asian countries, are now experiencing population stagnation or decline.[9]

Just as population momentum imposes a challenge on countries that have experienced high population growth, it also exposes countries with low or negative population growth to long-term effects. Unless low birth rates are offset by in-migration, these countries will have ever-smaller age cohorts, and even if their reproductive rates rise above

BOX 11.1

Up Next: The Middle East

Which path will the countries of the Middle East[a] take? They now face a "demographic window" where working-age populations will significantly outnumber the elderly and children in the country for the next several decades. That window could spur economic dividends or trigger political and social instability—depending on whether these countries make the right investments in their young populations today.

The outlook, generally speaking, is not bright. Too many Middle Eastern youth are excluded and marginalized by a lack of opportunity. For example, although Egypt's economy has improved in recent years, youth are at high risk for unemployment and job insecurity, especially young females. While female school enrollment has increased in the last few decades, women are still four times as likely to be unemployed as males.

In Syria, the situation is similar. In 2002, youth accounted for three out of four unemployed workers. While many young people pursue higher education in order to get public-sector jobs with security and benefits, they often face disappointment because those jobs are very scarce. Finally, the legal system in Syria creates rigid labor laws that make it difficult for the private sector to hire the young and also fosters difficulties in access to housing and credit.

But there is hope: A recent study found that youth unemployment, school dropouts, adolescent pregnancy, and youth migration in Middle Eastern countries could be reduced by as much as 60 percent by putting existing resources to better use, particularly by improving health and education for young people.[b]

....

[a] The term *Middle East* is often interchanged with *Near East* in UN definitions. Here, the term includes Iran, Syria, Morocco, Egypt, Algeria, Bahrain, Jordan, Kuwait, Lebanon, Qatar, Saudi Arabia, West Bank/Gaza, and Yemen, where data permits.

[b] Chaaban, Jad, 2008, The Costs of Youth Exclusion in the Middle East, Middle East Youth Initiative Working Paper Number 7, Wolfensohn Center for Development and Dubai School of Government, Washington, DC.

replacement levels, low or negative population growth will have economic consequences—the reverse of the "bonus" effects described above for high–population growth countries. These include a declining labor force, lower savings, and less investment. (Of course, as James Gustave Speth explains in Chapter 26, there are potential environmental and social benefits to slower population and economic growth.)

SUMMARY OF POPULATION POLICIES THAT
IMPROVE ECONOMIC CONDITIONS

For countries at every stage of the demographic transition, economic success depends on the ability to balance the needs of workers and non-workers, young and old:

- Employment opportunities must be plentiful enough to absorb large numbers of workers in the early stage of demographic transition; in later stages of the transition, capital investments must be sufficient to accommodate a smaller workforce.

- Changes over time in the proportion of young to old people in a country require mechanisms and institutions to transfer wealth across generations, for instance, to better support the creation and expansion of schools when a population is young and to better support old age pensions when a population is older.

- And finally, the choice and opportunity sets of the rich are dramatically greater than for the poor. Yet, if a country is to achieve a desired population growth rate, policy must step in to reduce that gap. This includes ensuring access to family planning, which will reduce the high rates of unmet need in high-fertility countries, and encouraging social conditions that enable women to have greater decision-making authority in the family over childbearing.

REFERENCES

1. Hobcraft, John, 2003, in Nancy Birdsall, Allen Kelley, and Steven W. Sinding, eds., *Population Matters*, University Press, Oxford.

2. Montgomery, Mark, and Cynthia Lloyd, 1997, Excess Fertility, Unintended Births, and Children's Schooling, Working Paper Number 100, Population Council, New York.

3. King, Elizabeth, 2003, in Birdsall, Kelley, and Sinding, *Population Matters*.

4. Demographic and Health Surveys, available online at http://www.measuredhs.com/topics/infant_child_mortality.cfm.

5. Joshi, Shareen, and T. Paul Schultz, 2007, *Family Planning as an Investment in Development: Evaluation of a Program's Consequences in Matlab, Bangladesh*, Economic Growth Center, Yale University, New Haven, http://www.econ.yale.edu/growth_pdf/cdp951.pdf, accessed on August 4, 2008.

6. Onwuka, Emmanuel C., 2006, Another look at the impact of Nigeria's growing population on the country's development, *African Population Studies* 21(1): 1–18; Population Reference Bureau, 2007, *World Population Datasheet 2007*, PRB, Washington, DC.

7. Population Reference Bureau, 2007, *World Population Datasheet 2007*, PRB, Washington, DC.

8. Bloom, David E., and Jeffrey G. Williamson, 1998, Demographic transitions and economic miracles in emerging Asia, *World Bank Economic Review* 12: 419–455.

9. Population Reference Bureau, 2007, *World Population Datasheet 2007*, PRB, Washington, DC.

CHAPTER 12

··········

Food

Will There Be Enough?

LESTER R. BROWN

In early 2008, soaring food prices sparked unrest in more than twenty countries.[1] Violent riots convulsed Haiti, killing four and unseating the prime minister.[2] In Cameroon, food price riots left forty people dead.[3] And in Egypt, after food prices doubled, eleven were killed in clashes while standing in line to buy subsidized bread.[4]

Some of the contributing factors to the 2008 food price spike were temporary; however, in the longer term, humanity's ability to feed itself is far from assured. Today farmers contend with shrinking cropland, shortages of irrigation water, diminishing returns from fertilizers, climate change, rising fuel costs, and a dwindling backlog of yield-raising technologies. At the same time, they face fast-growing demand from the more than 70 million people added to the world's population each year, the desire of billions to "move up the food chain" and consume more grain-based meat and dairy products, and now the millions of motorists turning to crop-based fuels to supplement tightening supplies of gasoline and diesel.[5] As a result of these converging trends, world grain production has fallen short of consumption in six of the last nine years, drawing down world grain stocks.[6]

Lester R. Brown is President of Earth Policy Institute and author of Plan B 3.0: Mobilizing to Save Civilization *(W. W. Norton, NY: 2008), from which this chapter was adapted.* Plan B 3.0 *is available online for free downloading at www .earthpolicy.org/Books/PB3/index.htm.*

Food crises are nothing new. Human history is replete with dire warnings of mass famine, and, frequently, the real thing. Still, despite persistent hunger, humanity has made dramatic progress toward food security in the last half century. Production of the "big three" grains—wheat, rice, and corn—has more than tripled since 1950, increasing from 630 million to 2 billion tons.[7] These dramatic gains were achieved in two ways: by *increasing the number of acres under cultivation* by clearing forests and grasslands, and by *boosting the productivity of each acre* with irrigation, high-yield crops, new farming practices, and synthetic fertilizer. But, the prospects for continuing gains on both fronts are limited.[8] As a result, farmers will be hard-pressed to maintain current levels of production, much less make the increases necessary to accommodate a growing population. Already, growth in world grain production has slowed. The rate of increase in the yields of the big three crops fell below the growth in world population after 1984.[9] And the number of hungry people in the world—which was greatly reduced from 1950 to 1984 and continued to decline until the late 1990s—has since turned upward.[10]

THE CLOSING FRONTIER

In the past, humans have confronted food shortages by finding new areas to farm. But today, the frontier is closed and arable land is increasingly scarce. Most of the world's readily farmable acres are already under the plow or covered with cities. Worse, a new UN study shows that some 20 percent of the world's cropland has been degraded by human activity—such as poor farming practices, pollution, and over-grazing—and is now unsuitable for crops.[11] One-quarter of humanity—1.5 billion people—depends directly on land that is being degraded.[12] And land degradation is only getting worse; because of poor land management, ecosystem health and productivity are declining.[13]

Water may pose an even more serious constraint on expanded food production. Indeed, lack of water already limits the amount of new land that can be farmed. Water for agriculture is supplied by rain, from surface water (including snowmelt and ice melt), or from underground aquifers. Given the finite amount of cropland that receives sufficient rain, most of the great expansion of farmland in the last fifty years was made possible by a near-tripling of irrigation—diverting water from rivers and drawing down aquifers.[14] This explosion of irrigation has made farms the largest global consumer of water, comprising 70 percent of annual freshwater use.[15]

The link between water and food is strong: It takes 1,000 to 2,000 liters of water to produce a kilogram of wheat and 13,000 to 15,000 liters to produce the same quantity of grain-fed beef.[16] But farms face growing competition for water from urban areas and industry.[17] Supplies of irrigation water are beginning to shrink in some countries as wells go dry and scarce water is diverted to cities. For the first time, harvests in large countries such as China are being reduced by water shortages.[18]

In many parts of the world, irrigation water sources are being used up much faster than they can be replenished. Scores of countries, including major grain producers such as India, China, and the United States, are overpumping aquifers as they struggle to satisfy their growing water needs. A groundwater survey released in Beijing in August 2001, for example, revealed that the water table under the North China Plain, an area that produces over half the country's wheat and a third of its corn, is falling fast.[19] In India, water tables are falling in almost every state.[20] And in the United States, the massive Ogallala aquifer, which supplies about 30 percent of the groundwater used for irrigation in the country, is so heavily overdrawn that water levels have dropped by more than 100 feet in some areas.[21] The Ogallala is a "fossil" aquifer, meaning that its waters were laid down millions of years ago, and it recharges slowly, if at all. Like fossil fuels, fossil water is a finite resource—when it's gone, it cannot be replaced. Farmers who lose their irrigation water have the option of returning to lower-yield dryland farming if rainfall permits. But in more-arid regions, such as in the U.S. Southwest or in the Middle East, the loss of irrigation water means the end of agriculture.

> The great expansion of farmland in the last fifty years was made possible by irrigation. But in many parts of the world, irrigation water sources are being used up much faster than they can be replenished.

GREEN REVOLUTION 2.0?

As prospects for new irrigated farmland shrink, it is time for fresh thinking on how to improve the productivity of existing cultivated land without harming the environment—time for a truly Green Revolution. Here, there are possibilities. For example, farmers can plant crop varieties that are more tolerant of drought. And, where soil moisture permits, farmers can increase the area of "multicropped" land that produces more than one crop per year. Indeed, the tripling in the world grain harvest since 1950 is due in some degree to impressive increases in multiple cropping in Asia.[22] With water shortages emerging as an important constraint on food production growth, the world needs a focused effort to raise water productivity similar to the one that nearly tripled land productivity during the last half of the twentieth century (see Chapter 14).

The first Green Revolution was made possible by the development of high-yield plant varieties. Average grain yields, under favorable conditions, have quadrupled over the last half century.[23] But now plant breeders are confronting biological limits to further productivity. As one agricultural expert has observed, "in the 1950s and 1960s, it was not difficult to anticipate the sources of the increase in agricultural production. . . . I find it much more difficult to tell a convincing story about the sources of increase in crop production over the next half-century."[24]

Synthetic fertilizer was another star of the first Green Revolution. In 1847, Justus von Liebig, a German chemist, discovered that all the nutrients that plants remove from the soil could be replaced in chemical form. This insight had little immediate impact on agriculture, partly because growth in world food production during the nineteenth century came primarily from expanding cultivated area.[25] But by the second half of that century, the use of nitrogen fertilizer increased exponentially, enabling huge increases in the production of relatively cheap food and improving the well-being of millions.[26] Norman Borlaug, the scientist who is often called "the father of the Green Revolution," notes the importance of fertilizer for feeding the world's people: "Without chemical fertilizer, forget it," he said. "The game is over."[27] Many farmers got a taste of life without fertilizer in 2008, when prices tripled and shortages were widespread, contributing to the food crisis that roiled many developing countries.[28] Such price increases and shortages are likely to recur, since nitrogen fertilizer is synthesized with natural gas, a fossil fuel in ever-shorter supply.

Synthetic fertilizer has had enormous benefits for humanity, but those gains have come at substantial cost to the ecosystems we depend on. Nitrogen in fertilizer not taken up by plants leaches into streams and groundwater or is released into the atmosphere—in both cases, severely disrupting ecological systems.[29] Some of the nitrogen returned to the atmosphere forms nitrous oxide (N_2O), which contributes to climate change and stratospheric ozone depletion.[30] And nitrogen in waterways nurtures the growth of aquatic plants, like algae, that deplete oxygen when they decompose. The resulting "dead zones" cannot support other aquatic life. There are now 405 such zones worldwide, up from 49 in the 1960s.[31] The number of dead zones is doubling every decade, contributing to the decline of fisheries—another critical source of food for a growing human population.[32]

CARS AND CLIMATE CHANGE: NEW THREATS TO FOOD SECURITY

The difficulties of finding new cropland and increasing productivity are as old as agriculture. Today, we face two new challenges. First, there is the massive diversion of crops for fuel. In a misguided effort to reduce its oil insecurity by converting grain into fuel for cars, the United States, the world's breadbasket, has helped drive up world grain prices. What happens to the harvest of the United States, the world's leading grain exporter, is obviously of concern to the entire world. Leading importers like Japan, Egypt, and Mexico will be particularly affected by any reduction in U.S. corn exports.

> The grain required to fill an SUV's 25-gallon tank with ethanol just once will feed one person for a whole year. . . . In 2008 a quarter of the U.S. grain harvest was turned into fuel ethanol.

From an agricultural vantage point, the world's appetite for crop-based fuels is insatiable. The grain required to fill an SUV's 25-gallon tank with ethanol just once will feed one person for a whole year. If the entire U.S. grain harvest (including corn, wheat, rice, and other field grains) were to be converted to ethanol, it would satisfy at most 18 percent of U.S. automotive fuel needs.[33]

In 2008 a quarter of the U.S. grain harvest was turned into fuel ethanol.[34] Historically the food and energy economies were separate, but with so many ethanol distilleries coming on line, the two are merging. As a result, the world price of grain is moving up toward its oil-equivalent value. If the fuel value of grain exceeds its food value, the market will simply move the commodity into the energy economy. The price of grain is now keyed to the price of oil.

And it is not only American corn that is being diverted for fuel. In Swaziland, where some 40 percent of the population is hungry, the government has dedicated thousands of acres of cassava, a staple crop, to ethanol production. This and other diversions of food to fuel prompted UN Special Rapporteur Jean Ziegler to call biofuels a "crime against humanity."[35] Indeed, the emerging competition between the owners of the world's 860 million automobiles and the 2 billion poorest people is uncharted

> The price of grain is now keyed to the price of oil.

territory.[36] Suddenly the world is facing a moral and political issue that has no precedent: Should we use grain to fuel cars or to feed people? If the aim is to eradicate hunger and stem the growing tide of state failure, we must choose the latter.

The connection between the food and energy economies does not end with ethanol, of course. From farm to plate, the modern food system relies heavily on cheap oil. That means threats to our oil supply are also threats to our food supply. As environmental activist Bill McKibben has said, the food we eat is practically "marinated in crude oil."[37]

All that oil consumption contributes to climate change, the second unprecedented challenge to agriculture. Crop yields suffer in warmer temperatures, falling on average 10 percent for each increase of 1 degree Celsius.[38] In addition, climate change is having a profound impact on irrigation water supplies. The effects may be most severe in Asia, where mountain glaciers in the Himalayas and on the Tibet-Qinghai Plateau are melting and could soon deprive the major rivers of India and China of the ice melt needed to sustain them during the dry season.[39] China and India are the world's leading producers of both wheat and rice—two of humanity's food staples. At issue is not just the future of mountain glaciers, but the future of world grain harvests, threatening food security for our growing world population.[40]

HUNGER GROWS

These trends are converging, creating a "perfect storm" of rising food prices and growing hunger. For nearly a billion people, hunger is not an abstract possibility; it is a painful daily reality. The UN Food and Agriculture Organization estimates that the number of undernourished people grew from 848 million to 963 million between 2004 and 2008, largely because of the food price crisis.[41] Food price hikes have also worsened micronutrient deficiencies, lowering resistance to disease and increasing risks during childbirth for both mothers and children. "Because good nutrition is crucial both for children's physical and cognitive development and for their productivity and earnings as adults," writes Joachim von Braun of the International Food Policy Research Institute, "the adverse consequences of this price shock will continue even after the shock ends."[42]

HOW MANY PEOPLE CAN THE EARTH SUPPORT?

Although nearly a billion people go to bed hungry each night, humanity has not yet surpassed its capacity to feed itself. When I am asked, "How many people can the Earth support?" I answer with another question: "At what level of food consumption?" Today, world population stands at 6.8 billion. If everyone on Earth ate like Americans—consuming roughly 800 kilograms of grain per person annually for food and animal feed—the 2-billion-ton annual world harvest of grain would support only 2.5 billion people. At the Italian level of consumption of close to 400 kilograms, the current harvest would support 5 billion. At the 200 kilograms of grain consumed by the average Indian, it would support a population of 10 billion.[43] In every society where incomes rise, people move up the food chain, eating more animal protein as beef, pork, poultry, milk, eggs, and seafood. The mix of animal products varies with geography and culture, but the shift to more livestock products as purchasing power increases appears to be universal.

> If everyone on Earth ate like Americans, the world's annual grain harvest would support only 2.5 billion people.

As consumption of livestock products, poultry, and farmed fish rises, grain use per person also rises. Of the 800 kilograms of grain consumed per person each year in the United States, only about 100 kilograms is eaten directly as bread, pasta, and breakfast cereals—the bulk of the grain is fed to livestock and consumed indirectly in the form of meat and dairy products. By contrast, in India, nearly all grain is eaten directly to satisfy basic food energy needs. Little is available for conversion into livestock products.[44]

For Americans and others in the developed world, there is much to be gained by moving down the food chain. People who live very low or very high on the food chain do not live as long as those in the middle. Of the United States, Italy, and India, life expectancy is highest in Italy, even though U.S. medical expenditures per person are much higher.[45]

For those who live in low-income countries like India, where a starchy staple such as rice can supply 60 percent or more of total caloric intake, eating more protein-rich foods can improve health and raise life expectancy. Those consuming a Mediterranean-type diet that includes meat, cheese, and seafood, but all in moderation, are healthier and live longer. The shift by affluent consumers down the food chain improves health and helps reduce the demand for land, water, and fertilizer.

TIME FOR PLAN B

Current food challenges could be alleviated by moving more of humanity to the middle of the food chain and by halting the conversion of crops to fuel. However, it remains an open question whether the world can produce enough food to feed our rapidly growing population, especially in the context of climate change and soaring energy prices, without irreversible damage to the environment.

UN projections show world population growth under three different assumptions about fertility levels. The medium projection, the one most commonly used, has world population reaching 9.2 billion by 2050. The high one reaches 10.8 billion. The low projection, which assumes that the world will quickly move below replacement-level fertility to 1.6 children per couple, has population peaking at just under 8 billion in 2041 and then declining.[46] If the goal is to eradicate poverty and hunger, we have little choice but to strive for the lower projection. Indeed, the alternative may be a halt in population growth because of rising mortality. More and more, those on the lower rungs of the global economic ladder are losing their tenuous grip. If we continue with business as usual, the number of hungry people will soar.

Stabilizing population growth—by ensuring access to family planning and reproductive health services, by educating girls and empowering women—may therefore be the most urgent item on the global agenda. The benefits are enormous and the costs are minimal. In an increasingly integrated world with a growing number of failing states, poverty and population growth have become national security issues. Slowing population growth helps eradicate poverty, and, conversely, eradicating poverty helps slow population growth.

In addition to stabilizing population, eradicating poverty requires ensuring long-term food security for the world's poor. This means stabilizing climate by abandoning business-as-usual energy policies and moving to cut carbon dioxide (CO_2) emissions 80 percent by 2020, as outlined in my book *Plan B 3.0: Mobilizing to Save Civilization*. At issue is whether we can mobilize to lower atmospheric CO_2 concentrations before higher temperatures dehydrate our crops and melt the mountain glaciers that feed major rivers.

This means that a nation's ministry of energy may have a greater influence on future food security than its ministry of agriculture. By emphasizing less-carbon-intensive forms of energy—such as solar, wind, and geothermal—the ministry of energy can help minimize crop-withering heat waves, prevent the melting of the glaciers, and prevent the ice sheet melting that would inundate the river deltas and floodplains that produce much of the Asian rice harvest.

In a world where cropland is scarce and becoming more so, decisions made in the ministry of transportation will also directly affect world food security. By replacing auto-based systems with more-diversified forms of transport that are less land- and energy-intensive, including light-rail, buses, and bicycles, and by restricting the use of grain to produce automotive fuel, the ministry of transportation can reduce fossil fuel use and help stabilize the climate.

And where water is a more serious constraint on expanding food production than land, it will be up to the ministry of water resources to do everything possible to raise the efficiency of water use and stabilize water tables and aquifers. With water, as with energy, the principal opportunities now are on the demand side—increasing water-use efficiency—not on expanding the supply.

Restoring a semblance of food security requires leading countries to collectively mobilize to slow population growth, eradicate poverty, restore the Earth's natural support systems, and stabilize climate. Although none of these goals can be achieved quickly, progress toward all is essential. If food security cannot be restored, social unrest and political instability will spread and the number of failing states will likely increase dramatically, threatening the stability of civilization itself. The food crisis of 2008 was not an anomaly. It was a warning of what is to come—unless we act now.

REFERENCES

1. Brown, Lester R., 2008, World facing huge new challenge on food front, *Plan B Update*, Earth Policy Institute, Washington, DC, April 16, www.earthpolicy.org/Updates/2008/Update72.htm.

2. Klarreich, Kathie, 2008, Food riots lead to Haitian meltdown, *Time*,

April 14; Food riots turn deadly in Haiti, 2008, April 5, http://news.bbc.co .uk/2/hi/americas/7331921.stm.

3. Adam, David, 2008, Food prices threaten global security—UN, *The Guardian,* April 9.

4. Egypt court convicts 22 for food riots, 2008, *International Herald Tribune,* December 15; Food price crisis bites in Egypt, 2008, *BBC News,* May 3, http:// news.bbc.co.uk/2/hi/middle_east/7381766.stm.

5. UN Population Division, World Population Prospects: The 2006 Revision Population Database, at esa.un.org/unpp, updated 2007.

6. U.S. Department of Agriculture, 2009, Production, Supply and Distribution, electronic database, at www.fas.usda.gov/psdonline, updated January 12.

7. Worldwatch Institute, 2001, *Signposts 2001,* CD-Rom, Washington, DC; U.S. Department of Agriculture (USDA), Production, Supply and Distribution, electronic database at www.fas.usda.gov/psdonline, updated September 12, 2007.

8. Brown, Lester R., 2004, *Outgrowing the Earth,* W. W. Norton & Company, New York, p. 60–69.

9. U.S. Department of Agriculture (USDA), 2007, Production, Supply and Distribution, electronic database at www.fas.usda.gov/psdonline, updated September 12; UN Population Division, World Population Prospects: The 2006 Revision Population Database, at www.esa.un.org/unpp, updated 2007.

10. UN Population Division, World Population Prospects: The 2006 Revision Population Database, at esa.un.org/unpp, updated 2007; FAO, FAO-STAT Food Security, electronic database, at www.fao.org/faostat, updated June 30, 2006.

11. Bai, Z. G., D. L. Dent, L. Olsson, and M. E. Schaepman, 2008, Global Assessment of Land Degradation and Improvement 1: Identification by Remote Sensing, Report 2008/01, FAO/ISRIC, Rome/Wageningen, http:// www.fao.org/nr/lada/dmdocuments/GLADA_international.pdf.

12. Ibid.

13. Ibid.

14. Brown, Lester R., 2004, *Outgrowing the Earth,* W. W. Norton & Company, New York, p. 60–69.

15. UN Water/Food and Agriculture Organization, 2007, *Coping with Water Scarcity: Challenge of the 21st Century,* FAO, Rome.

16. UN Water/Food and Agriculture Organization, FAO urges action to cope with increasing water scarcity, http://www.fao.org/newsroom/en/ news/2007/1000520/index.html.

17. Hassan, Rashid, Robert Scholes, Neville Ash, eds., 2005, *Ecosystems and Human Well-being: Current State and Trends,* volume 1, *Findings of the Condition and Trends Working Group of the Millennium Ecosystem Assessment,* Chapter 26, Cultivated Systems, Island Press, Washington, DC, accessed online at http:// www.millenniumassessment.org/documents/document.766.aspx.pdf.

18. U.S. Department of Agriculture (USDA), 2007, Production, Supply and Distribution, electronic database at www.fas.usda.gov/psdonline, updated September 12; UN Population Division, World Population Pros-

pects: The 2006 Revision Population Database, at esa.un.org/unpp, updated 2007; Ma, Michael, 2001, Northern cities sinking as water table falls, *South China Morning Post*, August 11.

19. Ma, Michael, 2001, Northern cities sinking as water table falls, *South China Morning Post*, August 11; share of China's grain harvest from the North China Plain based on Yang, Hong, and Alexander Zehnder, 2001, China's regional water scarcity and implications for grain supply and trade, *Environment and Planning A*, volume 33, and on grain production from U.S. Department of Agriculture (USDA), Production, Supply and Distribution, electronic database, at www.fas.usda.gov/psd/psdonline, updated June 11, 2007.

20. Roy, Aditi Deb, and Tushaar Shah, 2002, *Socio-Ecology of Groundwater Irrigation in India*, International Water Management Institute, Colombo, Sri Lanka, p. 25–26.

21. U.S. Geological Survey, *Ground Water Atlas of the United States*, http://pubs.usgs.gov/ha/ha730/ch_e/E-text5.html; Dennehy, K. F., 2000, High Plains Regional Ground-Water Study, Fact Sheet FS-091-00, U.S. Geological Survey, Reston, VA, August 18.

22. U.S. Department of Agriculture (USDA), 2001, Production, Supply and Distribution, electronic database, at www.fas.usda.gov/psdonline, updated August 10, 2007; Worldwatch Institute, *Signposts 2001*, CD-Rom, Washington, DC.

23. Ruttan, Vernon, 1999, The transition to agricultural sustainability, *Proceedings of the National Academy of Sciences*, May 25, http://www.pnas.org/content/96/11/5960.full#sec-4.

24. Ibid.

25. Brown, Lester R., 2004, *Outgrowing the Earth*, W. W. Norton & Company, New York, p. 64.

26. Fertilizer use data from UN Food and Agriculture Organization (FAO), Fertilizer Yearbook, FAO, Rome (various years); International Fertilizer Industry Association (IFA), Short Term Prospects for World Agriculture and Fertilizer Demand 2005/06–2007/08, IFA, Buenos Aires, December 2006; IFA, Fertilizer Consumption Statistics, electronic database at www.fertilizer.org/ifa/statistics, updated March 10, 2008; Hassan, Rashid, Robert Scholes, Neville Ash, eds., 2005, *Ecosystems and Human Well-being: Current State and Trends*, volume 1, *Findings of the Condition and Trends Working Group of the Millennium Ecosystem Assessment*, Island Press, Washington, DC, Chapter 26, Cultivated systems, accessed online at http://www.millenniumassessment.org/documents/document.766.aspx.pdf.

27. Bradsher, Keith, and Andrew Martin, 2008, Shortages threaten farmers' key tool: Fertilizer, *The New York Times*, April 30.

28. Ibid.

29. Hassan, Rashid, Robert Scholes, Neville Ash, eds., 2005, *Ecosystems and Human Well-being: Current State and Trends*, volume 1, *Findings of the Condition and Trends Working Group of the Millennium Ecosystem Assessment*, Island Press, Washington, DC, p. 28, 37, 40, accessed online at http://www.millenniumassessment.org/documents/document.766.aspx.pdf.

30. Hassan et al., 2005, *Ecosystems and Human Well-being*, p. 28.

31. Diaz, Robert J., and Rutger Rosenberg, 2008, Spreading dead zones and consequences for marine ecosystems, *Science* 321(5891): 926–929, August 15, http://www.sciencemag.org/cgi/content/abstract/321/5891/926.

32. Hassan et al., 2005, *Ecosystems and Human Well-being*, p. 17–18.

33. Brown, Lester R., 2007, Distillery demand for grain to fuel cars vastly understated: World may be facing highest grain prices in history, *Eco-Economy Update*, Earth Policy Institute, Washington, DC, January 4; using corn ethanol conversion based on Keith Collins, chief economist, USDA, statement before the U.S. Senate Committee on Environment and Public Works, September 6, 2006, p. 8; energy content of ethanol relative to gasoline from Oak Ridge National Laboratory (ORNL), Bioenergy Conversion Factors, at bioenergy.ornl.gov/papers/misc/energy_ conv.html, viewed August 3, 2007; U.S. gasoline consumption in 2007 from Table 2: Energy Consumption by Sector and Source, in DOE, EIA, 2007, *Annual Energy Outlook 2007*, DOE, EIA, Washington, DC, February; USDA, Production, Supply and Distribution, electronic database at www.fas.usda.gov/psdonline, updated September 12, 2007.

34. U.S. Department of Agriculture (USDA), Production, Supply, and Distribution, electronic database, at www.fas.usda.gov/psdonline, updated January 12, 2009; corn for ethanol from USDA, Feedgrains Database, electronic database at www.ers.usda.gov/Data/feedgrains, updated January 15, 2009.

35. Monbiot, George, 2007, The western appetite for biofuels is causing starvation in the poor world, *The Guardian*, November 6.

36. Number of automobiles from Ward's Communications, 2006, *Ward's World Motor Vehicle Data 2006*, Ward's, Southfield, MI, p. 240.

37. McKibben, Bill, 2007, *Deep Economy: The Wealth of Communities and the Durable Future*, Times Books, New York.

38. Peng, Shaobing, et al., 2004, Rice yields decline with higher night temperature from global warming, *Proceedings of the National Academy of Sciences*, July 6, p. 9971–9975.

39. UN Environment Programme (UNEP), 2007, *Global Outlook for Ice and Snow*, UNEP, Nairobi, Kenya, p. 139–140.

40. Food and Agriculture Organization, FAOSTAT, electronic database at faostat.fao.org, viewed March 18, 2008; U.S. Department of Agriculture (USDA), Production, Supply and Distribution, electronic database at www .fas.usda.gov/psdonline, updated March 11, 2008.

41. Food and Agriculture Organization, 2008, *Hunger on the rise: Soaring prices add 75 million people to global hunger rolls*, FAO, Rome; Food and Agriculture Organization, 2008, Number of hungry people rises to 963 million, FAO, Rome.

42. von Braun, Joachim, 2008, Food and Financial Crises: Implications for Agriculture and the Poor, International Food Policy Research Institute, Washington, DC, December.

43. Author's calculations from U.S. Department of Agriculture (USDA), Production, Supply and Distribution, electronic database, at www.fas.usda .gov/psdonline, updated August 10, 2007; UN Population Division, World

Population Prospects: The 2006 Revision Population Database, at esa.un
.org/unpp, updated 2007.

44. U.S. Department of Agriculture (USDA), Production, Supply and Distribution, electronic database, at www.fas.usda.gov/psdonline, updated
August 10, 2007; UN UNPopulation Division, World Population Prospects:
The 2006 Revision Population Database, at esa.un.org/unpp, updated 2007;
FAO, FAOSTAT, electronic database at faostat.fao.org, updated June 30,
2007.

45. Organisation for Economic Co-operation and Development, 2007,
Total Expenditure on Health per Capita, US$ PPP, table, in OECD Health
Data 2007—Frequently Requested Data, at www.oecd.org, July.

46. UN, Department of Economic and Social Affairs, Population Division,
World Population Prospects: The 2006 Revision and World Urbanization Prospects,
www.esa.un.org/unpp.

CHAPTER 13

··········

Understanding the Global Food Crisis

Malthusian Nightmare or Free-Trade Fiasco?

WALDEN BELLO

From Thomas Malthus to Paul Ehrlich, history echoes with warnings that population growth will inevitably outstrip agricultural capacity, causing famine and social collapse. While the worst predictions have proved unfounded, concerns are surfacing again in the context of the current food crisis, as commodity prices soar and hungry people take to the streets in Mexico, Haiti, and elsewhere. Even the *Wall Street Journal* has contemplated the clash between humanity and food, with a front-page article entitled "New Limits to Growth Revive Malthusian Fears."[1]

To really understand the current food crisis, one must go beyond simplistic unicausal theories about uncontrolled population growth, though the latter has a role. And the policy response must go beyond efforts to reduce fertility, though reproductive health must be championed as a necessary condition for the achievement of greater freedom and autonomy for women.

Although it is acute in developing countries where population growth rates are high, the food crisis is largely rooted in a free-market economic model that has eviscerated rural cultures and undermined food security around the globe—and sparked a grassroots resistance movement.

Walden Bello is Senior Analyst at and former Executive Director of the Bangkok-based research and advocacy institute Focus on the Global South. This chapter is adapted from an article that appeared in the June 2, 2008, edition of The Nation.

The roots of the crisis can be seen in Mexico, where in 2007 tens of thousands of people staged demonstrations to protest a 60 percent increase in the price of tortillas. The proximate cause of the price increase, as many analysts observed, was biofuel: Because of U.S. government subsidies, American farmers were devoting more and more acreage to corn for ethanol rather than for food, which sparked a steep rise in corn prices. However, an intriguing question escaped many observers: How on Earth did Mexicans, who live in the land where corn was domesticated, become dependent on U.S. imports in the first place?

> How on Earth did Mexicans, who live in the land where corn was domesticated, become dependent on U.S. imports in the first place?

ERODING MEXICAN AGRICULTURE

The Mexican food crisis cannot be fully understood without taking into account the fact that in the years preceding the tortilla crisis, the homeland of corn had been converted to a corn-importing economy by "free-market" policies promoted by the International Monetary Fund (IMF), the World Bank, and Washington. The process began with the early 1980s debt crisis. One of the two largest developing-country debtors, Mexico was forced to beg for money from the Bank and IMF to service its debt to international commercial banks. The quid pro quo for a multibillion-dollar bailout was what a member of the World Bank executive board described as "unprecedented thoroughgoing interventionism" designed to eliminate high tariffs, state regulations, and government support institutions, which neoliberal doctrine identified as barriers to economic efficiency.[2]

Interest payments rose from 19 percent of total government expenditures in 1982 to 57 percent in 1988, while capital expenditures dropped from an already low 19.3 percent to 4.4 percent.[3] The contraction of government spending translated into the dismantling of state credit, government-subsidized agricultural inputs, price supports, state marketing boards, and extension services. Unilateral liberalization of agricultural trade pushed by the IMF and World Bank also contributed to the destabilization of peasant producers.

This blow to peasant agriculture was followed by an even larger one in 1994, when the North American Free Trade Agreement (NAFTA) went into effect. Although NAFTA had a fifteen-year phaseout of tariff protection for agricultural products, including corn, highly subsidized U.S. corn quickly flooded in, reducing prices by half and plunging the corn sector into chronic crisis. Largely as a result of this agreement, Mexico's status as a net food importer was firmly established.

With the shutting down of the state marketing agency for corn, distribution of U.S. corn imports and Mexican grain has come to be monopolized by a few transnational traders, like U.S.-owned Cargill and partly-U.S.-owned Maseca, operating on both sides of the border. This has given them tremendous power to speculate on trade trends, so movements in biofuel demand can be manipulated and magnified many times over. At the same time, monopoly control of domestic trade has ensured that a rise in international corn prices does not translate into significantly higher prices paid to small producers.

It has become increasingly difficult for Mexican corn farmers to avoid the fate of many of their fellow corn cultivators and other smallholders in sectors such as rice, beef, poultry, and pork, who have gone under because of the advantages conferred by NAFTA on subsidized U.S. producers. According to a 2003 Carnegie Endowment report, imports of U.S. agricultural products threw at least 1.3 million Mexican farmers out of work—many of whom have since found their way to the United States.[4]

> Imports of U.S. agricultural products threw at least 1.3 million Mexican farmers out of work—many of whom have since found their way to the United States.

CREATING A RICE CRISIS IN THE PHILIPPINES

That the global food crisis stems mainly from neoliberal restructuring of agriculture is even clearer in the case of rice. Unlike corn, less than 10 percent of world rice production is traded. Moreover, there has been no diversion of rice from food consumption to biofuels. Yet in 2008 alone, prices nearly tripled, from $380 a ton in January to more than $1,000 in April. It is likely that the inflation stems partly from speculation by wholesaler cartels at a time of tightening supplies. However, as with Mexico and corn, the big puzzle is why a number of formerly self-sufficient rice-consuming countries have become severely dependent on imports.

The Philippines—now the world's largest importer of rice—is a case in point. Manila's desperate effort to secure supplies at any price was front-page news in the spring of 2008, and pictures of soldiers providing security for rice distribution in poor communities have become emblematic of the global crisis. Some observers were quick to blame population growth. One lawmaker noted, "The country's inordinately huge population growth rate threatens food security and aggravates the looming rice shortage."[5]

In fact, the Philippines provides a grim example of how neoliberal economic restructuring transforms a country from a net food exporter to a net food importer. The broad contours of the Philippines' story are similar to those of Mexico. Dictator Ferdinand Marcos was guilty of many crimes and misdeeds, including failure to follow through on land reform, but one thing he cannot be accused of is starving the agricultural sector of government funds. To head off peasant discontent, the regime provided farmers with subsidized fertilizer and seeds, launched credit schemes, and built rural infrastructure, with land under irrigation rising from 500,000 hectares in the mid-1960s to 1.5 million in the mid-1980s. Owing to these investments, the Philippines achieved self-sufficiency in rice for most of the Marcos period, though in its last full year, 1985, it had to import over 500,000 tons. When Marcos fled the country in 1986, there were reported to be 900,000 tons in government warehouses.[6]

Paradoxically, the next few years under the new democratic rule saw the gutting of government investment capacity. As in Mexico, the World Bank and IMF, working on behalf of international creditors, pressured the Corazon Aquino administration to make repayment of the $26 billion foreign debt a priority. Aquino acquiesced, though she was warned by the country's top economists that the "search for a recovery program that is consistent with a debt repayment schedule determined by our creditors is a futile one."[7] Thus, structural adjustment, which was already in effect in the last years of Marcos, was tightened under Aquino.

Between 1986 and 1993 the equivalent of 8 to 10 percent of GDP left the Philippines yearly in debt-service payments—roughly the same proportion as in Mexico. Interest payments as a percentage of expenditures rose from 7 percent in 1980 to 28 percent in 1994; capital expenditures plunged from 26 percent to 16 percent.[8] In short, debt servicing became the national budgetary priority.

Spending on agriculture fell by more than half, from 7.5 percent of total government spending in 1982 under Marcos to 3.3 percent in

1988 under Aquino.[9] Even before Marcos left the country in 1986, the government's "Masagana 99" rural credit program, to which many observers attributed the rise in rice yields, had already fallen victim to the IMF–World Bank adjustment program. But the Bank, Fund, and their local acolytes were not worried, since one purpose of the belt-tightening was to get the private sector to energize the countryside. But agricultural capacity quickly eroded. Irrigation coverage stagnated at 1.5 million hectares. By the end of the 1990s, only 17 percent of the Philippines' road network was paved, compared with 82 percent in Thailand and 75 percent in Malaysia. Crop yields were generally ane-mic; the average rice yield of 2.8 metric tons per hectare was way below those in China and Vietnam, where governments actively promoted rural production.[10] The post-Marcos agrarian reform program shriv-eled, deprived of funding for support services, which had been the key to successful reforms in Taiwan and South Korea. As in Mexico, Filipino peasants were confronted with full-scale retreat of the state as provider of comprehensive support—a role they had come to depend on.

And the cutback in agricultural programs was followed by trade lib-eralization, with the Philippines' 1995 entry into the World Trade Organization (WTO) having the same effect as Mexico's joining NAFTA. WTO membership required the Philippines to eliminate quo-tas on all agricultural imports except rice and allow a certain amount of each commodity to enter at low tariff rates. While the country was allowed to maintain a quota on rice imports, it nevertheless had to admit the equivalent of 1 to 4 percent of domestic consumption over the next 10 years. In fact, because of gravely weakened production resulting from lack of state support, the government imported much more than that to make up for possible shortfalls. These imports, which rose from 263,000 metric tons in 1995 to 2.1 million tons in 1998, depressed the price of rice, discouraging farmers and keeping growth in production at a rate far below that of the country's two top suppliers, Thailand and Vietnam.[11]

The consequences of the Philippines' joining the WTO barreled through the rest of its agriculture like a super-typhoon. Swamped by cheap imports—much of them subsidized U.S. grain—farmers reduced land devoted to corn from 3.1 million hectares in 1993 to 2.5 million in 2000.[12] Massive importation of chicken parts nearly killed that indus-try, while surges in imports destabilized the poultry, hog, and vegetable industries.[13]

During the 1994 campaign to ratify WTO membership, government economists, coached by their World Bank handlers, promised that losses in corn and other traditional crops would be more than compensated for by the new export industry of "high-value-added" crops like cut flowers, asparagus, and broccoli. Little of this materialized. Nor did many of the 500,000 agricultural jobs that were supposed to be created yearly by the magic of the market; instead, agricultural employment dropped from 11.2 million in 1994 to 10.8 million in 2001.[14]

The one-two punch of IMF-imposed adjustment and WTO-imposed trade liberalization swiftly transformed a largely self-sufficient agricultural economy into an import-dependent one as it steadily marginalized farmers. It was a wrenching process, the pain of which was captured by a Filipino government negotiator during a WTO session in Geneva. "Our small producers," he said, "are being slaughtered by the gross unfairness of the international trading environment."[15]

> "Our small producers are being slaughtered by the gross unfairness of the international trading environment."

THE GREAT TRANSFORMATION

The experience of Mexico and the Philippines was paralleled in one country after another subjected to the ministrations of the IMF and the WTO. A study of fourteen countries by the UN's Food and Agriculture Organization found that the levels of food imports in 1995–1998 exceeded those in 1990–1994.[16] This was not surprising, since one of the main goals of the WTO's Agreement on Agriculture was to open up markets in developing countries so they could absorb surplus production in the north. As U.S. Agriculture Secretary John Block put it in 1986, "The idea that developing countries should feed themselves is an anachronism from a bygone era. They could better ensure their food security by relying on U.S. agricultural products, which are available in most cases at lower cost."[17]

What Block did not say was that the lower cost of U.S. products stemmed from subsidies, which became more massive with each passing year despite the fact that the WTO was supposed to phase them out. From $367 billion in 1995, the total amount of agricultural subsidies provided by developed-country governments rose to $388 billion in 2004.[18] Since the late 1990s, subsidies have accounted for 40 percent of the value of agricultural production in the European Union and 25 percent in the United States.[19]

The apostles of the free market and the defenders of dumping may seem to be at different ends of the spectrum, but the policies they advocate are bringing about the same result: a globalized capitalist industrial agriculture. Developing countries are being integrated into a system where export-oriented production of meat and grain is dominated by large industrial farms like those run by the Thai multinational Charoen Pokphand Group and where technology is continually upgraded by advances in genetic engineering from firms like Monsanto. And the elimination of tariff and nontariff barriers is facilitating a global agricultural supermarket of elite and middle-class consumers serviced by grain-trading corporations like Cargill and Archer Daniels Midland and transnational food retailers like the British-owned Tesco and the French-owned Carrefour.

There is little room for the hundreds of millions of rural and urban poor in this integrated global market. They are confined to giant suburban favelas, where they contend with food prices that are often much higher than the supermarket prices, or to rural reservations, where they are trapped in marginal agricultural activities and are increasingly vulnerable to hunger. Indeed, within the same country, famine in the marginalized sector sometimes coexists with prosperity in the globalized sector.

This is not simply the erosion of national food self-sufficiency or food security but what Africanist Deborah Bryceson of Oxford calls "de-peasantization"—the phasing out of a mode of production to make the countryside a more congenial site for intensive capital accumulation (see Box 13.1).[20] This transformation is a traumatic one for hundreds of millions of people, since peasant production is not simply an economic activity. It is an ancient way of life, a culture, which is one reason displaced or marginalized peasants in India are committing suicide at unprecedented rates. In the state of Andhra Pradesh, farmer suicides rose from 233 in 1998 to 2,600 in 2002; in Maharashtra, suicides more than tripled, from 1,083 in 1995 to 3,926 in 2005.[21] One estimate is that some 150,000 Indian farmers took their lives between 1997 and 2005.[22]

> Within the same country, famine in the marginalized sector sometimes coexists with prosperity in the globalized sector.

Collapse of prices from trade liberalization and loss of control over seeds to biotech firms is part of a comprehensive problem, says global justice activist Vandana Shiva: "Under globalization, the farmer is

BOX 13.1

Yes, There Is an Alternative: Food Sovereignty

Margaret Thatcher famously observed that "there is no alternative" to the prevailing model of globalized, free-market capitalism. Applied to agriculture, that model has devastated small farmers and eroded food security for many. But now, an alternative paradigm is emerging. Peasant organizations around the world have become increasingly militant in their resistance to the globalization of industrial agriculture. Indeed, it is because of pressure from farmers' groups that the governments of the south have refused to grant wider access to their agricultural markets and demanded a massive slashing of U.S. and EU agricultural subsidies, which brought the WTO's Doha Round of negotiations to a standstill.

Farmers' groups have networked internationally; one of the most dynamic to emerge is *Via Campesina* (Peasant's Path). Via not only seeks to get "WTO out of agriculture" and opposes the paradigm of a globalized capitalist industrial agriculture; it also proposes an alternative: food sovereignty. Food sovereignty means, first of all, the right of a country to determine its production and consumption of food and the exemption of agriculture from global trade regimes like that of the WTO. It also means consolidation of a smallholder-centered agriculture via protection of the domestic market from low-priced imports; remunerative prices for farmers and fisherfolk; abolition of all direct and indirect export subsidies; and the phasing out of domestic subsidies that promote unsustainable agriculture. Via's platform also calls for an end to the Trade Related Aspects of Intellectual Property Rights regime, or TRIPs, which allows corporations to patent plant seeds; opposes agro-technology based on genetic engineering; and demands land reform.[a] In contrast to an integrated global monoculture, Via offers the vision of an international agricultural economy composed of diverse national agricultural economies trading with one another but focused primarily on domestic production.

Once regarded as relics of the preindustrial era, peasants are now leading the opposition to a capitalist industrial agriculture that would consign them to the dustbin of history. With the global food crisis, they are moving to center stage—and they have allies and supporters. For as peasants refuse to go gently into that good night and fight de-peasantization, developments in the twenty-first century are revealing the panacea of globalized capitalist industrial agriculture to be a nightmare. With environmental crises multiplying, the social dysfunctions of urban-industrial life piling up, and industrialized agriculture creating greater food insecurity, the farmers' movement increasingly has relevance not only to peasants but to everyone threatened by the catastrophic consequences of global capital's vision for organizing production, community, and life itself.

....

[a]Seragih, Henry, and Ahmad Ya'kub, 2004, The Impact of WTO and Alternatives to Agricultural Trade, paper presented at the Regional Conference on Agricultural Negotiations in the WTO: Implications for Trade and Agriculture in East Asia, Hong Kong, January 12–14.

losing her/his social, cultural, economic identity as a producer. A farmer is now a 'consumer' of costly seeds and costly chemicals sold by powerful global corporations through powerful landlords and money lenders locally."[23]

AFRICAN AGRICULTURE: FROM COMPLIANCE TO DEFIANCE

De-peasantization is at an advanced state in Latin America and Asia. And if the World Bank has its way, Africa will travel in the same direction. As Bryceson and her colleagues point out, the *World Development Report* for 2008, which touches extensively on agriculture in Africa, is practically a blueprint for the transformation of the continent's peasant-based agriculture into large-scale commercial farming.[24] However, as in many other places today, the Bank's wards are moving from sullen resentment to outright defiance.

At the time of decolonization, in the 1960s, Africa was actually a net food exporter. Today the continent imports 25 percent of its food; almost every country is a net importer.[25] Hunger and famine have become recurrent phenomena, with the past three years alone seeing food emergencies break out in the Horn of Africa, the Sahel, and southern and central Africa.[26]

Agriculture in Africa is in deep crisis, and the causes range from wars to bad governance, lack of agricultural technology, and the spread of HIV/AIDS. However, as in Mexico and the Philippines, an important part of the explanation is the phasing out of government controls and support mechanisms under the IMF and World Bank structural adjustment programs imposed as the price for assistance in servicing external debt.

Structural adjustment brought about declining investment, increased unemployment, reduced social spending, reduced consumption, and low output. Lifting price controls on fertilizers while simultaneously cutting back on agricultural credit systems simply led to reduced fertilizer use, lower yields, and lower investment. Moreover, reality refused to conform to the doctrinal expectation that withdrawal of the state would pave the way for the market to dynamize agriculture. Instead, the private sector, which correctly saw reduced state expenditures as creating more risk, failed to step into the breach. In country after country, the departure of the state "crowded out" rather than "crowded in" private investment. Where private traders have replaced the state, noted an Oxfam report, "they have sometimes done so on highly unfavorable terms for poor farmers," leaving "farmers more food insecure, and governments reliant on unpredictable international aid flows."[27] The usually pro–private sector *Economist* agreed, admitting that "many of the private firms brought in to replace state researchers turned out to be rent-seeking monopolists."[28]

The support that African governments were allowed to muster was channeled by the World Bank toward export agriculture to generate foreign exchange, which states needed to service debt. But, as in Ethiopia during the 1980s famine, this led to the dedication of good land to export crops, with food crops forced into less-suitable soil, thus exacerbating food insecurity. Moreover, the World Bank's encouragement of several economies to focus on the same export crops often led to overproduction, triggering price collapses in international markets. For instance, the very success of Ghana's expansion of cocoa production triggered a 48 percent drop in the international price between 1986 and 1989.[29] In 2002–2003 a collapse in coffee prices contributed to another food emergency in Ethiopia.[30]

> Good land is dedicated to export crops, and food crops are forced into less-suitable soil, thus exacerbating food insecurity.

As in Mexico and the Philippines, structural adjustment in Africa was not simply about underinvestment but state divestment. But there was one major difference. In Africa, the World Bank and IMF micromanaged, making decisions on how fast subsidies should be phased out, how many civil servants had to be fired, and even, as in the case of Malawi, how much of the country's grain reserve should be sold and to whom.[31] In other words, Bank and IMF resident proconsuls reached to the very innards of the state's involvement in the agricultural economy to rip it up.

Compounding the negative impact of adjustment were unfair EU and U.S. trade practices, which were a far cry from those countries' "free-trade" ideals. Liberalization allowed subsidized EU beef to drive many west African and south African cattle raisers to ruin. With their subsidies legitimized by the WTO, U.S. growers off-loaded cotton on world markets at 20 percent to 55 percent of production cost, thereby bankrupting western and central African farmers.[32]

According to Oxfam, the number of sub-Saharan Africans living on less than a dollar a day almost doubled, to 313 million, between 1981 and 2001—46 percent of the whole continent.[33] The role of structural adjustment in creating poverty was hard to deny. The World Bank's chief economist for Africa admitted, "We did not think that the human costs of these programs could be so great, and the economic gains would be so slow in coming."[34]

Malawi is representative of the African tragedy spawned by the IMF and the World Bank. In 1999 the government of Malawi initiated a program to give each smallholder family a starter pack of free fertilizers and seeds. The result was a national surplus of corn.[35] What came after is a story that should be enshrined as a classic case study of one of the greatest blunders of neoliberal economics.

The World Bank and other aid donors forced the scaling down and eventual scrapping of the program, arguing that the subsidy distorted trade.[36] Without the free packs, output plummeted. In the meantime, the IMF insisted that the government sell off a large portion of its grain reserves to enable the food reserve agency to settle its commercial debts. The government complied. When the food crisis turned into a famine in 2001–2002, there were hardly any reserves left. As many as 1,500 people perished.[37] The IMF was unrepentant; in fact, it suspended its disbursements on an adjustment program on the grounds that "the parastatal sector will continue to pose risks to the successful implementation of the 2002/03 budget. Government interventions in the food and other agricultural markets . . . [are] crowding out more productive spending."[38]

By the time an even worse food crisis developed in 2005, the government had had enough of World Bank/IMF stupidity. A new president reintroduced the fertilizer subsidy, enabling 2 million households to buy it at a third of the retail price, and seeds at a discount. The result: bumper harvests for 2 years, a million-ton maize surplus, and the country transformed into a supplier of corn to southern Africa.

Malawi's defiance of the World Bank would probably have been an act of heroic but futile resistance a decade ago. The environment is different today, since structural adjustment has been discredited throughout Africa. Even some donor governments and NGOs that used to subscribe to it have distanced themselves from the Bank. Perhaps the motivation is to prevent their influence in the continent from being further eroded by association with a failed approach

In country after country, a common theme emerges: Free-market restructuring of agriculture has devastated small farmers and eroded food security.

and unpopular institutions when Chinese aid is emerging as an alternative to World Bank, IMF, and western government aid programs.

The food crises roiling the globe have many causes, and increased demand—from rising affluence as well as population growth—may be part of the picture. And there are certainly environmental limits to agricultural capacity, especially given the ecological costs of industrial agriculture and the ravages of climate change. But, in country after country, a common theme emerges: Free-market restructuring of agriculture has devastated small farmers and eroded food security.

REFERENCES

1. Lahart, Justin, Patrick Barta, and Andrew Batson, 2008, New limits to growth revive Malthusian fears, *The Wall Street Journal,* March 24.

2. Miller, Morris, 1991, *Debt and the Environment: Converging Crisis,* UN, New York, p. 215.

3. Bello, Walden, Shea Cunningham, and Bill Rau, 1994, *Dark Victory: The United States, Structural Adjustment, and Global Poverty,* Food First, San Francisco, p. 39.

4. Cited in Pollan, Michael, 2004, *A Flood of U.S. Corn Rips at Mexico,* Common Dreams News Center, April 23, http://www.commondreams.org/cgi-bin/print.cgi?file=views04/0423-02.htm.

5. Burgonio, T. J., 2008, Runaway population growth factor in rice crisis, *Philippine Daily Inquirer,* March 30.

6. Carino, Conrad, 2008, Rice crisis "imminent" a long time ago, *Manila Times,* April 6, http://www.manilatimes.net/national/2008/apr/06/yehey/top_stories/20080406top3.html.

7. Alburo, Florian, et al., 1985, Towards recovery and sustainable growth, School of Economics, University of the Philippines, Diliman, Quezon City, September.

8. World Bank, 1997, World Development Indicators 1998, World Bank, Washington, DC, p. 199.

9. Government data from Riza Bernabe, personal communication, May 5, 2008.

10. Obanil, Rovik, 2002, Rice safety nets act: More of a burden than a shield, *Farm News and Views,* first quarter, p. 10.

11. Selected Agricultural Statistics, 1998 and 2002, Department of Agriculture, Quezon City; Obanil, Rovik, 2002, Rice safety nets act: More of a burden than a shield, *Farm News and Views,* first quarter, p. 10.

12. Selected Agricultural Statistics, 1998 and 2002, Department of Agriculture, Quezon City.

13. See Bello, Walden, et al., 2004, *The Anti-Development State: The Political Economy of Permanent Crisis in the Philippines,* University of the Philippines, Quezon City, p. 146–148.

14. Selected Agricultural Statistics, 1998 and 2002, Department of Agriculture, Quezon City.

15. Submission of Republic of the Philippines, World Trade Organization Committee on Agriculture, Geneva, July 1, 2003.

16. Food and Agriculture Organization (FAO), 2000, *Agriculture, Trade, and Food Security: Issues and Options in the WTO Negotiations from the Perspective of the Developing Countries*, volume 2, *Country Case Studies*, FAO, Rome, p. 25.

17. Quoted in Cakes and caviar: The Dunkel draft and third world agriculture, 1993, *Ecologist* 23(6): 220, November–December.

18. OECD Agricultural Trade Statistics, http://www.oecd.org/dataoecd/48/2/40010981.xls.

19. Oxfam International, 2002, *Rigged Rules and Double Standards*, Oxfam International, Oxford, p. 112.

20. Bryceson, Deborah, 2000, Disappearing peasantries? Rural labor redundancy in the neo-liberal era and beyond, p. 304–305, *Disappearing Peasantries? Rural Labor in Africa, Asia, and Latin America*, Deborah Bryceson, Cristobal Kay, and Jos Mooij, eds., London, cited in Davis, Mike, 2006, *Planet of Slums*, Veerso, London, p. 15.

21. Patnaik, Utsa, 2004, External Trade, Domestic Employment, and Food Security: Recent Outcomes of Trade Liberalization and Neo-Liberal Economic Reforms in India, paper presented at the International Workshop on Policies against Hunger III, Berlin, October 20–22, p. 1.

22. The Hindu, 2007, November 12, http://www.hindu.com/2007/11/12/stories/2007111257790100.htm.

23. Shiva, Vandana, 2004, *The Suicide Economy*, April, Znet, http://www.countercurrents.org/glo-shiva050404.htm.

24. Havnevik, Kjell, Deborah Bryceson, Lars-Erik Birgegard, Prosper Matandi, and Atakilte Beyene, 2008, African agriculture and the World Bank: Development or impoverishment? *Pambazuka News*, March 11, http://www.pambazuka.org/en/category/features/46564.

25. Africa's hunger: A systemic crisis, 2006, BBC News, Jan 21, http://news.bbc.co.uk/2/hi/afria/462232.stm; The Development of African Agriculture, http://www.africangreenrevolution.com/cgi-bin/african_green_rev/printer_friendly.cgi?f.

26. See, inter alia, Oxfam International, 2006, *Causing Hunger: An Overview of the Food Crisis in Africa*, Oxfam, Oxford, July.

27. Oxfam International, 2006, *Causing Hunger*, p. 18.

28. The new face of hunger, 2008, *Economist*, April 17, http://www.economist.com/world/internatiional/PrinterFriendly.cfm?story_id=11049284.

29. Abugre, Charles, 1993, *Behind Crowded Shelves: An Assessment of Ghana's Structural Adjustment Experiences, 1983–1991*, Food First, San Francisco, p. 87.

30. Oxfam International, 2006, *Causing Hunger*, p. 20

31. See Did the IMF Cause a Famine? 2008, Yingsakfoodnetwork.com, April 28, http://wwwyingsakfoodnetwork.com/did_the_imf.asp.

32. Trade talks round going nowhere sans progress in farm reform, 2003, *Business World* (Phil), Sept. 8, p. 15.

33. Oxfam International, 2006, *Causing Hunger*, p. 13.

34. Quoted in Miller, Morris, 1991, *Debt and the Environment*, p. 70.

35. Nolen, Stephanie, 2007, How Malawi went from a nation of famine to

a nation of feast, *Globe and Mail*, October 12; Starter Packs: A Strategy to Fight Hunger in Developing Countries: Lessons from Malawi, CAB Abstracts, http://www.cababstractsplus.org/google/abstract.asp?aspAcNo =20053142997.

36. Nolen, Stephanie, 2007, How Malawi went from a nation of famine to a nation of feast, *Globe and Mail*, October 12.

37. Ibid.

38. IMF statement, quoted in Famine in Malawi exposes IMF negligence, 2002, *Economic Justice News* 5(2), June, http://www.50years.org/cms/ejn/ story/89. This article summarizes a report by ActionAid, State of Disaster: Causes, Consequences, and Policy Lessons from Malawi, released on June 13, 2002.

CHAPTER 14

··········

How Much Is Left?

An Overview of the Water Crisis

ELEANOR STERLING and ERIN VINTINNER

The first full views of Earth from Apollo 8—a brilliant blue oasis in a vast black matrix—graphically showed how water distinguishes our world from other planets and underscored our planet's misnomer. Indeed, given that over 70 percent of its surface is covered in water, Earth should have been named Planet Ocean or the Blue Planet. Water defines our world, from shaping its surface to fostering the evolution and complexity of life as we know it. Access to freshwater has driven the emergence of societies, and loss of water has brought them crashing down.

All life on Earth is linked through water as it cycles at different scales across the planet, from wetlands to rain clouds to rivers to oceans. It is a finite and powerful force, and both its presence and absence have profound implications at every level, from global climate patterns to ecosystems to the processes within living cells.

Because we are completely dependent on freshwater, in a global sense, we are all downstream from someone else in the world's water cycle. Which is why, as we face a global water crisis, we must all work together—water-rich and water-poor regions alike. While we cannot control the uneven natural distribution of freshwater, we have direct responsibility for the ways in which we use water in agriculture, indus-

Eleanor Sterling is Director of the Center for Biodiversity and Conservation (CBC) at the American Museum of Natural History. Erin Vintinner is Biodiversity Specialist at CBC.

try, municipalities, and domestic capacities. We have the ability to turn this crisis around, for the sake of all life on our blue planet. But first, we must understand the scope of the problem. How much water do we have left, and which areas will be hardest hit? Why are we faced with shortages, and how can we equitably share what remains?

WHERE IN THE WORLD IS OUR WATER?

Even though we live on a planet covered by water, 97 percent is salty, and the majority of what's fresh is stored in glaciers or deep underground aquifers. That means that all freshwater-dependent life forms must share less than 1 percent of the water on Earth.

This precious fraction of freshwater is truly our most valuable resource and is renewable at the local scale. Yet as water passes through the water cycle, it is not evenly distributed across our landforms, rendering some regions water rich and others water poor. Six countries (Brazil, Canada, China, Colombia, Indonesia, and Russia) hold over half of the world's freshwater.[1] While Asia and the Middle East have about 60 percent of the world's population, they have only about 36 percent of river runoff (one main indicator of freshwater). In contrast, only 6 percent of the world's population lives in South America, yet the region has one-quarter of all of the world's runoff.[2]

> While Asia and the Middle East have about 60 percent of the world's population, they have only about 36 percent of river runoff. . . . Most of the world's megacities are in water-stressed regions.

These regional disparities translate into harsh conditions for people living in areas where freshwater supplies are scarce and populations are increasing. Most of the world's megacities are in water-stressed regions where periodic water shortages may occur. These include Mexico City, Calcutta, Cairo, Jakarta, Beijing, Lagos, and Manila,[3] whose residents can anticipate chronic water shortages that threaten food production, hinder development, and negatively impact species and ecosystems.[4]

By 2025, the United Nations estimates, 2.8 billion people in forty-eight countries will be living in areas facing water stress or scarcity.[5, 6, 7] These areas will be concentrated mainly in Africa, west Asia, and the Middle East—including the countries of Ethiopia, Kenya, Nigeria, and India and parts of China.

It is important to note that calculations of water stress and scarcity are based on estimates of a country's renewable freshwater supplies and do not include water withdrawn from nonrenewable groundwater. A country may temporarily alleviate the effects of water stress by mining ancient aquifers, which are underground reservoirs of water stored within interconnected spaces in the soil, sand, gravel, and rock. Groundwater plays a central role in the water cycle, because it is a major contributor to flow in streams, rivers, and wetlands. However, since groundwater can take a long time to recharge from surface waters draining into the soil, the practice of mining aquifers is often not sustainable, especially if population and freshwater demands increase.[8]

> By 2025, 2.8 billion people in forty-eight countries will be living in areas facing water stress or scarcity.

There are numerous water-stressed areas in the United States, particularly in the dry Southwest, where water shortages are now chronic. Current trends such as declining groundwater levels, increasing population, and growing demand for public water supplies indicate that the amount of freshwater in many areas of the country is reaching its limits. Alarmingly, water managers in thirty-six U.S. states anticipate that they will face local, regional, or statewide water shortages during the next ten years.[9] These water shortages would lead to severe restrictions in water usage in all areas (agriculture, industry, and domestic use) and prompt exploration into new water management strategies or freshwater sources that are more expensive and have a greater impact on the environment.

Already, the U.S. Geological Survey has reported groundwater depletion in the Southwest and Great Plains, which in some cases has caused the land to sink. Depletion of groundwater risks saltwater intrusion, which occurs when salt water creeps into the aquifer, rendering freshwater unusable for drinking or irrigation. This has already been reported in coastal areas of Florida, Georgia, South Carolina, and New Jersey and even in inland Arkansas.[10]

A 2003 report by the General Accounting Office anticipated water shortages that could result in severe economic, environmental, and social impacts for the United States. It is difficult to measure the nationwide economic costs of water shortages, but an analysis by the National Oceanic and Atmospheric Administration estimated that eight water shortages that occurred during the last twenty years cost $1 billion or more. Water shortages can harm water-dependent animal and plant species, wildlife habitat, and water quality. They also impact communities by creating conflict between water users over decisions on how water is consumed in homes and businesses, and who is subject to inequalities in levels of relief from water shortages.[11]

WHY IS SO LITTLE LEFT?

We make many choices in the allocation and use of our finite freshwater resources, and each decision has consequences for human societies, as well as other life on Earth. The ability to deliver water wherever it is wanted has been one of the most important factors in the growth of civilizations and cities. But over time our ingenuity has wrought larger-and-larger-scale schemes, often with heavy social and environmental costs that we are just beginning to understand.

Once delivered to where we want it, water is often wasted or degraded, undermining its renewable nature. Effective water management to ensure a sustainable supply of healthy water for all Earth's creatures (including ourselves) and ecosystems will demand a shift in the way we use, manage, and value our water wealth.

To begin to place a value on water, we need to understand how we are using it. Currently, our water crisis is being exacerbated by a number of factors. Climate change is altering precipitation patterns and affecting freshwater sources such as glaciers and snowpack. We are diverting and polluting surface waters, overmining groundwater, and deforesting, degrading, and paving over essential natural systems that help with water collection and purification.

Through hydropower dams and the withdrawal of water from surface and ground sources, humans appropriate over 50 percent of the Earth's renewable freshwater.[12] While this water is renewable at a local scale, it often carries heavy pollution loads following usage, affecting some of the most threatened freshwater ecosystems and species in the world. Scientists estimate that half the world's wetlands, which act as natural water treatment systems, may have been lost in the last hundred years. As our numbers continue to grow, competition for freshwater between humans and the negative consequences of our actions for other species will only intensify.

AGRICULTURE: OUR BIGGEST WATER USER

Of all the water that humans use, on a global scale the majority goes to large-scale agriculture—about 70 percent—while industry (mainly hydropower or nuclear power) accounts for 22 percent, and municipal and personal use is 8 percent. But these numbers vary greatly by country, with the United States allocating more (46 percent) to industry than agriculture (41 percent), while domestic use lies at 13 percent. In contrast, 96 percent of Bangladesh's water is used for agriculture, with just 1 percent used for industry and 3 percent for domestic use.[13]

Agriculture's large "water footprint" is the result of the huge acreage of irrigated cropland and the inefficient water delivery systems used for irrigation. Today, 40 percent of the world's food is grown in areas where irrigation is necessary—where crops wouldn't flourish with rainfall alone. By using irrigation, some farmers can raise two or even three crops each year instead of only one.

However, our water delivery systems are often inefficient, with at times up to 50 percent of irrigation water lost to leaks and evaporation.[14] Farmers are increasingly using more-efficient systems, including drip irrigation, which has been particularly effective in arid regions such as Israel, where a high-technology version was first invented in the 1950s, as well as in California's Central Valley. However, drip irrigation is not a panacea. Some drawbacks to this method include high installation costs and technical issues such as clogging. Other advances for irrigation efficiency include laser-leveling of fields to prevent pooling, increasing organic matter in the soil to foster its ability to absorb water, planting cover crops that slow down water movement, retiring land that has high rates of erosion, and, in particular, planting a diversity of crops and those most suited to the soil and rainfall of the region.

Unfortunately, this commonsense approach to agriculture is not yet in practice in many parts of the world. "Thirsty" crops—those needing large quantities of water to grow—are often grown in areas where water supplies are low. Rice, which is the main source of directly consumed calories for about half of the world's population, is one of the thirstiest crops, requiring more than 400 gallons of water to grow 1 pound.

Advances in technology have unlocked hidden sources of water for agriculture. Well drilling has afforded access to underground aquifers where water is stored in porous rock and soil. But at times these advances have outstripped the ability of water systems to renew themselves.

As one example, drilling into the Ogallala aquifer underlying much of the Great Plains region from Texas to South Dakota has turned the high plains from a dust bowl region, where farmer after farmer failed in the early twentieth century, to one of the most productive agricultural regions of the world. However, as smaller farmers are now learning, this productivity is not sustainable in the long term. The Ogallala aquifer was formed over 20 million years ago, and much of the water has built up and been held within it for millions of years. Deep wells in the Ogallala still bring up ancient water, laid down at the end of the last Ice Age. But we are currently consuming this "fossil" water at a rate of ten times the natural recharge rate. From the early 1900s, when the Ogallala was first tapped for irrigation, to 2005, the water table in parts of the high plains dropped by more than 100 feet.[15]

Elsewhere in the world, the overmining of groundwater is also taking its toll. Europe relies on groundwater for 65 percent of its drinking water,[16] and the practice is increasing across Asia, particularly in India, Pakistan, and China.

DAMS, POLLUTION, AND WASTEFUL CONSUMPTION

Every day we use products whose water footprint—the amount of water used in their production, packaging, and distribution—is unexpectedly large. It takes 74 gallons of water to produce a single cup of coffee and over 750 gallons to produce a fast-food meal, including a one-third-pound hamburger, french fries, and orange juice. Diets that include large amounts of meat and other animal products require more water than diets that consist mainly of vegetables and grains. Similarly, diets that are made up of highly processed foods require more water than those consisting of whole foods such as fruits and vegetables.

Not only food products but *everything*, including the cars we drive and the computers we use, takes water to produce. Manufacture of a single medium-sized cotton T-shirt consumes 700 gallons of water.

> Manufacture of a single medium-sized cotton T-shirt consumes 700 gallons of water.

By consciously thinking about our consumer practices and focusing on what we need more than what we want, we can each contribute to water sustainability.

To support our water use in agricultural, industrial, and domestic sectors, we rely on a massive water infrastructure anchored by dams. Built through the ages to control, contain, and divert water, dams have provided a wide range of significant benefits. But in taming the world's rivers, these massive marvels of engineering have also had negative impacts by blocking migration routes for many fish species and altering river habitat. The reservoirs behind large dams have also displaced millions of people. And large dams often don't live up to expectations—delivering less water and less profit than expected and, in the case of some hydropower dams, generating less power than predicted, too.

Significant water withdrawals for agricultural, industrial, and municipal use also have downstream effects. Ten of the world's major rivers, including the Colorado, Ganges, Jordan, Nile, Rio Grande, and Yellow, regularly run dry before reaching the sea.

The water crisis has two dimensions: quantity and quality. While we must consider our impacts on the supply of water, we must also take into account how we are polluting these same supplies. Water is polluted from industrial and domestic use, including inadequate sanitation. This combination of factors has had disastrous results worldwide. Currently, 90 percent of the groundwater under China's cities is contaminated;[17] 70 percent of India's rivers and lakes are unsafe for drinking or bathing;[18] 40 percent of the rivers and streams in the United States are too dangerous for drinking, fishing, and swimming;[19] 75 percent of people in Latin America and the Caribbean suffer from chronic dehydration because they don't have access to safe drinking water;[20] and all of Africa's 677 major lakes are now threatened to varying degrees by unsustainable use, pollution, and climate change.[21]

Today's sophisticated sewage systems treat biodegradable food, human waste, and metals, but they are not designed to capture the massive amounts of synthetic chemicals used to manufacture consumer products or chemical ingredients in agents used in our bathrooms and kitchens. Scientists have found levels of pollutants such as pesticides, organic pollutants such as PCBs (polychlorinated biphenyls), and heavy metals such as lead and mercury in the tissues of polar bears, seals, and seabirds living in remote areas of the Arctic where none of these pollutants are used. The pollutants travel via long-range atmospheric transport, ocean currents, and runoff passing through the Arctic drainage basin.

SHARING WHAT'S LEFT

Our overconsumption and increasing levels of pollution suggest that in some of the more water-rich parts of the world, the value we place on water is extremely low. But in other parts of the world, water is a luxury few can afford. In 2004, the World Health Organization and the United Nations Children's Fund estimated that more than one-sixth of the

BOX 14.1

The Missing Piece: A Water Ethic

Sandra Postel

Now for the million dollar questions: Why has so much modern water management gone awry? Why is it that ever more money and sophisticated engineering have not solved the world's water problems? Why are rivers drying up and water tables falling?

The answer, in part, is simple: We have been trying to meet insatiable demands by expanding a finite water supply. In the long run, that is a losing proposition: It is impossible to expand a finite supply indefinitely, and in many parts of the world the "long run" has arrived.

For sure, measures to improve water use have let many places contain their water demands and delay an ecological reckoning. Such measures as thrifty irrigation, water-saving plumbing, native landscaping, and wastewater recycling can cost-effectively reduce the water used to grow food, produce goods, and meet household needs. The conservation potential of these measures has barely been tapped.

Yet something is missing, something less tangible than low-flow showerheads but, in the final analysis, more important. In our technological world, we no longer grasp water's most fundamental role as the basis of life, or the need for the wild river or even for the diversity of species collectively performing nature's work. By and large, society views water as a "resource" valued only when it is put to use.

We have been quick to assume rights to use water but slow to recognize obligations to preserve and protect it. Better pricing and more open markets will assign water a higher economic value and breed healthy competition that will weed out unproductive uses. But this will not solve the deeper problem. What is needed is a set of principles that stops us from chipping away at natural systems until nothing is left of their life-sustaining functions.

In short, we need a water ethic—a guide to right conduct in the face of complex decisions we do not and cannot fully understand. The essence of such an ethic is to make the protection of freshwater ecosystems a central goal in all that we do. This may sound idealistic in light of our ever-more-crowded world. Yet it is no more radical than suggesting a building be given a solid foundation before thirty stories are added to it. Water is the foundation of every human enterprise, and if that foundation is insecure, everything built upon it will be insecure, too. Our stewardship of water will indeed determine the staying power of human societies.

The adoption of such a water ethic would be a historic shift away from strictly utilitarian water management and toward an integrated, holistic approach that views people and water as interconnected parts of a greater whole. Instead of asking how to manipulate rivers to meet our demands, we would ask how to satisfy human needs while accommodating the ecological requirements of freshwater ecosystems. It would lead us, as well, to deeper questions, in particular how to narrow the gap between the haves and have-nots.

An ethically grounded water policy must presuppose that all living things deserve enough water to survive before some get more than enough.

On paper, at least one government has grounded its water policy in such an ethic. South Africa's 1998 water law establishes a reserve consisting of two parts. The first is a nonnegotiable water allocation to meet the basic needs of all South Africans. (When the ANC came to power, 14 million South Africans lacked water for basic needs.) Part two is an allocation of water to support eco-system functions. The water constituting this two-part reserve has priority over uses such as irrigation, and only this water is guaranteed as a right.

At the core of South Africa's policy is an affirmation of the "public trust," which says governments hold certain rights and entitlements in trust for the people and are obliged to protect them for the common good. Another rule fast becoming essential is the "precautionary principle," which says that it is wise to err on the side of protecting too much rather than too little.

The utilitarian code that guides most water management may fit with prevail-ing market-based socioeconomic paradigms, but it is neither universal nor unchanging. Ethics are not static; they evolve with social consciousness. But that evolution is not automatic. As societies wrap their collective minds around the consequences of global environmental change, a new ethic may emerge that says it is not only right and good but necessary that all living things get enough water before some get more than enough. Because in the end, we're all in this together.

....

Sandra Postel is director of the Global Water Policy Project. This box is adapted with per-mission from an article that appeared in The American Prospect *in April 2008.*

world's population lacked access to safe drinking water, and more than twice that number lacked access to safe sanitation, exposing billions of people to potential waterborne diseases.[22]

On a daily basis, humans require a minimum of about 13 gallons per person per day to maintain an adequate quality of life: 10 percent for drinking, 40 percent for sanitation, 30 percent for bathing, and 20 percent for cooking. The average person in Kenya uses only 3 gallons per day, while the average UK resident uses about 30. These numbers stand in stark contrast to the 150 gallons per day used for domestic and municipal purposes by residents of the United States and Canada.[23]

The true nature of the water crisis is, not that there is too little freshwater on Earth, but that humans fail to respect water as a resource and fail to manage it so that every living being dependent on it has a safe and adequate supply. Therefore, while water scarcity can be a product of physical limitations due to competing demands, unequal water distribution, pollution, environmental stresses, and unpredictable access to water, many people experience water scarcity due to flawed management policies, poverty, inequality, and unequal institutional power structures.[24]

Access to adequate supplies of clean water is central to human development and many of the UN's Millennium Development Goals. One such goal is to reduce by half the proportion of people living without access to safe and sustainable drinking water by 2015. Yet the biosocial roots of the water crisis persist. Addressing the politics of water scarcity is an essential part of the effort to reduce the proportion of the world's population lacking clean water.

Benjamin Franklin noted in *Poor Richard's Almanac*, "When the well is dry, we learn the worth of water." Water—so unevenly spatially and temporally distributed—is fast becoming our most valuable, and most contested, resource. Even in water-rich areas, that resource should not be taken for granted. There is inherent uncertainty in climate patterns, and good conservation habits can help make water supplies last. By 2050, the U.S. population is expected to increase by more than 100 million people, and the global population is likely to increase by 30 percent to more than 9 billion. This population growth and corresponding increase in food and energy production and urban expansion have serious implications for freshwater resources.

> The average person in Kenya uses only 3 gallons of water per day; the average American or Canadian uses 150.

It is essential, therefore, to recognize our roles as stewards of the world's water, so we can begin to find solutions to the challenges presented by the water crisis. These solutions include exploring conservation measures; working with nature to manage water resources; incorporating social, environmental, and more-accurate economic costs into the value of water; reducing waste in our water-delivery infrastructure; addressing the social and political roots of water stress and scarcity; and investing in appropriate and efficient technological approaches to increase water recycling and reduce water use and contamination.

> The true nature of the water crisis is, not that there is too little freshwater on Earth, but that humans fail to respect water as a resource and fail to manage it so that every living being dependent on it has a safe and adequate supply.

Our blue planet is our only home, and it contains all the freshwater that humans and other species rely on to survive. We must raise our water consciousness by recognizing the vital role of water in our lives. We must resolve to value and protect the most valuable resource of all.

REFERENCES

1. Sterling, E. J., and M. D. Camhi, 2007, Sold down the river, *Natural History*, November.

2. Harrison, P., and F. Pearce, 2000, *AAAS Atlas of Population and Environment*, Victoria Dompka Markham, ed., American Association for the Advancement of Science and the University of California Press, p. 51–54 from http://atlas.aaas.org/index.php?part=2&sec=natres&sub=water.

3. WorldWatch Institute, 2008, Boosting water productivity, http://www.worldwatch.org/node/811, and Barlow, M., 2008, *Blue Covenant: The Global Water Crisis and the Coming Battle for the Right to Water*, The New Press, New York.

4. Hinrichsen D., B. Robey, and U. D. Upadhyay, 1997, Solutions for a water-short world, *Population Reports*, Series M, Number 14, Population Information Program, Johns Hopkins School of Public Health, Baltimore, December.

5. Hinrichsen, D., B. Robey, and U. D. Upadhyay, 1997, Solutions for a water-short world, *Population Reports*, Series M, Number 14, Population Information Program, Johns Hopkins School of Public Health, Baltimore, December, http://www.infoforhealth.org/pr/m14/m14table.shtml#table1.

6. UN Environment Programme, 2002, Vital Water Graphics: An Overview of the State of the World's Fresh and Marine Resources, UNEP, Nairobi, Kenya, ISBN: 92-807-2236-0 from http://www.unep.org/dewa/assessments/ecosystems/water/vitalwater/.

7. Postel, S., 1998, Water for food production: Will there be enough in 2025? *BioScience* 48: 629–637.

8. Hinrichsen, D., B. Robey, and U. D. Upadhyay, Solutions for a water-short world.

9. FitzHugh, T. W., and B. D. Richter, 2004, Quenching urban thirst: Growing cities and their impacts on freshwater ecosystems, *Bioscience* 54 (8): 741–754.

10. United States General Accounting Office, 2003, Freshwater Supply: States Views of How Federal Agencies Could Help Them Meet the Challenges of Expected Shortages, July, http://www.gao.gov/new.items/d03514.pdf.

11. Ibid.

12. Postel, Daily, and Ehrlich, 1999, Human appropriation of renewable freshwater, *Science* 271: 5250.

13. Gleick, P., et al., 2006, *The World's Water 2006–2007: The Biennial Report on Freshwater Resources*, Island Press, Washington, DC.

14. Seckler, D., 1996, The New Era of Water Resources Management, Research Report 1. International Irrigation Management Institute (IIMI), Colombo, Sri Lanka, http://www.iwmi.cgiar.org/pubs/pub001/body.htm.

15. United States Geological Survey, *Ground Water Atlas of the United States*, http://pubs.usgs.gov/ha/ha730/ch_e/E-text5.html; Dennehy, K. F., 2000, High Plains Regional Ground-water Study, Fact Sheet FS-091-00, U.S. Geological Survey, Reston, VA, http://co.water.usgs.gov/nawqa/hpgw/factsheets/DENNEHYFS1.html.

16. UN Environment Programme, 1999, Global Environmental Outlook 2000, Earthscan, London, http://www.unep.org/geo2000/english/0046.htm.

17. Lui, J., 2007, Yangtze's Decline Highlights China's Growing Water Problems, http://www.worldwatch.org/node/5050.

18. Development Alternatives, 2001, Troubled Waters, http://www.devalt.org/water/WaterinIndia/issues.htm.

19. United States Environmental Protection Agency, 2008, Polluted Runoff (Non-Point Source Pollution): Section 319 Success Stories, Volume II, http://www.epa.gov/owow/nps/Section319II/intro.html.

20. Pan American Health Organization, 2003, CEPIS: Water Crisis Threatens Health and Development, http://www.paho.org/english/dd/pin/ptoday12_mar04.htm.

21. UN Environment Programme, 2006, Africa's Lakes: Atlas of Our Changing Environment, http://na.unep.net/AfricaLakes/AtlasDownload/PDFs/Africas-Preface-Screen.pdf.

22. WHO and UNICEF, 2006, Meeting the MDG Drinking Water and Sanitation Target: The Urban and Rural Challenge of the Decade, retrieved from http://www.wssinfo.org/en/40_MDG2006.html.

23. Gleick, P., et al., 2006, *The World's Water 2006–2007*.

24. UN Development Programme, 2006, Human Development Report, http://hdr.undp.org/en/media/HDR06-complete.pdf.

CHAPTER 15

············

The Biggest Footprint

Population and Consumption in the United States

VICKY MARKHAM

From a global population-environment perspective, the United States is unique: It is the only large industrialized nation that is still experiencing significant population growth, and it leads the world in per capita resource consumption and associated pollution and waste. As a result, Americans have the largest environmental impact, or "footprint," of any people on Earth.

This dubious leading role is not matched with real leadership in addressing our environmental impact. Assuming such leadership will require a fundamental shift in how we think about and act on

> Americans have the largest environmental impact, or "footprint," of any people on Earth.

these issues—both as individuals and as a nation. And it will require a holistic approach: We must understand, and address, both population change and resource consumption, as they relate to one another, in order to achieve long-term sustainability.

····································

Vicky Markham is the Director of the Center for Environment and Population (CEP). This chapter is adapted from the U.S. National Report on Population and the Environment and the U.S. Population, Energy & Climate Change report (both available on www.cepnet.org).

THE POPULATION-ENVIRONMENT INTERACTION
IN THE UNITED STATES

To understand U.S. population dynamics and their environmental consequences, it is necessary to examine which population trends are important and why. First, we explore the demographic factors that are having the greatest impact on the environment in the United States today.

Population Size and Growth

Human numbers serve as a "multiplier" of the effects of resource consumption. It is important to consider the country's relatively high population numbers (the United States is by far the largest developed nation) and how fast they are growing and expected to grow into the foreseeable future.

- *The U.S. population is growing at just under 1 percent annually, adding about 3 million people per year, or 8,000 per day.*[1] At this rate, the nation's population will double about every seventy years.[2]

- *The nation's population growth is caused by natural increase (births minus deaths) plus immigration.* Today, natural increase accounts for about 60 percent of U.S. population growth, with the remaining 40 percent due to net international migration.[3]

- *The U.S. population is projected to reach about 420 million by 2050.*[4] Over the next several decades, the U.S. population growth rate is expected to decline slightly, to reach about 0.8 percent by the middle of the century.[5]

Population Distribution

Environmental impact is determined not just by the number of people but by where they live. For example, denser population can increase environmental pressures (for instance, along fragile coastlines) or relieve them (through "clustered" rather than "sprawling" development).

- *Today, the South and West are the most populous and fastest-growing regions in the nation.* More than half of the total U.S. population (180 million people) currently lives in the South and West.[6] In the past seven years, the South and West grew four times as fast as the Northeast and Midwest,[7] and this trend will continue into the near future.

- *More than 50 percent of all U.S. residents live within 50 miles of the coast, crowding into just 17 percent of the nation's land area.*[8] An additional 25 million people, accounting for about half the projected U.S. population increase, are expected to move to these areas in the next decade alone. Over 80 million people visit the shore for recreation every year.[9]

- *Americans are choosing to live near natural amenities* (near lakes, mountains, and coastal and scenic areas). The population of U.S. counties where income and employment are based largely on outdoor recreational and entertainment activities grew 20 percent during the 1990s, mostly due to migration from other parts of the country.[10]

- *Temporary migration (vacation and seasonal living) is on the rise.*[11] More than one-third of the vacant homes nationwide (nearly 4 million) are seasonal, recreational, or for occasional use.[12] Seasonal growth spurts can trigger environmental impacts, such as more land and resources used to build second homes, increased water and energy demands, and traffic congestion.[13] This environmental stress is most noticeable in high-tourism states (such as Florida and California) and in many of the nation's most popular national parks and other protected areas.[14]

- *Changes in population distribution can create population-environment "hot spots."* When populations grow in particularly environmentally sensitive areas, severe and sometimes irreversible consequences can result. For example, in the arid Southwest, population growth exacerbates the region's chronic water scarcity and destroys fragile desert ecosystems. In the South, many people have moved to coastal areas where development drains wetlands and coastal salt marshes, eliminating life-supporting habitat for a wealth of plants and wildlife and making the coastlines more vulnerable to storms and erosion.[15] Four of America's top ten fastest-growing states are coastal, while four more are in the driest part of the country.[16]

- *America has become a metropolitan nation.* More than 80 percent of Americans (226 million people) lived in metro areas (cities and their suburbs) in 2000.[17] Although there are many differences among regions, one thing they have in common is that nearly all of the nation's large metropolitan areas (with populations of at least 250,000) grew in the past decade.[18] Most metropolitan areas have been developing land faster than their populations have been growing. In the 1980s and 1990s, the U.S. population grew by 17 percent, and developed land grew by 47 percent.[19]

- *Growth outside cities in the suburban and surrounding "exurban" areas far outpaces growth within cities.*[20] Over the past decade, the population of America's metropolitan areas outside central cities increased by nearly 13 percent, while growth inside central cities grew by 4 percent.[21]

Population Composition

The composition, or makeup, of a population—its age, income, education, and other characteristics—can determine where and how people live, move, vacation, develop land, and consume natural resources.

- *A large youthful population will keep the U.S. population growing.* Today, 34 percent of Americans are age twenty-four or under, and another 28 percent are between twenty-five and forty-four.[22] This relatively large cohort of youth ensures that momentum for significant future population growth is already in place.

- *Today's older population is larger than it has ever been.* The overall median age in 2006 (36.4 years) was also higher than at any time in the nation's history.[23] The nation's aging baby boomers (born between 1946 and 1964) number 78 million, or 26 percent of the total U.S. population.[24] They are wealthier, spend more money, consume more resources, have more homes per capita, and move more often than any generation before them.[25] In addition, a substantial share of America's population aged sixty-five and older is settling in "retirement magnet" states such as Arizona, Florida, and Nevada, where pressure on natural resources (especially water) is already evident. Over the next quarter century, the proportion of elderly Americans is projected to double in at least fourteen states in the South and West.[26]

Households

Households are an important demographic unit for measuring a population's environmental impacts. Every household has a minimum number of possessions, occupies a certain amount of space, and emits certain waste and/or pollutants. The extent of environmental stress depends on household size (the number of people sharing a home), the size of homes, the amount of land surrounding and used to build homes, and, of course, the number of households.[27]

- *The average number of people per household in the United States has declined, and the number of households has multiplied.* The average household had 2.6 people in 2000, down from 3.1 in 1970 (or 1 fewer person for every two households).[28] In part, declining household size is due to a rising divorce rate: The proportion of households with divorced heads increased from 5 percent in 1970 to 15 percent in 2000.[29] Smaller household size, combined with robust population growth, has spurred a nationwide building boom: Between 1990 and 2000, 14 million new housing units were built.[30]

- *Living space in and around homes has gone up.* Between 1970 and 2000, the average size of new single-family homes increased by more than 700 square feet.[31] During that period, the proportion of homes nationwide that are 3,000 square feet or more nearly doubled (from 11 percent to 20 percent), while the proportion under 1,200 square feet declined (from 12 percent to 5 percent).[32] Although the average lot size of single-family homes has varied over time, 67,000 new homes on the largest lots (from 9,000 to over 22,000 square feet) were sold between 1999 and 2003.[33] In recent years, lot sizes have grown the most on the outskirts of metro areas, resulting in sprawl and the rapid development of already dwindling open spaces.

Resource Consumption and Affluence

Resource consumption is often associated with a population's level of income, or affluence. Evidence shows that, as a whole, more-affluent people consume more resources and generate more waste and pollution than their lower-income counterparts.

- *The United States is among the world's wealthiest nations.* America has a per capita income of nearly $40,000, compared with about $26,000 for other developed countries, $4,000 for developing countries, and $9,000 globally.[34] (Our environmental impact is commensurate with our relative affluence; see Box 15.1.)

ENVIRONMENTAL IMPACTS OF U.S. POPULATION GROWTH AND RESOURCE CONSUMPTION

The trends described above affect ecosystems, natural resources, and plant and animal species, both inside and outside U.S. borders. They have significant effects on land use, water, forests, biodiversity, fisheries and aquatic resources, agriculture, energy, climate change, and waste.

- *Land Use.* Today U.S. land is converted for development at about twice the rate of population growth.[35] Each American effectively occupies about 20 percent more developed land (for housing, schools, shopping, roads, and other uses) than he or she did twenty years ago.[36] The nation's most predominant form of land-use change is "sprawl": low-density development spread into suburban

BOX 15.1

Our Ecological Footprint

The "Ecological Footprint"[a] is a rough measure of humanity's demand on nature. The size of our footprint is calculated by determining how much land and water area a human population requires to produce the resources it consumes and to absorb its wastes, using prevailing technology.

Globally, humanity's Ecological Footprint now exceeds the planet's regenerative capacity by about 30 percent. It now takes the Earth one year and four months to regenerate what we use in a year. Turning resources into waste faster than waste can be turned back into resources puts us in global ecological "overshoot," depleting the very resources on which human life and biodiversity depend.

The result of overshoot is collapsing fisheries, diminishing forest cover, depletion of freshwater systems, and the buildup of pollution and waste, the sum of which creates problems like global climate change. Overshoot also contributes to resource conflicts and wars, mass migrations, famine, disease, and other human tragedies—and tends to have a disproportionate impact on the poor, who cannot buy their way out of the problem by getting resources from somewhere else.

Our demand on the planet has more than doubled over the past forty-five years as a result of population growth and increasing individual consumption. In 1961, almost all countries in the world had more than enough capacity to meet their own demand; by 2005, the situation had changed radically, with many countries able to meet their needs only by importing resources from other

and rural areas, with increased vehicle use and new houses, roads, shops, and other infrastructures. More than half (53 percent) of the nation's wetlands have been lost, mainly due to urban/suburban development and land-use change for agriculture.[37]

- *Water.* U.S. public freshwater supply withdrawals and the human population both grew by 8 percent from 1995 to 2000.[38] About 40 percent of the nation's rivers, 46 percent of lakes, and 50 percent of estuaries are now too polluted for fishing and swimming.[39] Only 2 percent of U.S. rivers and streams remain free-flowing; the rest have been dammed or diverted.[40] America is among the world's top ten in per capita water withdrawal, with each American using three times the world average.[41]

nations and by using the global atmosphere as a dumping ground for carbon dioxide and other greenhouse gases.

More than three-quarters of the world's people live in nations that are "ecological debtors"—their national consumption has outstripped their country's biocapacity. Thus, most of us are propping up our current lifestyles, and our economic growth, by drawing upon (and increasingly overdrawing) the ecological capital of other parts of the world.

National footprints vary widely. In 2005, the United States and China had the largest total footprints, each using 21 percent of the planet's biocapacity. China had a much smaller per person footprint than the United States but a population more than four times as large. Per capita, the United Sates has the biggest footprint of any country on Earth except the United Arab Emirates—about 9.5 global hectares per person. At the other end of the spectrum, citizens of Malawi used less than 1 hectare each. If everyone in the world lived as Americans do, we would need five Earths to support us.

If we continue with business as usual, by the early 2030s we will need two planets to keep up with humanity's demand for goods and services. But there are many effective ways to change course.

....

[a]Conceived by Mathis Wackernagel and William Rees at the University of British Columbia, the Ecological Footprint is now in wide use by scientists, governments, and others working to monitor ecological resource use and advance sustainable development. This box is adapted from the 2008 *Living Planet Report*, published by the World Wildlife Fund, the Zoological Society of London, and Global Footprint Network.

- *Forests.* The United States is the world's largest consumer of forest products—in the last four decades alone, U.S. wood consumption overall grew by 50 percent.[42] In 2000, per capita sawn-wood consumption was nearly twice that of developing countries and ten times the world's average.[43] Today nearly twice as many people are being supported by the same forested area that existed a hundred years ago.[44]

- *Biodiversity.* About 6,700 known plant and animal species are considered at risk of extinction in the United States.[45] Almost 1,000 species are listed by the U.S. government as endangered, and 300 as threatened (over twice the number listed a decade ago), mainly (85 percent) from habitat loss and alteration.[46] Half of the continental United States can no longer support its original vegetation.[47]

- *Fisheries and Aquatic Resources.* Thirty percent of assessed fish populations in U.S. coastal waters are either overfished or fished unsustainably.[48] A third of all U.S. lakes, a quarter of the rivers, all of the Great Lakes, and two-thirds of the nation's coastline were under a fish consumption advisory from pollutants in 2004,[49] many related to mercury contamination.[50] Nearly 40 percent of North American freshwater animal species are in serious decline and at risk of disappearing.[51]

- *Agriculture.* Nearly 3,000 acres of U.S. farmland are lost every day to development, with the rate of loss increasing.[52] Because cities often arose near fertile cropland, that land is now being overtaken by suburban sprawl. America's prime farmland was developed 30 percent faster than other rural land in the past two decades.[53]

- *Energy.* With only 5 percent of the global population, the United States consumes almost 25 percent of the world's energy.[54] America consumes more oil than any other nation.[55] The American residential sector uses more energy than that of any other country, with household appliances the fastest-growing energy consumers in the United States, next to cars. The commercial and transportation sectors are projected to be the nation's fastest-growing energy-use sectors from 2005 to 2030.[56]

- *Climate Change.* The United States uses more energy than any other country and is the single largest emitter of carbon dioxide (CO_2) greenhouse gas among industrialized nations, accounting for nearly a quarter of all global emissions.[57] The nation's average temperature increase over the next hundred years is projected to be 5 to 9 degrees Fahrenheit.[58] Predictions of sea level rise and more severe weather events that will impact coastal areas are predicted, particularly in the U.S. Mid-Atlantic and Gulf coasts.

- *Waste.* Each American produces about 5 pounds of trash daily, up from less than 3 in 1960, five times the average amount in developing countries.[59] Indeed, the United States is the largest per capita municipal waste producer in the world.[60] Our municipal and industrial waste threatens other resources: Nearly half the nation's 1,300 Superfund toxic waste sites are contaminating or threatening drinking-water supplies.[61]

CONCLUSION: CHOICES MATTER

To reverse the trends described above, we must make environmental sustainability a national priority—a standard that informs decision-making at all levels. Because Americans have such an outsize environmental footprint, our choices—both public and private—have enormous impact. That is why the goal of a sustainable environment must guide our public choices, from school curricula to the investments and policies of business and government at all levels. At the same time, we must each consider the environment in our personal choices—decisions about what we buy, where and how we live, and even deliberations about family size—because it is all linked.

As we have seen, human numbers alone do not determine environmental impact. That impact results from the complex interplay of resource use and population dynamics. For example, 20,000 people in a sprawling suburban development in the arid West may do more environmental damage than the same population living more compactly on less-fragile land. But, while resource use is belatedly receiving some of the attention it deserves, population dynamics are rarely considered in current environmental discourse. If we are to tackle the monumental environmental threats we face today, that must change.

Climate change provides a case in point. Forestalling further human-induced warming will require a multifaceted approach: Americans must choose to reduce energy use and greenhouse gas emissions *and* address the population dynamics that serve to magnify the impact of our consumption.

On the energy side, this includes changes in our everyday consumer choices—walking instead of driving, eating less meat, replacing incandescent lightbulbs with energy-efficient compact fluorescents. It also includes local, state, and federal governments' commitment to and financial incentives for energy efficiency and greenhouse gas reduction. These might include local-national plans and tax/financial incentives for energy-efficient mass transport, household appliances, heating and cooling, and building codes. Federal research for new energy-efficient technologies should be a priority, and such products should be made available to all Americans at every price point.

On the population side, this includes acknowledgment that population dynamics are part and parcel of climate-change challenges and solutions. But what does that mean, exactly? Again, choices matter. In the United States, a large youthful population guarantees that our numbers will continue to grow, and the choices made by young Americans will determine the magnitude of that growth. But too many young people in the United States lack the means and knowledge to make thoughtful choices about childbearing. Despite recent decreases, U.S. rates of teen pregnancy, childbirth, abortion, and sexually transmitted diseases are among the highest of all industrialized nations.[62, 63] Nearly half of all pregnancies among American women are unintended—a far higher percentage than in other developed countries.[64] By making high-quality reproductive health care and family planning more accessible and affordable, we can improve the health and well-being of American women and families. At the same time, reducing unintended pregnancy would slow population growth and shrink our collective footprint.

> Despite recent decreases, U.S. rates of teen pregnancy, childbirth, abortion, and sexually transmitted diseases are among the highest of all industrialized nations.

We know more than ever, with scientific certainty, about the role of population dynamics in environmental change. It is time to put that knowledge into action to maintain a healthy, sustainable planet for all generations.

REFERENCES

1. Population Reference Bureau, 2005, *2005 World Population Data Sheet*, www.prb.org, and CEP calculation.

2. Doubling times are calculated using the formula provided (69/annual growth rate) by Population Reference Bureau, Human Population: Fundamentals of Growth, Population Growth and Distribution, http://www.prb .org/Content/NavigationMenu/PRB/Educators/Human_Population/ Population_Growth/Population_Growth.htm, accessed 2005.

3. U.S. Census Bureau, 2005, *Population Estimates,* www.census.gov, Tables 11 and 12.

4. U.S. Census Bureau, 2005, *Statistical Abstract of the United States,* Table 3, Resident Population Projections: 2004 to 2050, and Table 2, Population: 1960 to 2003.

5. U.S. Census Bureau, 2008, *National Population Projections,* Table 1, Projections of the Population and Components of Change for the United States: 2010 to 2050.

6. U.S. Census Bureau, *2007 Population Estimates,* www.census.gov, Table GCT-T1, Population Estimates, accessed 2008.

7. Ibid.

8. Beach, Dana, 2002, *Coastal Sprawl: The Effects of Urban Design on Aquatic Ecosystems in the United States,* and Beach, Dana, *America's Oceans in Crisis, 2006.*

9. Ibid.

10. Johnson, Kenneth M., and Calvin L. Beale, 2002, Nonmetro recreation counties: Their identification and rapid growth, *Rural America* 17 (4).

11. Markham, Victoria D., et al., 2003, *U.S. State Reports on Population and*

the Environment: New Hampshire, Center for Environment and Population (CEP) with National Wildlife Federation, at www.cepnet.org.

12. U.S. Census Bureau, Census 2000 Summary File 1: Vacant Housing Units, Vacancy Status, search by region and state, http://factfinder.census.gov/servlet/DTGeoSearchByListServlet?ds_name=DEC_2000_SF1_U&state=dt&_ts=106518818243, accessed 2004.

13. Markham, Victoria D., et al., 2003, *U.S. State Reports on Population and the Environment: New Hampshire,* Center for Environment and Population (CEP) with National Wildlife Federation, at www.cepnet.org.

14. National Park Service, Public Use Statistics Office, *Statistical Abstract 2003,* Table 3: 2003 Recreation Visits/Visitor Days by Region, and Table 2: 2003 Recreation Visits by State, and Table 4: 2003 Percent of Recreation Visits by Population Center, http://www2.nature.nps.gov/stats/abst2003.pdf, accessed 2004.

15. U.S. Census Bureau, 2003, *Statistical Abstract of the United States,* Table 24, Population in Coastal Communities: 1970 to 2002.

16. U.S. Census Bureau, 2007, *Population Estimates,* Table: Population Change: April 1, 2000 to July 1, 2007.

17. Perry, Marc J., and Paul J. Mackun, 2001, *Population Change and Distribution, 1990 to 2000,* U.S. Census Bureau.

18. U.S. Census Bureau, 2004, *Statistical Abstract of the United States,* Table 24: Large Metropolitan Statistical Areas–Population: 1990 to 2003.

19. Fulton, William, et al., 2001, *Who Sprawls Most? How Growth Patterns Differ Across the U.S.,* Brookings Institution Center on Urban and Metropolitan Policy.

20. Ibid.

21. Op. cit., Perry, Marc J., 2001.

22. U.S. Census Bureau, *2007 Population Estimates,* Table Dp-1, General Demographic Characteristics, http://www.census.gov/, accessed 2008.

23. Ibid.

24. U.S. Census Bureau, 2006, Oldest baby boomers turn 60 (Facts for Features Press Release), January 3, http://www.census.gov/Press-Release/www/releases/archives/facts_for_features_special_editions/006105.html.

25. Gillion, Steven, 2006, *Boomer Nation: The Largest and Richest Generation Ever, and How It Changed America.*

26. Population Reference Bureau, Ameristat, 2000, *Baby-Boomer Retirees Changing the U.S. Landscape.*

27. Liu, J., et al., 2003, Effects of household dynamics on resource consumption and biodiversity, *Nature* 421: 530–533.

28. Population Reference Bureau, Ameristat, 2003, *While U.S. Households Contract, Homes Expand.*

29. Yu, E., and J. Liu, 2007, Environmental impacts of divorce, *Proceedings of the National Academy of Sciences* 104(51): 20629–20634.

30. U.S. Census Bureau, 1990 Census, Table ST-98-47: Housing Units, Households, Households by Age of Householder, and Persons per Household, http://eire.census.gov/popest/archives/household/sthuhh2.txt, accessed June 10, 2004; 2000 Census, State Housing Unit Estimates: April 1,

2000 to July 1, 2002, http://eire.census.gov/popest/data/household/HU-EST2002-01.php, accessed 2004.

31. Population Reference Bureau, Ameristat, 2003, *While U.S. Households Contract, Homes Expand.*

32. U.S. Census Bureau, Manufacturing, Mining, and Construction Statistics: Square Feet of Floor Area in New One-Family Houses Completed, http://www.census.gov/const/C25Ann/sftotalsqft.pdf, accessed 2004.

33. U.S. Census Bureau, Manufacturing, Mining, and Construction Statistics: Characteristics of New Housing, Lot Sizes of New One-Family Housing Sold, http://www.census.gov/const/C25Ann/lotsizesold.pdf, accessed 2005.

34. Population Reference Bureau, 2005, *2005 World Population Data Sheet.*

35. Sierra Club, A Complex Relationship: Population Growth and Suburban Sprawl, http://www.sierraclub.org/sprawl/population/factsheet.asp#2, accessed 2004.

36. Calculation based on population figures from U.S. Census Bureau, *Statistical Abstract* (1982 and 2003) and developed land acreage figures from U.S. Department of Agriculture, Natural Resources Conservation Service, *National Resources Inventory: 2002 Annual NRI Highlights.*

37. New England Coastal Basins Program, U.S. Geological Survey, 2005, Mercury in Water, Sediment, and Fish, http://nh.water.usgs.gov/projects/nawqa/sw_merc.htm.

38. Hutson, Susan S., et al., 2000 (updated 2005), Estimated Use of Water in the U.S., 2000, U.S. Geological Survey Circular Number 268.

39. American Rivers, 2004, *America's Most Endangered Rivers of 2004: Ten Rivers Reaching the Crossroads in the Next 12 Months.*

40. The Nature Conservancy, 2003, Sustainable Waters and Sustainable Rivers program description, http://www.freshwaters.org/eswm/sustrivs/, accessed 2005.

41. NationMaster.com, accessed 2006.

42. U.S. Forest Service, 2000, *State of Forestry in the United States of America: An Overview,* U.S. Department of Agriculture.

43. U.S. Forest Service, 2000, *State of Forestry in the United States of America: An Overview,* U.S. Department of Agriculture; U.S. Department of Agriculture, *National Report of Sustainable Forests, 2003,* USDA, Forest Service.

44. Society of American Foresters, http://www.safnet.org/aboutforestry/facts.cfm, accessed 2006.

45. Cohn, Jeffrey P., and Jeffrey A. Lerner, 2003, Integrating Land Use Planning and Biodiversity, Defenders of Wildlife.

46. U.S. Fish and Wildlife Service, Summary of Listed Species as of 11/01/2005, http://ecos.fws.gov/tess_public/servlet/gov.doi.tess_public .servlets.TESSBoxscore?format=display&type=archive&sysdate=11/01/2005, accessed 2005.

47. Biodiversity Project, 2000, *Getting on Message: Making the Biodiversity-Sprawl Connection,* Biodiversity Project, Madison, WI.

48. Pew Oceans Commission, 2003, *America's Oceans in Crisis.*

49. U.S. Environmental Protection Agency, Office of Water, 2005, National Listing of Fish Advisories, Fact Sheet EPA-823-F-05-004.

50. Natural Resources Defense Council, 2008, 700 freshwater fish species

at risk, survey says, http://www.nrdc.org/news/newsDetails.asp?nID=3019, accessed 2008.

51. Union of Concerned Scientists, The Arctic National Wildlife Refuge: Is Loss of a Pristine Wilderness Worth the Oil That Might Be Gained? http://www.ucsusa.org/global_environment/archive/page.cfm?pageID=780 #eco, accessed 2005.

52. American Farmland Trust, 2002, *Farming on the Edge: Sprawling Development Threatens America's Best Farmland.*

53. Ibid.

54. Markham, Victoria D., (ed.), Paul Harrison, and Fred Pearce, 2000, *AAAS Atlas of Population and Environment*, American Association for the Advancement of Science (AAAS), at www.cepnet.org and www.aaas.org.

55. BP Statistical Review of World Energy, 2005 (U.S. percent share of total global oil consumption).

56. U.S. Energy Information Administration, 2008, *Annual Energy Outlook, 2008* (revised early release), Table 2, Energy Consumption by Sector and Source, http://www.eia.doe.gov/oiaf/aeo/index.html, accessed 2008.

57. U.S. Energy Information Administration, 2005, *U.S. Emissions of Greenhouse Gases in the United States 2004.*

58. National Assessment Synthesis Team, 2000, *U.S. National Assessment: The Potential Consequences of Climate Variability and Change*, Summary, U.S. Global Change Research Program.

59. U.S. Environmental Protection Agency, Basic Facts: Municipal Solid Waste, http://www.epa.gov/epaoswer/non-hw/muncpl/facts.htm.

60. OECD Environmental Data Compendium, Nationmaster.com, accessed 2006.

61. Environmental Defense, Environmental Scorecard, Superfund Report: Entire United States, http://www.scorecard.org/env-releases/land/us.tcl#trends accessed 2006.

62. U.S. Centers for Disease Control, NCHS Data on Teen Pregnancy, http://www.cdc.gov/nchs/data/infosheets/infosheet_teen_preg.htm, accessed 2008.

63. Guttmacher Institute, 2001, *Teenage Sexual and Reproductive Behavior in Developed Countries: Can More Progress Be Made?* Guttmacher Institute, Washington, DC.

64. Guttmacher Institute, An Overview of Abortion in the U.S., http://www.guttmacher.org/media/presskits/2005/06/28/abortionoverview.html, accessed 2008.

CHAPTER 16

....

Population Growth, Reproductive Health, and the Future of Africa

DR. FRED T. SAI

Africa's colonial legacy—exploitation, artificial borders, too many small and unviable countries after independence—is the chief source of a catalogue of misfortunes that is by now familiar: civil and regional conflict, famine and hunger, land degradation, corruption, and ill health. Rapid population growth is not the cause of these problems, but it does make them more difficult to solve. Better access to reproductive health services would help slow population growth, while measurably improving the health and well-being of Africa's people.

Those improvements are badly needed. Africa has seen considerable economic gains—overall, growth in sub-Saharan economies has topped 5 percent a year since 2005. But the continent still lags behind all other developing regions. In 2006, the nations of sub-Saharan Africa, with 782 million people, had a collective gross national income of $648 billion—a per capita average of well under $2.50 per day. And of that $648 billion, South Africa accounted for the largest share ($255 billion), followed by Nigeria ($90 billion), leaving only $303 billion for the remaining 590 million Africans—a per capita average of only $1.40 per day.[1]

Dr. Fred T. Sai, a Ghanaian family health physician and Presidential Advisor on Reproductive Health and HIV/AIDS, served as Chairman of the United Nations International Conference on Population and Development in 1994 and as President of the International Planned Parenthood Federation from 1989 to 1995.

The human condition in Africa is daunting. Of the twenty-two countries on the UN 2007–2008 list of low human development indicators, all are in Africa. In these countries, illiteracy rates are high, infrastructure is inadequate, and health services are rudimentary. For decades, a staggering debt burden sapped funds that could have been used to boost human development in Africa, though recent debt relief efforts by donor countries—the Heavily Indebted Poor Countries Initiative and the Multilateral Debt Relief Initiative—have begun to alleviate that burden.[2] And while some progress has been made in reducing political instability and civil unrest, much more must be done to sustain economic growth, durable peace, and equitable income distribution.

> Better access to reproductive health services would help slow population growth, while measurably improving the health and well-being of Africa's people.

POPULATION AND PROGRESS

Rapid population growth frustrates Africa's progress toward economic and human development. It has often been said that the one thing Africa has is plenty of land and that consequently population growth is not a bad thing. It is true that Africa's population density, at 65 people per square mile, is low compared with the world average of 115. But this simplistic view does not take into account that the natural carrying capacity of much of the land in Africa is low and subject to the vagaries of a capricious climate. It also ignores the fact that it is the *rapidity* of population growth that causes so much stress and suffering. Indeed, even after peaking in the late 1970s, Africa's population growth rates remain the highest in the world—far higher than those experienced in the past by the now-developed world, or by other developing regions today.

That rapid growth is increasingly a concern for African governments. Today, three-quarters of Africans live in countries with governments that view their population growth rates as too high. This is one of the most significant population policy developments of recent years: In 2005, two-thirds of African governments enacted policies to slow population growth, up from 25 percent in 1976.[3] Concern about population growth is both social and economic: How will governments provide schools, health care, and jobs for the next generation?

There is also the fundamental question of food: In many regions, particularly the Sahel, the human population is growing faster than the food supply. Although a net food exporter before 1960, Africa has become more dependent on food imports and aid over the last four decades. Agricultural production has slightly improved in recent years but is struggling to keep pace with population growth. Even before the latest food crisis began in 2008, the number of undernourished people in Africa had more than doubled since the late 1960s to more than 200 million.[4] In the summer of 2008, the Africa Progress Panel, led by former UN Secretary-General Kofi Annan, warned of "a significant increase in hunger, malnutrition, and infant and child mortality" and noted that "many countries are already experiencing the reversal of decades of economic progress and 100 million people are being pushed back into absolute poverty."[5]

> Three-quarters of Africans live in countries with governments that view their population growth rates as too high.

The food crisis that began in 2008 was triggered by a rapid rise in food prices; many of its underlying causes are described in Chapters 12 and 13. Population growth and pressure on the environment also play significant roles.[6] Although most Africans depend on the land for their livelihood, the land's capacity to produce is ebbing away under the pressure of growing numbers of people who do not have the wherewithal to put back into the land what they are forced to take from it. Africa is losing nearly 10 million acres of forest every year—twice the world's average deforestation rate—according to the United Nations Environment Programme (UNEP). At the same time, land degradation is a major worry for thirty-two countries in Africa. Erosion and chemical and physical damage have degraded about 65 percent of the continent's farmlands.[7]

As the land's vegetative cover shrinks, its already fragile soils lose the capacity to nourish crops and retain moisture. Agricultural yields fall, and the land becomes steadily more vulnerable to variable rainfall, turning dry spells into drought and periods of food shortage into famines. In most parts of Africa, farmers say that it is more difficult to make ends meet, that plots are much smaller and farther away, that fallow periods are shorter. All these trends impose extra strains on women, who are usually responsible for growing the family vegetables, fetching water, and gathering fuelwood.

Another serious problem is water. Africa is the world's second driest continent, after Australia. Over 300 million people on the continent already face water scarcity, and areas experiencing water shortages in sub-Saharan Africa are expected to increase by almost a third by 2050.[8] Water shortages will increasingly become a constraint on economic and social development, especially in countries with limited water supplies, rapid population growth, and/or fast-expanding industry and agriculture. (See Chapter 14.)

WHY AFRICA'S POPULATION IS GROWING SO RAPIDLY

Africa's population is growing, first, because death rates have fallen as a result of immunization and improvements in health care. Second, fertility remains high in most African countries. Total fertility rates (the average number of children each woman will have in her lifetime) vary from a low of 2.7 in South Africa to 5.9 in Nigeria and 7.1 in Mali and Guinea-Bissau.[9] For most of the African countries in the high-fertility group, projections indicate a doubling or tripling of population size, even with an assumed decline in fertility. If fertility were to remain constant at current levels, there would be a fivefold increase in population.[10]

Theories abound about why high fertility persists in many African countries. Some contend that high fertility simply reflects a preference for large families: According to a 2007 paper by the World Bank, "The main reason for not using contraception in many high-fertility countries in Sub-Saharan Africa is a desire to have more children, rather than unawareness about fertility control or lack of access to contraception."[11] But this is arguable. While many Africans want large families—especially where poverty is severe and the status of women is low—there is also considerable unmet need for family planning and other reproductive health services, as evidenced by the very high rate of unsafe abortion. The World Health Organization estimates that 4.2 million unsafe abortions occur in Africa each year, resulting in about 30,000 deaths annually.[12] This can only be seen as evidence of women's frustrated desire to control their fertility.

In sub-Saharan Africa as a whole, nearly a quarter of women of reproductive age are believed to lack access to family planning services; that is, they report that they want no more children or want to delay the next pregnancy by two years or more but are not using contraception.[13] Contraceptive use remains low in sub-Saharan Africa, where only 24 percent of married women use contraceptives, compared with 53 percent in north Africa and the Middle East, 46 percent in south Asia, 79 percent in east Asia and the Pacific, and 71 percent in Latin America and the Caribbean.[14] Satisfying unmet need for contraception would greatly diminish the percentage of unplanned, unwanted, or mistimed pregnancies.

High fertility is also related to the low status of African girls and women. Many girls are born prematurely or at low birth weight because their own mothers were malnourished, ill, or overworked. If she survives infancy, an African girl will most likely grow up on a diet that does not meet her minimum nutritional requirements. As a child, she will have a heavy burden of household chores and will probably receive less education than her brothers. She is likely to be married off young, especially if a good bride price is available, and taught that her main role in life is to bear and rear as many children "as God brings." During pregnancy, her needs for adequate rest, good nutrition, and health care will too often be ignored—so the burden of poor health will be passed to the next generation.

REPRODUCTIVE HEALTH IN AFRICA: AN URGENT NEED

Low social status and a lack of reproductive health services are a deadly combination for African women. Today, pregnancy and unsafe abortion are the leading causes of death among women of reproductive age in most African countries. The maternal mortality rate is higher in Africa than on any other continent: 820 deaths per 100,000 live births for the continent as a whole in 2005, and an average of 900 for sub-Saharan Africa. In some countries it is believed to be as high as 2,000 deaths per 100,000 live births.[15] This amounts to 270,000 deaths per year—over 700 deaths every day. The high mortality rate masks an even higher morbidity rate: The same afflictions that kill hundreds of thousands of women maim and render sterile many millions more of their sisters. For every woman who dies, fifty to a hundred others suffer short-, medium-, or long-term debilities from their pregnancies and deliveries.

> Pregnancy and unsafe abortion are the leading causes of death among women of reproductive age in most African countries.

Lack of reproductive health services also fuels the epidemic of sexually transmitted infections (STIs), including HIV/AIDS. According to the World Health Organization, STIs have become the most common group of notifiable diseases (those which must be reported to the public health and/or police authorities) in most countries worldwide, but prevalence rates are particularly high in Africa. WHO estimated in 2001 that there are 69 million new cases of curable STIs (syphilis, gonorrhea, chlamydia, and trichomoniasis) every year in sub-Saharan Africa—the highest rate of new cases in any region.[16]

Africa is more seriously affected by HIV/AIDS than any other region. In 2007, over two-thirds (67 percent) of all persons infected with HIV were living in sub-Saharan Africa—22 million. An estimated 11.9 million adults and children became infected with HIV in 2007, more than in all other regions of the world combined. The 1.5 million AIDS deaths in sub-Saharan Africa in 2007 represent three-quarters of global AIDS deaths. There are 11.6 million AIDS orphans in Africa—out of a global total of 15 million.[17]

Across Africa, women bear a disproportionate part of the AIDS burden: Not only are they more likely than men to be infected with HIV, but in most countries they are also more likely to care for those sickened by AIDS. Like other STIs, AIDS affects more women than men; over fourteen African women are currently infected for every ten African men.[18] Factors such as inferior health and social status, polygamy, other STIs, malnutrition, ear piercing, genital mutilation, and menstruation all render women more vulnerable to HIV transmission. When women are sick with AIDS or die, their children suffer, even if they were not infected themselves at birth.

A GROWING POLITICAL COMMITMENT

The most important factor in containing the spread of HIV is political commitment. Increasingly, African leaders are speaking out loudly, clearly, and repeatedly about AIDS, seeking to destigmatize it and encouraging discussion about safe sex everywhere from the classroom to the boardroom. As a result, the outlook for Africa is improving. In 2001 the members of Africa Unity (AU) agreed to give more emphasis to HIV/AIDS and recommended the allocation of 15 percent of national budgets to health and the integration of HIV and sexual and reproductive health programs. NGOs are increasingly accepted as legitimate partners of governments in the field of sexual and reproductive health. The participation of communities is either being actively promoted or being considered.

Political commitment, openness, and determination have even proved that AIDS can be reversed. There might even be said to be a thin silver lining to the HIV epidemic. It has been a much-needed wake-up call to many African leaders to address the many problems of the poor, and the result, encouraged as well by pressures from outside, has been more support for poverty alleviation and reproductive health. And there is a genuine effort to push for more responsive and democratic governance, a truly basic requirement of people-oriented development.

At the same time, there is revitalized political commitment to other aspects of reproductive health. Certainly most African governments have now come to accept that fewer, better-spaced births lead to healthier children and lower maternal mortality and morbidity. Meeting in Dakar in 2004, African ministers agreed that the Millennium Development Goals cannot be achieved unless the Cairo Program of reproductive health and gender equality is fully implemented. Apart from being important ends in themselves, gender equality and the empowerment of women were "key to breaking the cycle of poverty and improving the quality of life of the people of the continent," the leaders declared. The ministers also pledged to increase efforts to prevent, diagnose, and treat HIV/AIDS and other sexually transmitted infections, within the context of sexual and reproductive health.

In April 2007, the African Union Conference of African Health Ministers, meeting in Johannesburg, South Africa, adopted the Africa Health Strategy: 2007–2015, which significantly incorporated the African Charter's Protocol on the Rights of Women. This protocol is groundbreaking in a number of respects, not least in the sphere of reproductive rights, and abortion rights in particular: It is the first international human rights instrument to explicitly provide for abortion as a right in specific situations.[19]

The Africa Health Strategy also acknowledges in its introduction that, despite some progress, "Africa is still not on track to meet the health Millennium Declaration targets and the prevailing population trends could undermine progress made." Hopefully the milestone Africa Health Strategy will lead to international recognition that family planning and reproductive health programs are critical for health, demographic, and general development reasons—and encourage more investments in this area.

Unfortunately, as the nations of Africa have solidified their support for reproductive health, the priorities of donor countries and development agencies have shifted toward other issues. As global funds and initiatives bypass funding of family planning, less attention is being focused on the consequences of high fertility. Indeed, according to the World Bank, "less assistance is now available to

> As the nations of Africa have solidified their support for reproductive health, the priorities of donor countries and development agencies have shifted toward other issues.

these countries to meet the growing demand for family planning and related services than in previous decades, indicating an urgent need to refocus the attention of donors on completing what has been called a 'development success story.'"[20]

If Africa is to expand its recent economic gains and achieve a sustainable balance between resource use, population growth, and human development, donor nations must ramp up their investment in reproductive health. Given Africa's recent successes and new commitment, that investment could greatly improve the lives of African people.

REFERENCES

1. *World Development Indicators 2008*, World Bank, Washington, DC.

2. *Africa's Development: Promises and Prospects*, 2008, Report of the Africa Progress Panel.

3. *World Population Policies 2005*, UN, New York.

4. *State of Food & Agriculture 2006*, FAO, Rome.

5. *Africa's Development: Promises and Prospects*, 2008.

6. Soaring food prices put further pressure on African agriculture, 2008, FAO Newsroom, June 19, Nairobi/Rome, accessed at http://www.fao.org/newsroom/en/news/2008/1000868/index.html.

7. UNEP, 2008, *Africa: Atlas of Our Changing Environment*, http://na.unep.net/AfricaAtlas/AfricaAtlas/.

8. Ibid.

9. Population Reference Bureau Datafinder, http://www.prb.org, accessed December 2008.

10. Ibid.

11. *Population Issues in the 21st Century: The Role of the World Bank*, 2007, World Bank, Washington, DC.

12. Guttmacher Institute and Ipas Convene Consultation on Unsafe Abortion in Africa, 2006, March 23, accessed at http://www.guttmacher.org/media/inthenews/2006/03/23/index.html.

13. *World Contraceptive Use, 2003*, UN, New York.

14. *State of the World's Children 2007*, UNICEF, New York.

15. *Maternal Mortality in 2005: Estimates developed by WHO, UNICEF, UNFPA and the World Bank*, 2007, WHO, Geneva.

16. *Global Prevalence and Incidence of Selected Sexually Transmitted Infections*, 2001, WHO, Geneva.

17. *2008 Report on the Global AIDS Epidemic*, WHO/UNAIDS, Geneva.

18. Joint UN Programme on HIV/AIDS (UNAIDS), 2006, *2006 Report on the Global AIDS Epidemic*, UNAIDS, Geneva and New York.

19. A briefing paper on the AU Protocol on the Rights of Women in Africa is available from the Center for Reproductive Rights at http://www .reproductiverights.org/pdf/pub_bp_africa.pdf.

20. *Population Issues in the 21st Century: The Role of the World Bank*, 2007, World Bank, Washington, DC.

CHAPTER 17

....

Cancún

Paradise Lost

ADRIANA VARILLAS

Cancún is a paradise. So the magazines, television commercials, and posters that hang on travel agents' walls tell us. And so believe the many tourists who arrive here and stay in the luxurious *zona hotelera*, or hotel district, where they fall asleep listening to the ocean waves caressing the beach and awake to see God reflected in the indescribable blue of the ocean.

But a few feet away there is another Cancún: a place of desperate poverty and social decay, a place far from the magnificent natural beauty that draws tourists. This is the dwelling place of those who make the dream possible.

> Cancún is a microcosm of the intersection between population and the environment. But this is not a simple tale of too many people degrading scarce resources.

This book is about the connections between population and the environment, and Cancún is, in many ways, a microcosm of that intersection. Indeed, this tourist center has one of the highest population growth rates in all of Latin America, and it faces an ever-growing inventory of environmental problems.

Adriana Varillas is a journalist based in Cancún, Mexico, where she specializes in writing about the environment, gender, human rights, politics, and other issues. Translated from Spanish by Josie Ramos.

However, the story of this natural spa in the Mexican Caribbean is not a simple tale of too many people degrading scarce resources.

Instead, Cancún represents the multiple challenges of so many regions of the world, which face domestic and international migration, population growth, and poverty; where economic development enriches the lives of the few at the expense of the many; where short-sighted policies destroy the natural resources upon which prosperity and well-being depend; and where basic access to health services—including critical reproductive health services—is available only to the wealthy and well connected. As we strive to build a sustainable, equitable future for the billions of individuals who comprise our growing world, we have much to learn from the story of the real Cancún.

DREAMS OF PROSPERITY

Cancún was born from a dream—a dream of a beautiful, prosperous city built on the dollars of Mexican and foreign tourists. This city would attract wealthy visitors, generate employment, and alleviate the country's poverty. In 1969, a group of specialists from the Banco de México designed a master plan that gave life to the Cancún project. Financed by the Inter-American Development Bank, the dream was made real by the labor of thousands of migrants, who came to Cancún in search of a better life.

At that time, the area we now know as the *zona hotelera* was an unspoiled barrier island 14 miles long, with lush rain forests and the Caribbean Sea on one side and an enormous lagoon system on the other. It was a land of extraordinary beauty: vast virgin beaches with fine white sand and coral reefs that comprise a critical part of the longest barrier reef in the Atlantic, the Mesoamerican Coral Reef System. And there were the *cenotes*, the "pools of the gods"—underground wells filled with crystal clear freshwater connected by a system of rivers that form part of the world's largest subterranean river system. Wildlife was abundant: The island was an important nesting site for seabirds, and a number of sea turtles still nest there today. The mangrove-lined lagoons were home to manatees and many species of fish and shellfish.

In this paradise, the city's founders constructed the second largest international airport in Mexico. They built more than a hundred hotels, many of them of five-star quality, and dozens of restaurants featuring international cuisine. And so *Kan Kun*—which in Maya means "a nest of vipers"—became the first planned tourist center in Mexico and one of the top ten most popular destinations in the world. In 2008, its international airport welcomed 12.7 million visitors, who generated $3 billion in revenue, up 12 percent from the previous year.[1]

TOWARD RUIN . . .

But dreams can become nightmares. Sigfrido Paz Paredes, who collaborated in the city's planning, has stated that "within the formula for success was hidden the seed to bring it to ruin."[2]

As migrants poured into Cancún, a city that was originally planned for 250,000 inhabitants[3] stretched to accommodate a million.[4] In fact, Cancún's annual population increase is the second highest in Mexico (5.5 percent),[5] surpassed only by Playa del Carmen, just 42 miles to the south, which has the highest population growth rate in Latin America— 14 percent, according to the local municipal government.

Human presence, the fruit of this growth, has had its consequences. Today, the Cancún Lagoon's perimeter is filled with buildings; the original tropical forest is long gone. The lagoon itself is fouled with sewage and gasoline, heavy metals, and other chemicals.

Hotels and condominiums were built over the years by filling in or destroying valuable mangrove forests, a practice that continues today and greatly diminishes the area's natural defenses against tropical storms and hurricanes. Indeed, by 2004, 140,000 acres of mangrove and rain forest had been destroyed to make way for development in Cancún.[6] The beaches are increasingly eroded by hurricanes, which are more prevalent and destructive thanks to the effects of climate change. Tourists and residents alike produce 800 tons of trash every day in a land where there is no more space for landfills.

Even the ocean has not been spared. Obsolete drainage and sewage systems sometimes discharge their effluent into the sea. Once-abundant marine species are in crisis due to overfishing for consumption by tourists. And large swaths of the precious coral reef have been killed or damaged by pollution, rising water temperature, and cruise ships, as well as snorkelers and scuba divers who ignore instructions to leave the reef's ecosystems untouched.[7]

THE OTHER CANCÚN

While Cancún's tourist infrastructure was thoroughly planned, there was no corresponding plan to accommodate the tens of thousands of men, women, and youth who cook for, serve, and clean up after the tourists. The result has been a surge of informal settlements that have sprung up along the outskirts of the city. These settlements are home to tens of thousands of families who lack potable water and most other public services.[8]

Residents of these outlying settlements get up before dawn to take buses that take them from their *palapas*—shelters made out of metal sheeting, palm branches, and a mud called *huano*—to the opulent resorts of the tourism zone. The hardest part is leaving their work after long hours to return to their own realities—from the grand hotel back to the small *palapa*.

They return to unpaved streets without lighting, and drainage systems that flood whenever it rains. They return to settlements where people bathe in the underground wells—the *cenotes*. And, because the settlements lack sewage systems, residents also install latrines or defecate directly into the *cenotes*. Such practices represent a serious health hazard and the root of water contamination of a much larger area because the *cenotes* are connected by an underground river system to other wells that supply drinking water.[9]

A DREAM DEFERRED

In the *zona hotelera,* many are living the dream built on tourist-generated wealth. But the people who make the tourist experience possible each and every day meet only their most basic needs. These men and women make beds where they will never sleep; they scrape away leftover food that would have been welcomed at their own tables; they drive the taxis that take the tourists to the airplanes on which they will never fly.

> The people who make the tourist experience possible meet only their most basic needs.

Many work 12-hour days, six days a week, all to make sure that others have an enjoyable stay in "paradise."

The dream of a decent job with a living wage remains elusive for most. Workers in the tourism industry earn less than those in comparable occupations[10]—typically $250 per month. Most work on temporary contracts with the hotels, under which their employers are not obligated to provide benefits or salary increases based on seniority. At the same time, life here is expensive because tourism drives up local prices of food and other necessities.

Low wages compel some parents to send their children to work. Almost 500 children between the ages of six and fourteen get up at

dawn, not to run to school, but to work at crosswalks, stoplights, mar-
ketplaces, dumps, commercial centers, or outside restaurants and
bars.[11] They sell flowers, gum, cigarettes, and bracelets—and they are
vulnerable to pedophiles, who videotape and sexually abuse them in
exchange for small amounts of money.[12] Others work as butchers' assis-
tants, tortilla packers, upholsterers, or servants in the wealthy houses of
the majestic tourist center.[13]

Unprotected at home and adopted by the streets, many of Cancún's
children have become "soldiers" in the 115 gangs that have flourished
in the city's working-class neighborhoods.[14] One of those soldiers is
Gerardo Tec Luna, who was eleven years old the first time he was
arrested by local police. Today, eighteen-year-old Gerardo has two tat-
tooed tears on his cheek—which means, in gang symbolism, that he
has killed two people. There is no proof of the murders, but what is
certain is that in these past seven years, Gerardo has been arrested
seventy-three times.

AND STILL THE CITY GROWS

Cancún remains unable—or unwilling—to provide services for its current residents; however, its population continues to grow. While the
bulk of Cancún's population growth results from migration, the city
also grows from within. Many low-income residents of Cancún lack
access to the information and the means to plan their pregnancies.

In spite of a prevalent open attitude in the presence of "spring
break," there still exist many stigmas and barriers regarding sexuality,
leading many to reject the use of condoms and other contraceptive
methods. As a result, Cancún holds the lead in Mexico for pregnancies
among women under the age of seventeen.[15]

Lack of access to sexuality education and services results in not only
new lives but also new illnesses and deaths. The lack or misuse of condoms is fueling the rise of sexually transmitted infections (STIs),
including HIV/AIDS, in youth. The state of Quintana Roo, which
includes Cancún, currently has 600 known cases of HIV, with Cancún
itself ranked seventh in the nation in HIV infection rates. There were
440 recorded AIDS-related deaths between 2000 and 2006.[16]

PARADISE DESTROYED

Cancún is ceasing to be a paradise. The magazines, television ads, and
travel agents do not advertise it. The tourists are beginning to see it,

when they arrive to see a coastline with less sand, with more street vendors, with drug-selling taxi drivers, and with greater air, water, and noise pollution. They see a paradise destroyed in the form of pregnant teens and in child performers on street corners swallowing fire or doing tricks for money.

The affluent residents of Cancún also see it, as they face, resentfully, the declining quality of *their* life. Their paradise crumbles bit by bit in the face of rampant organized crime, traffic congestion, and overall social deterioration. And for the residents of the other Cancún—those who live in the distant *palapas* and work long hours for little pay—this city was never a paradise.

Although population growth has played an important role in Cancún's problems, it has been steadily ignored. The story of Cancún is, indeed, a story about population growth and social and environmental ruin. But it is more than that. Fundamentally, it is a story about a dream: a model of development that promises prosperity but ignores the natural world on which prosperity depends; that delivers wealth for a few, but grinding, inescapable poverty for many more.

If we are to build a sustainable, equitable future, we must heed the lessons of Cancún—the Cancún of the tourists and the Cancún of the migrants—of the posters and the slogans but also of the outlying settlements, devastated jungles, and sandless beaches. We must remember Cancún's hopes and dreams, and also its realities.

REFERENCES

1. Southeastern Airport Group (ASUR) and *Turismo y Ocio International Magazine*, http://turismoyocio.net/index.php?option=com_content&task =view&id=1442&Itemid=2.

2. Sigfrido Paz Paredes, personal communication.

3. Master plan of Cancún, designed by Infratur and the Banco de México, in 1970. According to the plan, in the year 2000, Cancún would have 250,000 inhabitants, even though at the time the plan was designed, Cancún already had 416,000, according to the National Institute of Statistics, Geography and Information (INEGI).

4. Officially, the municipality of Benito Juárez, with Cancún as its capital, has 526,108 inhabitants, according to the 2005 Census of Population and Housing conducted by the INEGI. The census was interrupted and altered in October 2005 by the impact of Hurricane Wilma. The government is sure that many households were not counted, so the actual number of inhabitants is between 800,000 and 1 million people, not counting the floating population that comes and goes to the city.

5. Census by INEGI, 2005.

6. Leadership for Environment and Development, 2004, The Case of Cancún: Impacts of Intensive Tourism Development, December, accessed online at http://casestudies.lead.org/index.php?cscid=157.

7. Neto, Frederico, 2002, Sustainable Tourism, Environmental Protection and Natural Resource Management: Paradise on Earth? paper presented at the International Colloquium on Regional Governance and Sustainable Development in Tourism-driven Economies, Cancun, Quintana Roo, Mexico, February 20–22, 2002, available at UN Web site http://unpan1.un.org/intradoc/groups/public/documents/un/unpan002109.pdf.

8. Potable Water and Drainage Commission of Quintana Roo (Capa).

9. Leadership for Environment and Development, 2004, The Case of Cancún.

10. Neto, 2002, Sustainable Tourism, Environmental Protection and Natural Resource Management: Paradise on Earth?

11. Census of Child Laborers, 2006, System for the Integral Development of the Family (DIF).

12. Enrique Quiroz, Director of the Program for the Integral Attention to At-Risk Minors and Adolescents, personal communication.

13. Census of Child Laborers, 2006, System for the Integral Development of the Family (DIF).

14. Headquarters of the State Municipality of Cancún, 2007.

15. Program for Integral Pregnancy Services (PAIDEA) of the System for the Integral Development of the Family (DIF) on the municipal level. Source found in the *El Periódico de Quintana Roo*, article by Wendy Sánchez.

16. State Department of Health (SESA).

CHAPTER 18

....

The Flip Side

How the Environment Impacts Our Reproductive Health

CHARLOTTE BRODY and JULIA VARSHAVSKY

The human footprint on our planet can be calculated in acres of forest destroyed, tons of greenhouse gases emitted, and species gone extinct. Now environmental health science has added a new measure of our collective impact: soaring rates of disease, defects, and disorders linked to exposure to environmental contaminants.

Infertility, miscarriage, birth defects, and premature puberty are among the health problems tied to chemical contaminants. These chemicals are everywhere—in our food, air, and water and in the products we use every day. While other chapters in this volume examine the impact of human reproduction—and rapid population growth—on the environment, this chapter looks at the flip side of that coin: how degradation of the environment affects our ability to bring healthy children into the world.

ALARMING TRENDS

Infertility and related reproductive health problems are on the rise. In the United States, health care providers report that more patients are

Charlotte Brody is the National Field Director for Safer Chemicals, Healthy Families; Julia Varshavsky is the Reproductive Health Program Coordinator for the Collaborative on Health and the Environment (CHE).

seeking treatment for infertility.[1, 2] A national survey administered by the Centers for Disease Control and Prevention (CDC) showed that the number of couples in the United States who experienced problems conceiving and/or carrying a pregnancy to term grew from 6.1 million in 1995 to 7.3 million in 2002. This means that one in every eight American couples experienced fertility problems in 2002.[3]

The rise in reported infertility is often explained as the result of people's choosing to begin families later in life, when fertility declines. But the CDC survey reported a surprising finding: The greatest increase in fertility challenges was seen in young adults—in women under the age of twenty-five, who are thought to be at the peak of their reproductive years.

Increased infertility can be traced to a sharp increase in conditions that affect reproductive capacity in both men and women. For example, researchers have found that sperm health and testosterone levels have decreased and birth defects of the reproductive system have increased in some countries.[4] The National Cancer Institute reported that in Massachusetts, men's testosterone levels dropped 17 percent from 1987 to 2004,[5] a decline that is not related to normal aging or to health and lifestyle factors known to influence hormone levels. Studies in Denmark, Finland, and several U.S. cities have produced similar results.[6, 7, 8]

> One in eight American couples experienced fertility problems in 2002. . . . The greatest increase in fertility challenges was seen in . . . women under the age of twenty-five.

Other male reproductive health disorders are on the rise. Testicular cancer is increasing worldwide in males under age fifty, and in European countries, researchers now expect that one in a hundred men will be diagnosed with the disease in their lifetimes.[9] According to the CDC, the incidence of hypospadias—a birth defect in which the opening of the penis is located on the shaft rather than at the tip—has doubled since the 1970s.[10] Cryptorchidism, or undescended testes, is estimated to have doubled between 1950 and 2000.[11]

THE GROWING THREAT OF ENDOCRINE DISRUPTORS

Although less is known about physical or medical conditions that impact female reproductive health, we do know that hormones play an extremely important role in regulating healthy ovarian development, puberty onset, menstruation, pregnancy, and menopause. So environmental contaminants that mimic, block, or disrupt normal hormonal activities can have profound effects on women's reproductive health. These ubiquitous chemicals are called "endocrine disruptors," because they interfere with the ability of the endocrine system to regulate body functions.

Endocrine disruptors are increasingly thought to play a role in the rise in premature puberty. Over the last forty to fifty years, the United States and several other countries have seen a drop in the age at which breast development occurs and menstruation begins in young girls.[12] These statistics are troubling because early puberty increases the risk of reproductive health problems, including breast cancer—a disease that now affects one in eight women in the United States.[13, 14] Early puberty also elevates the risk for depression, sexual victimization, obesity, and experimentation with sex, alcohol, and drugs at younger ages.[15, 16, 17, 18] And endocrine disruptors may affect gender determination: In some populations there has been a significant decline in the birth ratio of male to female newborns.[19, 20, 21]

> Some chemicals that are only mildly toxic to adults can be hazardous to a growing fetus. . . . With many toxic exposures, "the timing makes the poison."

Recently, scientists have learned more about the insidious effects of endocrine disruptors. In June 2007, twenty-four eminent scientists issued *The Faroes Statement: Human Health Effects of Developmental Exposure to Chemicals in Our Environment.*[22] The consensus in this important document is that many disorders—including reproductive abnormalities, diabetes, and cancer—can begin with exposures to endocrine-disrupting chemicals during the prenatal and early postnatal period. The *Faroes Statement* notes that some chemicals that are only mildly toxic to adults can be hazardous to a growing fetus. That finding led the authors to reformulate the centuries-old axiom, "the dose makes the poison." With many toxic exposures, they note, "the timing makes the poison."

Our understanding of adult disease from prenatal exposures to chemicals began with a drug called diethylstilbestrol (DES), which mimics the effect of estrogen in the human body. DES was frequently prescribed in the United States to prevent miscarriages from 1947 until it was deemed unsafe to do so in 1971. The daughters of women who took the drug while pregnant have much higher risks for a variety of reproductive tract disorders in their childbearing years, including a rare form of cervical-vaginal cancer, structural reproductive tract abnormalities, impaired fertility, and adverse birth and pregnancy outcomes. DES sons are more likely to have genital abnormalities, in addition to a possible increased risk of testicular and prostate cancer.[23, 24] DES effects may even be seen in the grandchildren of those exposed, leading scientists to believe that endocrine disruptors could have lasting impacts on generations to come.[25, 26, 27, 28]

While DES is no longer prescribed, the *Faroes Statement* describes how other chemicals that are ubiquitous in our environment are causing DES-like harms. These include bisphenol A (BPA), which is found in many common products—including baby bottles, water bottles, dental sealants, and the linings of food cans. When young mice are exposed to BPA, they have a greater risk of developing reproductive disorders such as uterine fibroids in adulthood.[29] These health impacts occurred at environmentally relevant levels of BPA exposure, or levels to which American women are currently exposed.[30]

WOMEN AND CHILDREN AT RISK

Women face particular reproductive risks from environmental contaminants, in part because they store higher amounts of toxic chemicals in their bodies than do men. One explanation for this difference is that women have an average of 2 to 10 percent more body fat than men, and many contaminants accumulate in fat cells.

> An average of 200 industrial chemicals and pollutants were found in the umbilical cord blood taken from ten newborn babies across the United States.

Also, women who become pregnant are "the first environment" for the next generation, and they may pass many of these chemicals on to their developing children during pregnancy and breast-feeding. In 2004, an average of 200 industrial chemicals and pollutants were found in the umbilical cord blood taken from ten newborn babies across the United States.[31]

The fetus and infant are extraordinarily vulnerable to environmental contaminants that can both cause immediate health problems and lead to many other health impacts in later life. For example, the National Research Council estimates that 60,000 children are born every year in the United States with neurological problems that could lead to poor school performance, because of exposure to mercury in utero.[32] A Baltimore study connected cord blood concentrations of per-fluorinated compounds used in Teflon and water- and stain-repelling textiles to low birth weight in babies.[33] Low birth weight is the leading cause of infant death and is linked to higher incidence of childhood infections, younger age at puberty, adult diabetes, and other health problems.

CONCLUSION

Although uncertainty remains and more collaborative research is needed, the weight of the scientific evidence linking environmental contaminants to reproductive health and fertility is strong. We have enough information now to take precautionary measures to reduce the negative impacts of chemicals on the environment and on our health. To that end, we must shift toward new paradigms in the way we think about environmental contaminants and human health, the way we study them, and the way we create policies and guidelines that reduce their impact and truly protect public health.

The traditional way of thinking about chemicals—the higher the exposure, the greater the impact—is not accurate for every person or every disease. Small amounts of hormonally active contaminants can have serious effects on reproductive health, and the timing of exposure is critical. Windows of susceptibility occur in the womb, in early life, and during other developmental stages of our lives such as puberty. As with prescription drugs, some people are more sensitive to contaminants than others, and combinations of chemicals can harm us much more than any one single chemical exposure.

The connection between environmental health and fertility is of vital interest to advocates for reproductive rights and health. At the 1994 UN Conference on Population and Development, in Cairo, the world's nations affirmed the "right of all couples and individuals to decide freely and responsibly the number, spacing and timing of their children, and to have the information and means to do so." That right, of course, embraces conception as well as contraception. But—as we have seen—that right can be rendered meaningless when exposure to toxic chemicals compromises fertility. For the promise of Cairo to be realized, all men and women must have the right to live in a society that protects public health and provides them with the freedom to have a safe and healthy pregnancy if or when they choose to have children.

REFERENCES

1. Luoma, Jon R., 2005, *Challenged Conceptions: Environmental Chemicals and Fertility* (prepared following the workshop Understanding Environmental Contaminants and Human Fertility: Science and Strategy, Vallombrosa Retreat Center, Menlo Park, CA), Stanford University School of Medicine, Department of Women's Health and the Collaborative on Health and the Environment, available at www.healthandenvironment.org/working_ groups/fertility.

2. Giudice, L. C., 2006, Infertility and the environment: The medical context, *Seminars in Reproductive Medicine* 24: 129–133.

3. U.S. Centers for Disease Control (CDC), National Center for Health Statistics National Survey of Family Growth (NSFG), www.cdc.gov/nchs/nsfg .htm.

4. Skakkebaek, N. E., E. Rajpert-De Meyts, and K. M. Main, 2001, Testicular dysgenesis syndrome: An increasingly common developmental disorder with environmental aspects, *Human Reproduction* 16: 972–978.

5. Travison, Thomas G., et al., 2006, A population-level decline in serum testosterone levels in American men, *The Journal of Clinical Endocrinology & Metabolism* 92(1): 196–202, doi:10.1210/jc.2006-1375.

6. Andersson, et al., 2005, Trends in Leydig cell function in Danish men, *Human Reproduction* 20: 26–27.

7. Perheentupa, A., et al., 2006, Clear Birth Cohort Effect in Serum Testosterone and SHBG Levels in Finnish Men, Endocrine Society Meeting 2006.

8. Travison, T. G., A. B. Araujo, A. B. O'Donnell, V. Kupelian, J. B. McKinlay, 2007, A population-level decline in serum testosterone levels in American men, *Journal of Clinical Endocrinology & Metabolism* 92: 196–202.

9. Bray, F., L. Richiardi, A. Ekbom, E. Pukkala, M. Cuninkova, and H. Moller, 2006, Trends in testicular cancer incidence and mortality in 22 European countries: Continuing increases in incidence and declines in mortality, *International Journal of Cancer* 118: 3099–3111.

10. CDC report: cdc.gov, http://www.ourstolenfuture.org/NewScience/ reproduction/hypospadias.htm.

11. Toppari, J., M. Kaleva, and H. E. Virtanen, 2001, Trends in the incidence of cryptorchidism and hypospadias, and methodological limitations of registry-based data, *Human Reproduction Update* 7(3): 282–286.

12. Herman-Giddens, M. E., 2006, Recent data on pubertal milestones in United States children: The secular trend toward earlier development, *International Journal of Andrology* 29: 241–246; Euling, S. Y., M. E. Herman-Giddens, P. A. Lee, S. G. Selevan, A. Juul, T. I. Sorensen, et al., 2008, Examination of U.S. puberty-timing data from 1940 to 1994 for secular trends: Panel findings, *Pediatrics* 121, Supplement 3: S172–191.

13. Evans, Nancy, ed., 2006, *State of the Evidence: What Is the Connection between the Environment and Breast Cancer?* fourth edition, Breast Cancer Fund and Breast Cancer Action.

14. American Cancer Society, *Breast Cancer Facts & Figures 2005–2006*, American Cancer Society, Inc., Atlanta.

15. Ouyang, F., M. J. Perry, S. A Venners, C. Chen, B. Wang, F. Yang, et al., 2005, Serum DDT, age at menarche, and abnormal menstrual cycle length, *Occupational and Environmental Medicine* 62: 878–884.

16. Krstevska-Konstantinova, M., C. Charlier, M. Craen, M. Du Caju, C. Heinrichs, C. de Beaufort, et al., 2001, Sexual precocity after immigration from developing countries to Belgium: Evidence of previous exposure to organochlorine pesticides, *Human Reproduction* 16: 1020–1026.

17. Denham, M., L. M. Schell, G. Deane, M. V. Gallo, J. Ravenscroft, and A. P. DeCaprio, 2005, Relationship of lead, mercury, mirex, dichlorodiphenyl-dichloroethylene, hexachlorobenzene, and polychlorinated biphenyls to timing of menarche among Akwesasne Mohawk girls, *Pediatrics* 115: e127–134.

18. Colon, I., D. Caro, C. J. Bourdony, and O. Rosario, 2000, Identification of phthalate esters in the serum of young Puerto Rican girls with premature breast development, *Environmental Health Perspectives* 108: 895–900.

19. UN, 2004, Live births by age of mother, sex of the child and urban/rural residence: Latest available year, 1995–2004, Table 10, *Demographic Yearbook*, available at http://unstats.un.org/unsd/demographic/products/dyb/DYB2004/Table10.pdf, accessed September 10, 2007.

20. Davis, D. L., M. G. Gottlieb, and J. R. Stampnitzky, 1998, Reduced ratio of male to female births in several industrial countries: A sentinel health indicator? *Journal of the American Medical Association* 279: 1018–1023.

21. UN, 2004, Live births by age of mother, sex of the child and urban/rural residence: Latest available year, 1995–2004.

22. International Conference on Fetal Programming and Developmental Toxicity, http://www.pptox.dk/.

23. Herbst, A. L., H. Ulfelder, D. C. Poskanzer, 1971, Adenocarcinoma of the vagina: Association of maternal stilbestrol therapy with tumor appearance in young women, *New England Journal of Medicine* 284: 878–881.

24. Schrager, S., and B. E. Potter, 2004, Diethylstilbestrol exposure, *American Family Physician* 69: 2395–2400.

25. Titus-Ernstoff, L., R. Troisi, E. E. Hatch, L. A. Wise, J. Palmer, M. Hyer, et al., 2006, Menstrual and reproductive characteristics of women whose mothers were exposed in utero to diethylstilbestrol (DES), *International Journal of Epidemiology* 35: 862–868.

26. Blatt, J., L. Van Le, T. Weiner, and S. Sailer, 2003, Ovarian carcinoma in an adolescent with transgenerational exposure to diethylstilbestrol, *Journal of Pediatric Hematology/Oncology* 25: 635–636.

27. Brouwers, M. M., W. F. Feitz, L. A. Roelofs, L. A. Kiemeney, R. P. de Gier, and N. Roeleveld, 2006, Hypospadias: A transgenerational effect of diethylstilbestrol? *Human Reproduction* 21: 666–669.

28. Klip, H., J. Verloop, J. D. van Gool, M. E. Koster, C. W. Burger, and F. E. van Leeuwen, 2002, Hypospadias in sons of women exposed to diethyl-stilbestrol in utero: A cohort study, *Lancet* 359: 1102–1107.

29. Sugiura-Ogasawara, M., Y. Ozaki, S. Sonta, T. Makino, and K. Suzu-

mori, 2005, Exposure to bisphenol A is associated with recurrent miscarriage, *Human Reproduction* 20: 2325–2329.

30. Newbold, R. R., W. R. Jefferson, and E. P. Banks, 2007, Long-term adverse effects of neonatal exposure to bisphenol A on the murine female reproductive tract, *Reproductive Toxicology*.

31. Environmental Working Group, 2005, Body Burden—The Pollution in Newborns, www.ewg.org/reports/bodyburden2/execsumm.php.

32. Toxicological Effects of Methylmercury, 2000, Commission on Life Sciences (CLS), National Academies Press.

33. Apelberg, B. J., F. R. Witter, J. B. Herbstman, A. M. Calafat, R. U. Halden, L. L. Needham, and L. R. Goldman, 2007, Cord serum concentrations of perfluorooctane sulfonate (PFOS) and perfluorooctanoate (PFOA) in relation to weight and size at birth, *Environmental Health Perspectives*.

PART III

Looking Back,
Moving Forward

CHAPTER 19

....

Cairo

The Unfinished Revolution

CARMEN BARROSO and STEVEN W. SINDING

THE SIGNIFICANCE OF CAIRO

The International Conference on Population and Development (ICPD), held in Cairo in 1994, was a watershed event. At that meeting, the world's nations recognized that their most pressing challenges—poverty, ill health, and environmental destruction—could be eased by addressing the needs of girls and women, particularly their reproductive health and rights. Some 179 countries approved the ICPD Program of Action, a framework for population policies based on individual rights to health, education, and control of sexuality and reproduction (see Box 19.1).

This unprecedented agreement by countries so disparate in their economic, political, and cultural situations was made possible by a combination of factors that were unique to that historical moment. Of particular significance was the strengthening of civil society in many countries, which brought new actors to advocate for and monitor public policies. At the same time, globalization—the growing awareness of

Carmen Barroso is Regional Director of the International Planned Parenthood Federation/Western Hemisphere Region (IPPF/WHR). Steven W. Sinding, a senior fellow at the Guttmacher Institute, served as Director-General of the International Planned Parenthood Federation from 2002 to 2006.

BOX 19.1

The Cairo Program of Action

Cairo's Program of Action (PoA) is ambitious: It contains more than 200 recommendations within five twenty-year goals in the areas of health, development, and social welfare. A central feature of the PoA is the recommendation to provide comprehensive reproductive health care, which includes family planning; safe pregnancy and delivery services; abortion where legal; prevention and treatment of sexually transmitted infections (including HIV/AIDS); information and counseling on sexuality; and elimination of harmful practices against women (such as genital cutting and forced marriage).

The Cairo PoA also defined reproductive health for the first time in an international policy document. The definition states that "reproductive health is a state of complete physical, mental, and social well-being and not merely the absence of disease or infirmity, in all matters relating to the reproductive system."

The PoA also says that reproductive health care should enhance individual rights, including the "right to decide freely and responsibly" the number and spacing of one's children, and the right to a "satisfying and safe sex life." This definition goes beyond traditional notions of health care as preventing illness and death, and it promotes a more holistic vision of a healthy individual.

Adapted from two Population Reference Bureau publications by Lori S. Ashford: New Population Policies: Advancing Women's Health and Rights *(March 2001) and* New Perspectives on Population: Lessons from Cairo *(March 1995). Both reports are available at www.prb.org.*

interdependence among countries and the rapid development and lower costs of communication technologies and international travel—fostered the development of transnational networks of activists. The new power and visibility of those activists represent what some have called a "global civil society."

Global civil society played a major role in all of the UN conferences of the 1990s, including those on the environment (1992), human rights (1993), and women (1995). At the Cairo conference, civil society won fundamental changes in the population policy paradigm. Women's rights activists criticized prevailing population policies, which they perceived as dominated by demographic concerns and as giving insufficient attention to individual rights. Perhaps surprisingly, they found key allies among those they were criticizing: advocates for population stabilization.

Indeed, Cairo saw unprecedented cooperation between population and women's rights advocates. This alliance has been interpreted as a marriage of convenience, because both partners were fiercely attacked by the Vatican and its conservative allies, which had intensified their historical opposition to contraception and abortion. But the alliance was not merely tactical. It reflected an underlying acknowledgment of a common agenda: Women's empowerment and slower population growth are usually mutually reinforcing.

Civil society produced new ideas and energy, but the ultimate decision makers were governments. Here, too, Cairo saw new cooperation and diminished rancor. At previous UN population conferences, the governments of the industrialized countries and of the developing world were often at odds. At the 1974 meeting in Bucharest, for example, the industrialized countries pressed the developing world to adopt demographic goals and promote family planning. Rejecting that appeal, the head of the Indian delegation stated in an oft-repeated aphorism that "development is the best contraceptive." By the time of the Mexico City conference in 1984, most developing countries were ready to tackle population growth and expand family planning. But at that meeting, the Reagan administration baffled the world by downplaying the significance of population growth and unveiling the infamous "Mexico City policy," which denies funds for any foreign nongovernmental organization that provides abortion services, referrals, or advocacy.

> Population and women's rights advocates recognized a common agenda: Women's empowerment and slower population growth are usually mutually reinforcing.

At the Cairo conference, in contrast, industrialized and developing countries agreed on the importance of population growth and the means to address it. The Clinton administration played a significant role in forging that consensus, in part by appointing a delegation with wide representation of population experts and women's health advocates who played leadership roles in the United States and around the world.

PROGRESS AFTER CAIRO

Cairo prompted many countries to address sexual and reproductive health issues, but progress remains uneven. A 2003 United Nations Population Fund (UNFPA) survey of 151 developing countries found that many have added elements that were called for in the Cairo Program of Action, but most still lack comprehensive sexual and reproductive health (SRH) programs (see Box 19.2 on ICPD objectives). About half the countries reported that primary health care services had been expanded to include youth-friendly reproductive health services; emergency contraception; voluntary counseling and testing for HIV; sexually transmitted infection prevention, treatment, and management; and family planning services. However, less than a quarter of the countries had free reproductive health services in all public health facilities, maternal health services for vulnerable groups in remote areas, or facilities for the prevention and management of obstructed labor or post-abortion complications.[1]

India is an interesting example of the evolution of government population policies. Until the early 1990s, the state-sponsored family planning program was driven by centrally determined targets for contraceptive use, with a heavy emphasis on sterilization. Women's groups voiced concerns about the coerciveness of this approach, as well as the poor quality of care in family planning clinics. After the Cairo conference, a new "target-free" approach replaced the old top-down system, with community-level planning focusing on the needs of clients, provision of a variety of contraceptive methods, and the integration of basic reproductive and child health care services.

Since Cairo, the global fertility rate has continued to decline, and access to modern contraceptive methods has improved, although unevenly and—because of the funding shortfalls described below—at a slower pace than in the 1980s and 1990s. Today 37 percent of countries have fertility levels below the "replacement level" of 2.1 children per woman; 27 percent of countries have levels between 2.1 and 4.0; 15 percent between 3.0 and 5; and 21 percent above 5.0.[2] A recent study shows that contraceptive use has increased in Asia, North Africa, Latin America, and sub-Saharan Africa since the early 1990s, but in sub-Saharan Africa the rise has slowed since 1997.[3] Indeed a

> One-fifth of countries still have fertility rates of five or more children per woman.

BOX 19.2

ICPD's Twenty-Year Goals, 1995–2015

- Provide universal access to a full range of safe and reliable family planning methods and related reproductive health services.
- Reduce infant mortality rates to below 35 deaths per 1,000 live births, and reduce under-five mortality rates to below 45 deaths per 1,000 live births.
- Close the gap in maternal mortality between developing and developed countries. Aim to achieve a maternal mortality rate below 60 deaths per 100,000 live births.
- Increase life expectancy at birth to more than seventy-five years. In countries with the highest mortality, aim to increase life expectancy at birth to more than seventy years.
- Achieve universal access to and completion of primary education; ensure the widest and earliest possible access by girls and women to secondary and higher levels of education.

recent study suggests that fertility declines have stalled altogether in six sub-Saharan states—a matter of particular concern given the paucity of SRH services and the large proportion of need for contraception that is unmet.[4]

THE AGENDA IS UNFINISHED

By a number of measures, we are still far from fulfilling the promise of Cairo. As we shall see in the next section, developing countries and donors alike have failed to make good on the financial commitments they made at the ICPD. Funding for reproductive health services has declined as a percentage of overall health spending, and in many cases in real terms as well.

Equally worrisome is the fact that health systems still lack the resources and the outreach capacity to meet the Cairo goal of universal access to SRH services. In many countries the special community-based outreach systems that were established to deliver family planning and other reproductive health services during the 1970s, 1980s, and 1990s have disappeared, leaving a void that has not been filled by any other form of health care. Examples include Bangladesh and Kenya, two countries that were highly successful in creating outreach systems to reach nearly their entire populations with family planning services. Neither country today has in place a rural health system capable of delivering SRH, much less other primary care services.

Sadly, when services are curtailed or health systems fail to reach out to entire populations, it is inevitably the poor, the marginalized, and the geographically remote who suffer the most. These are also, of course, the people most in need of publicly provided services. For the most vulnerable sectors of society, sexual and reproductive health conditions have not sufficiently improved. Appalling inequities persist, both between and within countries. In every country for which we have data, women in the highest income quintile have much better access to contraceptives when compared with the lowest quintile. In parts of sub-Saharan Africa, women have a 1 in 6 chance of dying in childbirth, while in parts of North America and Europe, lifetime risk is as low as 1 in 8,700.[5]

While we are not aware of any comprehensive analysis that shows the gap between universal access to SRH services and current reality, it is clear from recent funding analyses, as well as many country case studies that have been carried out by, among others, UNFPA and the World Bank, that the world remains far short of the target of universal access to SRH.

There are signs that countries are belatedly realizing that they have fallen behind in their commitments to SRH. For example, as Fred T. Sai describes in Chapter 16, in 2007 Africa's ministers of health convened a special meeting devoted to SRH, where they committed to give increased priority to sexual and reproductive health services and to more effectively link those services to HIV/AIDS prevention and treatment.

> Appalling inequities persist: In parts of sub-Saharan Africa, women have a 1 in 6 chance of dying in childbirth, while in parts of North America and Europe, lifetime risk is as low as 1 in 8,700.

But overall progress has been very disappointing. Maternal mortality has declined little if at all since the Safe Motherhood Initiative was launched in Nairobi twenty years ago. Likewise, although there has been considerable progress in helping AIDS victims remain alive as a result of the provision of low-cost antiretroviral drugs (ARVs) in many countries, the overall prevalence of HIV infection, leading to full-blown AIDS, has continued to rise year after year, as has AIDS mortality. In other words, while progress is being made on the treatment side of the equation, prevention efforts, especially in Africa, are lagging badly.

THE BIGGEST CHALLENGES

The Funding Shortfall

As concern about population growth has diminished, countries have shifted resources away from population activities, including family planning services. The resulting shortfall in funding is among the most serious of obstacles to full implementation of the Cairo Program of Action.

The Program of Action called for developing countries and donors to meet specific funding targets at five-year intervals following the ICPD, from 2000 up to 2015. Developing countries were to contribute two-thirds of the total and donors one-third, and the targets by year were $17 billion by 2000, $18.5 billion by 2005, $20.5 billion by 2010, and $21.7 billion by 2015. These figures were broken down into four categories: family planning (including delivery system costs); other reproductive health services; STD prevention (including HIV/AIDS); and research, data collection, and policy analysis.[6]

After a brief upturn immediately following the ICPD, funding—especially by donors—has stayed flat or declined. However, measuring expenditures on the Program of Action is complicated by the fact that there is no uniform, agreed-upon set of accounting categories that all countries use. Moreover, funding for HIV/AIDS programs has exploded since 2000. And while most countries and donors exclude HIV/AIDS expenditures from the SRH category, some do not. UNFPA generally counts AIDS funding toward a country's fulfillment of its Cairo pledge. So, depending on whether or not AIDS is included, one could say that funding the ICPD pledge has been met or even exceeded as of 2005, or one could say that SRH has been woefully underfunded.

> Funding for treatment of people living with AIDS has increased enormously since 2000 while all other aspects of funding for sexual and reproductive health, including AIDS prevention, have languished.

The truth is that funding for treatment of people living with AIDS has increased enormously since 2000 while all other aspects of funding for sexual and reproductive health, including AIDS prevention, have languished. Hardest hit have been family planning services, where the Cairo pledge has practically been forgotten.

Developing countries are doing twice as well as donors at meeting the funding commitments made in Cairo—meeting 46 percent of their targets overall, as opposed to the donors' 23 percent. Still, it is clear that the community of nations is falling well short of the funding needed to ensure universal access to reproductive health services.[7]

Conservative U.S. Policies

During the Clinton administration, the United States not only supported the Cairo Program of Action, it provided strong political leadership in both crafting and adopting the landmark international consensus. However, when Republican President George W. Bush assumed office in 2001, he immediately announced his opposition to much of what the Clinton administration had supported and enacted. During two terms in office, the Bush administration systematically and vigorously sought to overturn the Cairo consensus, using nearly every available UN conference and consultation to attempt to redefine agreed language and impose its own ideological agenda on the international community. (The Bush administration's efforts to thwart the Cairo agenda are described in detail in Chapter 21.)

The international political implications of the conservative U.S. stance are not altogether clear. Most efforts to roll back the Cairo consensus have failed. In international conference after conference, U.S. efforts to amend or change language were roundly rejected, and the United States often stood in a very small minority of countries that were allied with the Holy See.

On the other hand, because the United States has long been far and away the largest funder of international reproductive health programs, including efforts to control the spread of HIV/AIDS, the imposition of ideological restrictions on the use of U.S. funds (e.g., "abstinence-only" restrictions, limitations on the provision of condoms, defunding of large and effective service delivery NGOs) has had a profound impact on the nature and effectiveness of those programs. And, the U.S. preeminence in international politics has meant that the administration's hostility toward the Cairo agenda almost certainly had a "chilling effect" on the extent to which other countries—particularly aid-dependent countries—have chosen to embrace that agenda. In a word, alongside the Vatican, the Bush administration succeeded in making sexual and reproductive health and rights *controversial*—a subject that many countries would prefer to avoid in discussions, for example, of the Millennium

Development Goals or other international development issues. Bush successfully broke any credible claim to a global "Cairo consensus."

The election of Barack Obama portends a positive sea change on SRH policy by the United States. Obama has already reversed the Bush era restrictions to funding family planning organizations that did not agree to sign the infamous global gag rule, which forbade them to engage in any abortion-related activity, including giving women information about legal safe abortion services and advocating for changes in restrictive abortion laws. Obama's nomination of Hillary Clinton, long a champion of women's rights and SRH within development cooperation, to be his secretary of state strongly reinforces the likelihood that the United States will be an important champion of the Cairo agenda for the next several years.

REVERSING THE TREND

Integrating Poverty and SRH

A major blow was struck to the implementation of the Cairo Program of Action when any reference to reproductive health and rights was excluded from the UN's Millennium Declaration in 2000 and the eight Millennium Development Goals (MDGs) that were meant to prescribe international action to alleviate, and ultimately eradicate, extreme poverty. According to former *New York Times* reporter Barbara Crossette,[8] UN insiders, fearful that inclusion of the ICPD goal of universal access to reproductive health would inject a note of controversy that could imperil overall approval of the MDG package, excluded the Cairo goal from the MDGs—themselves essentially a restatement of the goals of a series of UN-sponsored international development conferences of the late 1980s and 1990s.

This exclusion has had profoundly negative implications for sexual and reproductive health programs, because the MDGs have been embraced by most developing countries and donor governments as the defining agenda for development and development cooperation programs. To be excluded from the MDGs was tantamount to being excluded from the international development discourse and the priorities, including the funding priorities, of most countries. Accordingly, shortly after the MDGs were first announced, the SRH community began a concerted campaign to restore the Cairo goal to a prominent place within the MDGs. This campaign culminated in 2006 with approval of the Cairo goal of universal access to SRH as an additional "target" under the MDGs, linked specifically to MDGs 5 and 6, relating to maternal health and HIV/AIDS, respectively. With this target in place, the UN has now developed specific indicators to be employed by countries and international agencies to monitor progress on the MDGs.

> Empowering individuals to manage their own fertility, to protect themselves against sexually transmitted diseases, and to bear children safely is essential to the achievement of every one of the MDGs.

The partial victory represented by the inclusion of the Cairo target was possible because many political leaders and development professionals have come to recognize that the alleviation of poverty and, indeed, the achievement of most of the specific MDGs, will not be possible unless individuals have the information and the means to manage their reproductive and sexual health. Whether one is talking about increasing school enrollments; gender equality; reducing infant, young child, or maternal mortality; combating AIDS; increasing caloric intakes; protecting the environment; or, indeed, lifting the billion poorest people out of poverty, ensuring sexual and reproductive health and rights is fundamental. Empowering individuals to manage their own fertility, to protect themselves against sexually transmitted diseases, and to bear children safely is essential to the achievement of every one of the MDGs.

Thanks to the largely successful campaign to raise the profile and priority of SRH in development circles, recent years have seen promising signs of restored commitment to the Cairo agenda. Increasingly, HIV/AIDS programs are looking to incorporate SRH interventions among their services, particularly voluntary counseling and testing and contraceptive services for HIV-positive women. Programs to reduce maternal mortality are giving increased attention not only to emergency obstetrical care, but also to ensuring adequate contraceptive supplies to prevent unintended and unwanted pregnancies. And advocates of gender equity are increasingly calling for programs that empower women both to successfully negotiate safe sex and to protect themselves from unwanted pregnancies. Today, SRH is not itself an MDG, but it is firmly embedded within many of the strategies countries are employing to implement the Millennium Development Goals.

Empowering Women

Gender inequalities—for example in income, housing, and education—have a strong impact on sexual and reproductive health. It is virtually impossible to bridge the gaps in sexual and reproductive health conditions when these other gaps remain. A woman may be unable to exercise her sexual and reproductive rights if she lacks access to a decent health care system, if she hasn't had the chance to attend school, if she is unemployed, or if she is employed but receives a meager salary. Disparities in sexual and reproductive health follow an underlying pattern of social disadvantages.

Gender inequalities compound and aggravate social and economic inequalities. Unequal power relationships between women and men, as well as differences in education and poverty levels, effectively block access to sexual and reproductive health care for millions of women worldwide—with profoundly negative impacts on their overall health. In many parts of the world, women are at risk of contracting HIV because social norms tolerate husbands' having multiple partners while wives are discouraged from insisting on condom use. Gender stereotypes and limited opportunities for girls also have an impact on teenage pregnancy rates, which are persistently high and even increasing in many countries, even as fertility rates among adult women are declining. The most egregious manifestation of gender inequality, gender-based violence, is widespread and has multiple consequences for sexual and reproductive health. The World Health Organization estimates that one in three women worldwide has experienced domestic violence.

Empowering women is increasingly recognized as key to combating poverty and promoting public health. The World Bank has called women's education the single most influential investment that can be made in the developing world. The third Millennium Development Goal is to promote gender equality and empower women. The task force of the Millennium Project analyzed practical solutions to achieve this goal and selected seven strategic priorities, including economic empowerment, guaranteeing sexual and reproductive rights, improving education opportunities, combating violence against girls and women, and increasing political representation in decision-making bodies.[9]

Integrating SRH and AIDS

For varied and complex reasons, since Cairo, the fields of AIDS prevention and treatment and SRH have moved on separate tracks. This divergence happened in part because—notwithstanding Cairo's identification of AIDS as a key reproductive health issue—most actors in the reproductive health field were slow to incorporate HIV/AIDS prevention and treatment in their programs. Recognizing the urgency of the challenge, AIDS activists forged ahead with free-standing programs.

Another factor was a decision made at the 2000 G8 summit to establish a Global Fund to Fight AIDS, Tuberculosis and Malaria, which linked AIDS with other communicable diseases rather than with reproductive and sexual health. This decision was reinforced when the World Health Organization (WHO) moved its HIV/AIDS unit out of the maternal and reproductive health section and into the communicable diseases area, where tuberculosis and malaria control were also located. These decisions by the most powerful donor countries and the preeminent international health body sent a signal to ministries of health around the world that HIV/AIDS should be handled primarily as a communicable disease rather than as a sexually transmitted one—despite the fact that sex between men and women was fast becoming the primary mode of transmission.

Finally, far more of the rapidly growing international assistance to combat AIDS is spent on treatment of those infected with HIV/AIDS than on the prevention of new infections. As a result, each year since the pandemic began, there have been more new cases than there were the year before.

Today there is increasing awareness of the need for a closer link between SRH programs and HIV/AIDS. The prevention of heterosexual transmission is unlikely unless there are effective programs of behavior-change communication, voluntary counseling and testing for HIV, and strong programs to protect couples from transmitting the virus to each other during sexual intercourse and to prevent HIV-positive mothers from transmitting it to their newborns.

Sexual and reproductive health programs are ideally situated to meet these challenges. Few areas of public health have seen more effective behavior-change communication than family planning programs, which effectively revolutionized reproductive behavior. Reproductive health programs in many countries and communities have long-established relationships of trust and policies of protecting client anonymity. These are essential conditions for encouraging women, in particular, to get tested for HIV and referred for treatment if they are HIV-positive, treated to protect the fetus if they are pregnant, and provided barrier methods to prevent further infection in either case. In an environment of widespread stigma and discrimination surrounding HIV and AIDS, SRH clinics can serve as much-needed safe spaces.

Fortunately, there appears to be a growing awareness that the gap

between HIV/AIDS and SRH programs must be bridged, if not eliminated altogether. As we noted above, African ministers of health have
agreed to move in this direction, both in program organization and
budgeting. The last several international HIV/AIDS conferences have
highlighted the importance of the linkage. Four major international
organizations—the Joint United Nations Program on HIV/AIDS
(UNAIDS), WHO, UNFPA, and the International Planned Parenthood
Federation (IPPF)—have issued guidelines for the practical implementation of integration. And UNAIDS has called for far more emphasis on
prevention, including much closer cooperation with SRH programs.

The decrease of international funding for family planning and sexual and reproductive health services has been coupled with an increase
in funding for HIV/AIDS. This phenomenon should not be viewed as a
threat but as an opportunity to advance both fields. In fact, HIV/AIDS
prevention and sexual and reproductive health care are complementary. The majority of new cases of HIV infection are contracted through
sex, birth, or breastfeeding—which all fall under the purview of SRH
programs. Also, HIV/AIDS as
well as other problems related to
sexual and reproductive health
are affected by a set of common
causes, including gender inequality, poverty, stigma, and discrimi-
nation. For these reasons, it is both logical and practical to mount an
integrated response that joins the struggle against HIV/AIDS with promotion of sexual and reproductive health.

> The gap between HIV/AIDS and sexual
> and reproductive health programs
> must be bridged, if not eliminated.

Focus on the National Level

There are important opportunities for change at the national level.
The sexual and reproductive health community, which has been so
effective at the international level, now must focus on the policies and
political negotiations taking place in each country. Bilateral and multilateral aid agencies are directing more funds through national governments, and decisions on sexual and reproductive health budgets are
increasingly made by recipient governments instead of donor institutions. All this makes advocacy at the country level more important.
Funding for international programs is still needed because international networks and alliances offer much-needed support for local
NGOs and advocates struggling to hold their national governments
accountable for meeting sexual and reproductive health needs.

CONCLUSIONS

Inequality and poverty undermine our dreams as a global community. Environmental degradation and global warming cast a dark cloud over our common future. Both have intrinsic links to the millions of youth living in poverty, with no future to aspire to and no alternatives to early childbearing. This situation further contributes to the vicious cycle of poverty. The Cairo agenda is central to addressing these interrelated problems.

Six years remain to make sure that the promises made in Cairo become a reality:

- six more years to build bridges with the HIV/AIDS movement and make sure that HIV/AIDS and sexual and reproductive health services go hand in hand;
- six more years to mobilize supporters at the national level to create the political will needed to elevate sexual and reproductive health on each country's public agenda;
- six more years to combat poverty and inequality, while promoting sexual and reproductive health and strengthening health systems.

This requires substantial investment—and mobilization of political will—at all levels of government and civil society.

REFERENCES

1. UNFPA, 2004, Investing in People: National Progress in Implementing the ICPD Programme of Action 1994–2004, UNFPA, New York.

2. UN Population Division, World Population Prospects: The 2006 Revision.

3. Bongaarts, John, Fertility Transition in the Developing World: Progress or Stagnation? PAA, 2007.

4. Garenne, M. Michel, 2008, Fertility Changes in Sub-Saharan Africa, DHS Comparative Reports Number 18, Macro International, Inc., Calverton, MD.

5. Who's Got the Power? Transforming Health Systems for Women and Children, Millennium Project, Task Force on Child Health and Maternal Mortality.

6. These estimates were subsequently deemed to be underestimations because of the rapid spread of the AIDS pandemic. However, paradoxically, HIV/AIDS programs are not usually included in calculations of SRH funding.

7. Speidel, J. J., 2005, Population donor landscape analysis, for Review of

Packard Foundation International Grantmaking in Population, Sexual and Reproductive Health and Rights (unpublished report), September 6.

8. Crossette, Barbara, 2004, Reproductive Health and the Millennium Development Goals: The Missing Link, Population Program of the William and Flora Hewlett Foundation, Palo Alto, CA.

9. Grown, Caren, Geeta Rao Gupta, and Aslihan Kes, et al., 2005, Taking Action: Achieving Gender Equality and Empowering Women, Earthscan, London.

CHAPTER 20

....

The New
Population Challenge

JUDITH BRUCE and JOHN BONGAARTS

The past four decades have witnessed unprecedented demographic change—and an intense, often polarized debate about how to slow population growth. The poles of the debate were staked out at the first global population conference, held in Bucharest in 1974. Participants at that meeting shared a concern about the adverse socioeconomic and environmental consequences of rapid population growth, but they disagreed about policy. "Supply-siders" emphasized the need to enhance the supply of contraceptives and improve family planning services, while the "demand" camp argued that improvements in social and economic conditions, reductions in mortality, and greater gender and economic equality would be needed to stimulate demand for smaller families.

The debate continued in the years following the conference, but, for the most part, it was the supply-siders' vision that prevailed. Spurred by documentation of a large unmet need for contraception, donors expanded funding and technical assistance for voluntary family planning programs in developing countries. These programs, which provide contraceptive information and services, enable women and men to take control of their reproductive lives and avoid unwanted childbearing. Newly available contraceptive methods, such as the pill and IUD,

Judith Bruce is a Senior Associate and Policy Analyst with the Population Council's Poverty, Gender, and Youth program. John Bongaarts is a Population Council vice president and Distinguished Scholar.

greatly facilitated the delivery of family planning services. A few notable success stories (e.g., in Taiwan and Korea) encouraged other governments to follow this approach.

By the time of the third international population conference, held in Cairo in 1994, access to family planning services had expanded substantially and fertility had declined in much of the developing world. However, fertility remained stubbornly high in some areas where poverty and women's status had not improved. This raised doubts about the effectiveness of supply-side policies that focused purely on family planning services to reach demographic goals. In addition, many raised strong objections to coercive measures. Although coercion was—and is—rare in population and family planning programs, notable instances such as China's one-child policy generated considerable controversy. And emerging evidence indicated that the quality, as well as quantity, of services is essential to attract and retain clients. This confluence of factors led participants in the Cairo conference to recommend a more comprehensive approach to population policy, one that emphasizes reproductive health, women's rights, and human development.

The Cairo process healed many of the rifts since Bucharest. Rather than continuing the debate over supply or demand, the Cairo agreement embraced both. It elaborated what *kind* of supply (not just services and technologies, but also their manner of provision, their client orientation, and the human rights that must underpin them) and the *kind* of development (investments in girls' education, gender equality, and efforts to alleviate poverty and early childhood mortality) that would support demand for lower fertility. The Cairo conference explicitly recognized that there are a number of synergistically linked solutions, all of which must be embraced. This process laid the foundation for nonideological, evidence-based population policy.

Unfortunately, in the years since Cairo, fundamentalist religious forces and their allies have ignited an old and destructive debate about individuals' (specifically women's) right to use birth control and to choose the number and timing of children, and also about the role of government in promoting population and reproductive health policies and programs. The "yes, and" of Cairo was replaced by a "no, but" divisiveness. In this contentious environment, advocates have focused on maintaining sup-

> New demographic realities have greatly increased the need for a more comprehensive approach to population policy.

port for hard-won reproductive health service systems and a steady stream of contraceptive supplies. Investments in the "demand" side of population—progressive development policies—have received relatively less attention.

If that neglect continues, we can expect diminishing returns from population and family planning programs. This is because new demographic realities—as outlined in Box 20.1 and as we will show below—have greatly increased the need for a more comprehensive approach to population policy.

ELEMENTS OF GROWTH: THE NEW DEMOGRAPHIC LANDSCAPE

According to the 2007 UN medium projection, world population will continue to grow at least until 2050, adding 3.7 billion to the 2005 population of 6.5 billion. Why does population continue to grow despite largely favorable trends in fertility and contraceptive use? A partial answer is provided by a population decomposition exercise that divides future population growth into four key segments:[1]

- *High fertility*—Fertility above the replacement level (about 2.1 births per woman) is a key cause of further growth. In countries with information on fertility preferences, it is possible to further divide this factor into its subcomponents: *unwanted childbearing* and *the desire for large families.*

- *Declining mortality*—Declines in death rates—historically the main cause of population growth—will almost certainly continue. Higher standards of living will ensure longer and healthier lives in most countries. The main exceptions are countries heavily affected by the AIDS epidemic.

- *Young age structure*—Because the largest generation of adolescents in history is now entering the childbearing years, this component is responsible for a substantial part of future growth in the developing world. A large cohort of young people generates "population momentum"—even if each of these young women has only two children, they will produce more than enough births to maintain significant growth over the next few decades.

- *Migration*—In-migration raises population size and out-migration reduces it. In most developing regions and countries, migration is a relatively minor demographic factor.

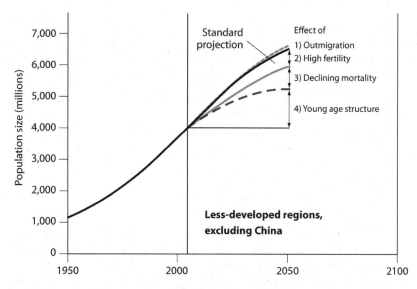

FIGURE 20.1 Alternative population projections 2005–2050 and effects of components of population growth.

Source: Based on United Nations 2007

TABLE 20.1 Effects of Components on Population Growth: 2005–2050

	Components of population growth (%)				
	High fertility	*Declining mortality*	*Young age structure*	*Migration*	*Combined (multiplicative)*
Africa	44	18	28	−1	117
Sub-Saharan	52	20	26	−1	129
Asia	−3	11	26	−1	34
China	−13	9	15	−1	7
Latin America	−1	7	37	−5	38
South (excluding China)	12	14	32	−2	64
South	6	13	28	−2	50

Source: Based on United Nations 2007

Quantitative estimates of the impacts of these four components of population growth are provided in Table 20.1. In the developing world as a whole, population is expected to grow by 50 percent between 2005 and 2050. This growth is the result of the combined effects of three growth factors: 6 percent from high fertility, 13 percent from declining mortality, and 28 percent from young age structure.

BOX 20.1

The Developing World's New Demographic Landscape

Population

- From 1950 to 2005, the developing world's population tripled from 1.7 to 5.3 billion. Growth varied widely by region, from 179 percent in Asia to 327 percent in sub-Saharan Africa.[a]
- The rate of growth has declined to 1.4 percent as a whole but is virtually unchanged, at 2.5 percent, in the fifty least developed countries.[b]

Reproductive Health and Fertility

- The average total fertility rate has fallen from 6 or more children per woman to near 3. Between the early 1960s and early 2000s, the largest declines were in Asia (−56 percent) and Latin America (−58 percent), with the smallest in sub-Saharan Africa (−19 percent). Some countries (mostly in Asia) completed the transition to replacement-level fertility in record time, while others (mostly in sub-Saharan Africa) have seen little change.[c]
- Average ideal family size for women twenty to twenty-nine is now less than three in most countries in Asia and Latin America but remains around five in sub-Saharan Africa.[d]
- Contraceptive use, once rare, is now widespread; prevalence among married women exceeds 60 percent.[e]
- Today, a far higher proportion of young couples are sexually active before marriage.[f]
- Unwanted fertility has not declined much. The rise in contraceptive use has been offset by more years of exposure to unintended pregnancy.[g]

---

[a]UN, 2007, *World Population Prospects: The 2006 Revision*, Population Division Department of Economic and Social Affairs, UN, New York.

[b]Ibid.

[c]Ibid.

[d]Westoff, Charles F., and Akinrinola Bankole, 2002, *Reproductive Preferences in Developing Countries at the Turn of the Century*, DHS Comparative Reports Number 2, ORC Macro, Calverton, MD.

[e]UN, 2006, *Levels and Trends of Contraceptive Use as Assessed in 2002*, Population Division, Department of Economic and Social Affairs, ST/ESA/SER.A/239, UN, New York.

[f]National Research Council and Institute of Medicine, 2005, Growing up global: The changing transitions to adulthood in developing countries, *Panel on Transitions to Adulthood in Developing Countries*, Cynthia B. Lloyd, ed., Committee on Population and Board on Children, Youth, and Families, Division of Behavioral and Social Sciences and Education, National Academies Press, Washington, DC.

[g]Westoff and Bankole, 2002, *Reproductive Preferences in Developing Countries at the Turn of the Century*.

Maternal and Child Heath

- Due to public health measures, improved nutrition, and rising standards of living, infant mortality has fallen to 6 percent, compared with 18 percent in the 1950s.[h]
- Maternal mortality has declined but remains unacceptably high among the poorest populations, especially the youngest first-time mothers. Some regions have shown some progress, but ratios in sub-Saharan Africa have remained at 905 per 100,000 live births.[i]

HIV/AIDS

- AIDS has killed 25 million people, and 33 million are currently infected. About 2.5 million new HIV infections occur each year. The epidemic appears to have stabilized in most countries, and the number of new infections is declining in many.[j]
- The impact of AIDS on population growth and age structure is smaller than was expected: Population in sub-Saharan Africa is expected to grow to 1 billion by 2050, despite the epidemic, and growth is expected to remain positive even where the epidemics are most severe.[k]
- In sub-Saharan Africa, 75 percent of the 6.2 million people aged fifteen to twenty-four with HIV are female, up from 62 percent a few years earlier. The female-to-male ratio is generally three to one throughout Africa; in some communities, it reaches eight to one.[l]

Early Marriage/Sexual Initiation

- Although child marriage is declining in many regions, more than a third of women currently aged twenty to twenty-four were married before age eighteen. Rates of child marriage are highest in western/middle Africa (45 percent) and south Asia (42 percent).[m]
- Most sexually active adolescent girls are married, but a rising share of sexual initiation is outside of marriage.[n]

....

[h]UN, 2007, *World Population Prospects*.

[i]Kenneth Hill, et al., 2007, Estimates of maternal mortality worldwide between 1990 and 2005: An assessment of available data, *The Lancet*, 370(9595): 1311–1319.

[j]UNAIDS, 2007, *AIDS Epidemic Update: December 2007*, Joint UN Program on HIV/AIDS (UNAIDS) and World Health Organization (WHO), Geneva.

[k]Bongaarts, John, Thomas Buettner, Gerard Heilig, and Francois Pelletier, 2008, Has the AIDS epidemic peaked? *Population and Development Review* 14(2): 199–224.

[l]UNAIDS, 2007, AIDS Epidemic Update—Regional Summary, accessed April 30, 2008 at http://www.unaids.org/en/CountryResponses/Regions/SubSaharanAfrica.asp.

[m]Mensch, B., S. Singh, and J. Casterline, 2005, Trends in the Timing of First Marriage among Men and Women in the Developing World, Policy Research Division Working Paper Number 202, Population Council, New York.

[n]Ibid.

The roles of the different factors vary considerably among regions in the developing world. For example, the high-fertility component is largest in sub-Saharan Africa (52 percent) and smallest in Asia and Latin America (a negative value indicates that fertility is expected to fall below replacement in the future). The impact of declining mortality is also highest in sub-Saharan Africa, reflecting its projected large improvement in life expectancy from current low levels. The young-age-structure component is an important driving force of population growth in all regions and the dominant one in Asia and Latin America. Migration effects are relatively minor, diminishing future population growth by an average of –2 percent, except in Latin America where it reaches –5 percent.

TWO CRUCIAL POLICY LEVERS:
HIGH FERTILITY AND YOUNG AGE STRUCTURE

Effective population policies for the twenty-first century must respond to the shifting sources of population growth outlined above. Of course, declining mortality is entirely beneficial, and migration is a relatively minor factor in population growth. So, for policy purposes, the two key elements of growth are *high fertility* and a *young age structure*.

High Fertility

In addressing high fertility, the distinction between wanted and unwanted fertility is crucial. Despite gains in family planning, unwanted fertility remains significant: About one in five births is unwanted by the woman, and a larger proportion is mistimed. Unwanted fertility can result from limited access to family planning services (geographic or social distance, prohibitive costs), inherent limitations of technology (contraceptive failure, side effects), poor communication between partners, and women's relatively limited ability to negotiate the terms of sexual relations, including contraceptive protection.

> The desire for large families is closely associated with poverty and inequality.

High fertility is also driven by the desire for large families. In most developing countries, women still want more than two children; in some areas, such as sub-Saharan Africa, desired family size is typically around five children. It is important to remember that the desire for large families is closely associated with poverty and inequality. Where child mortality rates are high and social safety nets are frayed or non-existent, poor couples will have many children to ensure that some will survive to work and support their parents in old age. And where women lack access to education and economic opportunity, they must rely on childbearing as a source of status and security. Of course, even in affluent societies where women enjoy relatively equal rights, some couples will choose to have large families—but average fertility in those societies is invariably at or below replacement level.

In much of Asia and Latin America, wanted fertility is not far from the replacement level, so high fertility consists mostly of unwanted childbearing. In contrast, in sub-Saharan Africa, the high fertility component is not only very large (Table 20.1) but mostly caused by the desire for large families.

The two components of high fertility can be addressed by a combination of supply- and demand-side approaches. Quality reproductive health services, geared to meet the specific needs of clients, are vitally important in reducing unwanted childbearing. And where desired family size remains high, population policy must address underlying socioeconomic conditions, especially where poverty and gender inequality are most severe. Where investments are made in human development—such as child survival programs and educating girls through their adolescent years—couples prefer smaller families. This holds true even in very poor societies such as Sri Lanka and the state of Kerala in India. Although poor, these societies have achieved high levels of literacy and female empowerment, low infant and child mortality—and dramatic declines in fertility rates.

> Increasing the age at marriage by five years could directly reduce 15 to 20 percent of future population growth.

Young Age Structure

As the leading driver of population growth in the developing world, young population age structures merit increased attention in population policy. Of course, a young age structure is in itself not amenable to modification. However, the onset and pace of childbearing can be altered to offset population momentum. For example, increasing the age at marriage and associated childbearing by five years could directly reduce 15 to 20 percent of future population growth. Many interventions that promote later marriage and childbearing—including girls' education and enforcement of child marriage laws—are beneficial in their own right and have also been shown to slow population momentum.

POPULATION POLICIES FOR THE TWENTIETH CENTURY

The demographic landscape has changed dramatically in the last forty years; now it is time for population policy to change in response. Early population programs focused on the "low-hanging fruit" of contraceptive supply—a strategy that made sense when few women in the developing world had access to family planning services. Today, we must reach into the higher branches. The population policy options outlined below address the two crucial levers of rapid population growth—high fertility and young age structure—with an array of supply and demand approaches. Moreover, these policies are vitally important to advance public health, gender equity, and social justice.

Investing in Girls

Investing in girls has always been a good idea—it is now even better in light of the shifting composition of future population growth. Investments in girls through adolescence provide a demographic "three-for": reducing population momentum by delaying marriage and childbearing, thereby increasing the space between generations; lowering desired family size as more educationally accomplished girls are less reliant on multiple children for security; and decreasing the age and power differential between partners, thus positively affecting women's ability to meet their fertility goals. Benefits also extend to the next generation, because those who marry later and with more authority are likely to invest in their children (especially their girl children) in ways that establish a virtuous cycle of improved health and education. Specifically, we must (1) help girls stay in school through adolescence, (2) provide social and economic alternatives to early marriage and childbearing, (3) end child marriage and support married girls, and (4) focus on the youngest first-time mothers.

> The new challenge is not just getting girls (and boys) to school on time but keeping them there through adolescence.

Help Girls Stay in School Through Adolescence. Girls' education has long been associated with positive developments in health and population outcomes. Attending school during adolescence is crucial to girls' reproductive health.[2] Girls who stay in school during adolescence have later sexual initiation and are less likely to be sexually active than their same-age peers who are out of school. Those who are sexually active as students are also more likely to use contraceptives than their peers.

In recent decades, governments of developing countries have emphasized primary school attendance. Great progress has been made: Primary school attendance rates are up, and the gap between boys and girls is narrowing. Thus the new challenge is not just getting girls (and boys)

> This vicious cycle of intergenerational poverty, high fertility, and poor health can be broken by concerted investments in the poorest girls in the poorest communities.

to school on time but keeping them there through adolescence—minimizing dropout, especially around the time of puberty for girls. It is also important to increase the proportion of girls in the appropriate grade for their age: In most countries of sub-Saharan Africa, for example, the majority of adolescent girls in school are in primary grades.[3]

Provide Social and Economic Alternatives to Early Marriage and Childbearing. In poor communities, early marriage, followed by early childbearing, is a vital security and survival strategy for girls. Girls (and their parents) may view marriage as the only possible economic choice and accepted social role. For young married girls, the failure to produce a child in a short period of time may make them vulnerable to unilateral divorce or abandonment. Thus, the opportunity costs of repeated childbearing outweigh any other possible choices they have.

The vicious cycle of intergenerational poverty, high fertility, and poor health can be broken by concerted investments in the poorest girls in the poorest communities. In these communities, many young people, especially girls, are permanently diverted in early adolescence from the path of school and access to decent work. The absence of sufficient investments in girls during the crucial ages around puberty—ages ten to fourteen—limits their prospects and encourages dependence on marital and sexual relationships and childbearing for social and economic security.

The promotion of girls' education to the secondary level and their inclusion in community development efforts, functional financial literacy, and publicly visible/civic activities offer girls the beginnings of autonomy and lays the foundations for shifting gender norms. Girls must have social power and economic authority to counter pressures for sexual relations as a livelihood strategy—both inside and outside of marriage.

End Child Marriage, Support Married Girls.   Child marriage—legally, marriage of a girl or boy before her or his eighteenth birthday—is still with us. Demographic and health survey data indicate that in fifty less-developed countries, about 38 percent of young women aged twenty to twenty-four were married before age eighteen.[4] If present patterns continue, over 100 million girls will be married as children in the next decade. Although the proportion of women who marry early is declining in most, but not all, parts of the world, it is vital to underscore that millions of girls are currently married, with many more in the pipeline. Child marriage is a practice that mainly affects girls (56 percent of women aged twenty to twenty-four versus 14 percent of men the same age were married by age twenty).[5]

> Married girls have a higher risk for sexually transmitted infections and HIV than their sexually active unmarried peers.

Marriage transforms virtually all aspects of girls' lives. Typically, a girl who marries is moved from her familiar home and village, loses contact with friends, is initiated into sexual activity with someone she barely knows, and soon becomes a mother. The implications for health and well-being are striking. Married girls have a higher risk for sexually transmitted infections and HIV than their sexually active unmarried peers.[*] And the youngest (age sixteen and under) first-time mothers face an increased risk of mortality and morbidity—for themselves, and for their children.

Measures to eliminate child marriage must be context-specific and engage parents and community members. Essential elements of any strategy will be schooling in some form and giving girls better access to their peers. This can be achieved either through programs that promote primary to secondary school progression or that work with girls who have not been in school, creating group structures—assets in and of themselves and platforms through which the girls can receive functional literacy training, mentoring, health information, and either direct access to or referral to services.

At the same time, it is essential to invest in the 50-million-plus mar-

---

[*]This is true for a complexity of reasons; see, for example, Clark, Shelley, Judith Bruce, and Annie Dude, 2006, Protecting young women from HIV/AIDS: The case against child and adolescent marriage, *International Family Planning Perspectives* 32(2): 79–88.

ried girls in the developing world, whose social, economic, and reproductive lives are still ahead of them. We have had successful experiments with adolescent girls in a number of settings (Asia, Middle East, and Africa). "Married girls clubs" have been established to reduce social isolation, increase girls' agency and negotiating power with partners, and provide a venue for learning. (See Box 20.2, Berhane Hewan.) These groups empower and connect married girls—a goal important in its own right and crucial to improving their health and well-being and that of their children.

Focus on the Youngest First-Time Mothers. Special measures need to be taken to make first-time pregnancy safer. First births carry special risks for both mother and child, regardless of the age of the mother. The issue is of particular relevance to married girls because the vast majority of births to adolescent girls are first births that occur within marriage.[6] Adverse outcomes associated with first birth include obstructed labor (which can result in obstetric fistula in settings where access to care is limited),[7] preeclampsia/eclampsia, malaria, and infant mortality.[8] Disentangling the age and parity effects at young ages is difficult. But it is clear that the youngest first-time mothers—those under fifteen, owing to physical factors, and those under eighteen, owing to social factors—face special risks. Girls who give birth during adolescence require special attention because they are less mature and are simultaneously coping with their own and their baby's physiological, emotional, and economic needs.[9]

Programs engaging the youngest first-time mothers are strategic from three perspectives. First, demographically, as fertility falls, a rising proportion of births will be first births. Second, habits formed around the first birth, such as infant feeding choices and, crucially, spacing through contraception, tend to be carried forward across the life cycle. Third, focusing the health system on the social needs of these first-time mothers will generally raise the quality of service-site interpersonal relations (caring about the context of the mother, developing support and communication with the father). These habits, embedded in the system, also tend to carry over to the ways in which women across all ages and parities are treated.

Improving Access to High-Quality, Appropriate Reproductive Health Services

When many family planning programs were launched in the 1960s and 1970s, there was a great unmet need for contraception, and most clients

BOX 20.2

Berhane Hewan ("Light for Eve"):
Forming Girls' Groups in Rural Ethiopia

In the Amhara region of Ethiopia, rates of child marriage are among the highest in the world. Almost half of all girls in Amhara are married before their fifteenth birthday. A Population Council study in Amhara and Addis Ababa found that for many young girls, marriage may mean having forced sexual relations with a virtual stranger. Some 95 percent of the girls surveyed did not know their husbands before marriage, and 85 percent were given no warning that they were about to be married. Sexual initiation was often unwanted and traumatic: More than two-thirds of the married girls had not yet begun menstruating when they had sex for the first time, and 81 percent said their sexual initiation was physically forced.

Berhane Hewan (meaning "light for Eve" in Amharic) is a program designed to discourage early marriage and to help young married girls. Berhane Hewan creates a community dialogue about the negative consequences of child marriage, offers social and economic support to girls and their families, and encourages school attendance and delayed marriage. Unmarried girls in the program meet five times per week in small groups under the guidance of female mentors, all of whom are either women leaders from the community or local women with a background in teaching. For married girls, the program convenes clubs that offer reproductive health information, life and livelihood skills, and links to social services. And Berhane Hewan provides an economic incentive to delay marriage: Unmarried girls who participate regularly and remain unmarried for the duration of the project are presented with a goat at graduation.

The program has had a dramatic impact on girls' lives and community norms. Girls in the program are three times as likely to be in school, and younger adolescents are 90 percent less likely to be married, than their counterparts in another village. Moreover, married girls who participate in the program are nearly three times more likely to use family planning than their peers. In the villages where Berhane Hewan was launched, there have been no marriages of girls under the age of fifteen since the program began—and a greater proportion of girls are marrying later. The project is currently being expanded to reach 10,000 girls by the end of 2009.[a]

....

[a] Erulkar, Annabel S., and Eunice Muthengi, 2007, Evaluation of Berhane Hewan, a Pilot Program to Promote Education and Delay Marriage in Rural Ethiopia, Population Council, New York; Amin, Sajeda, Erica Chong, and Nicole Haberland, Programs to address child marriage: Framing the problem, *Promoting Healthy, Safe, and Productive Transitions to Adulthood Brief*, Number 14, Population Council, New York, updated January 2008.

were married women. Clients and their needs have changed since those early years, and family planning programs must change with them, by providing a diverse array of services, including temporary contraception, and responding to social and emotional realities—including sexual coercion.

Providing a Diverse Array of Services, Including Temporary Contraception. Today, there is still a substantial unmet need for contraception and reproductive health care, in part because desired family size has declined substantially in a majority of developing countries. As women want fewer children, the number of years in which they must avoid pregnancy increases—and now approaches two decades, on average. But over the course of those decades, women and their partners will need a wider range of contraceptive and reproductive health services.

For example, a typical client is no longer a married woman seeking to end her childbearing years. More often, clinic waiting rooms are filled with young women or married couples who want to delay their first birth or ensure space between pregnancies. This means there is a need for technologies and services that allow couples to safely avoid childbearing without a loss of fertility. But some service providers still emphasize permanent methods of birth control, which are inappropriate for many young clients.

A key related trend is that more young women are being sexually initiated *outside* of marriage, long before they have established permanent partnerships. Young men and women need more and better information and services, and reversible methods—especially those that protect against sexually transmitted diseases, including HIV.

Responding to Social and Emotional Realities—Including Sexual Coercion. Effective reproductive health services must also understand their clients' social and emotional needs. For example, sexual coercion is a serious problem, especially for young girls. In some settings, a very high proportion of first sexual relations may be tricked or forced. Thus, the social support and information at service points must build girls' agency and negotiating power. As the HIV epidemic becomes increasingly younger and female, service points and treatment alone are not sufficient to alter negative reproductive health outcomes that stem from HIV infection. This must be complemented by the expansion of single-sex safe and supportive spaces for girls. Such spaces can be established in youth clubs, community centers, religious institutions, and school facilities. Safe spaces are a primary asset to girls and young women, offering them access to peers and mentors and a base from which to offer or refer to services.

CONCLUSION

For decades, population policy has focused on unmet need for contraceptive services. Those services are as important as ever, though they must adapt to the needs of a new generation of clients. But today's new population challenges require accelerated investment in human development—in particular, investments in girls' education. They also demand fresh attention to the inequities that prevent women—and girls—from freely choosing the number and timing of their children. The next wave of population policies must be far better targeted, emphasize social inclusion, assure the observance of human rights, and, crucially, address the opportunity structures and capacities of young populations—especially young girls. The "unmet need" we must fill now is for justice.

REFERENCES

1. Bongaarts, J., 1994, Population policy options in the developing world, *Science* 263(5148): 771–776; Bongaarts, J., and R. A. Bulatao, 1999, Completing the demographic transition, *Population and Development Review* 25(3): 515–529.

2. Lloyd, Cynthia, B., 2008, The role of schools in promoting sexual and reproductive health among adolescents in developing countries, *Social Determinants of Sexual and Reproductive Health: Informing Future Research and Programme Needs,* S. Malarcher, ed., World Health Organization, Geneva.

3. Lloyd, Cynthia B., 2006, Schooling and Adolescent Reproductive Behavior in Developing Countries, paper commissioned by the UN Millennium Project for the report *Public Choices, Private Decisions: Sexual and Reproductive Health and the Millennium Development Goals,* UN Millennium Project, New York, available at http://www.unmillenniumproject.org/documents/CBLloyd-final.pdf.

4. Mensch, B., S. Singh, and J. Casterline, 2005, Trends in the Timing of First Marriage among Men and Women in the Developing World, Policy Research Division Working Paper Number 202, Population Council, New York.

5. DHS data from fifty-one countries as reported in NRC and IOM 2005.

6. Seventy-eight percent of births that occur before age eighteen are first births, and 90 percent of first births that occur before age eighteen occur within marriage. DHS data analyzed by Monica Grant, Policy Research Division, Population Council. DHS surveys covered 60 percent of developing-country populations.

7. For a discussion of a programmatic response, see Chong, Erica, 2004, Healing wounds, instilling hope: The Tanzanian partnership against obstetric fistula, *Quality/Calidad/Qualité*, volume 16, Population Council, New York.

8. Kiely, Michele, ed., 1991, *Reproductive and Perinatal Epidemiology*, CRC Press, Boca Raton, FL.

9. Haberland, Nicole, Erica L. Chong, and Hillary J. Bracken, 2004, A World Apart: The Disadvantage and Social Isolation of Married Adolescent Girls, brief based on background paper prepared for the WHO/UNFPA/Population Council Technical Consultation on Married Adolescents, Population Council, New York.

CHAPTER 21

············

Rethinking U.S. Population Policy

SUZANNE PETRONI

For more than forty years, the United States has been a leader in the international population arena. While this era began with two decades of remarkable unity and bipartisanship, U.S. population policy has since been characterized by the increasing influence of the Religious Right, political divisiveness, and ideologically driven restrictions on funding and policies. These restrictions have in turn impeded access by millions of women, men, and youth around the world to critical, lifesaving reproductive health services.

> For more than forty years, the United States has been a leader in the international population arena. While this era began with unity and bipartisanship, U.S. population policy has since been characterized by political divisiveness and ideologically driven restrictions on funding and policies.

This chapter follows U.S. policy and funding trends over the four decades in which this country has served as the world's largest donor to international family planning programs. It describes the origins of U.S. support and details some of the key influences, policies, and regulations that have guided—and

Suzanne Petroni is a senior program officer at The Summit Foundation in Washington, DC, where she manages the Global Population and Youth Leadership program.

increasingly circumscribed—U.S. assistance in this area. Ultimately, it argues for a policy that uncouples fundamentalist morality from the health and welfare of the impoverished people of the world.

THE GENESIS OF A POLICY

In the United States, birth control advocate Margaret Sanger first called attention to the need for global family planning in the 1920s.* In 1952 Sanger, alongside public health advocates and supporters of women's rights, helped found the International Planned Parenthood Federation (IPPF) to further this cause.† The establishment of IPPF coincided with a growing awareness and acceptance of birth control across the United States. It also reflected a post–World War II commitment to positive U.S. engagement in the international arena. Development assistance programs were launched in the 1950s, for both humanitarian and national security purposes—chiefly to stop the spread of communism. Indeed, it was largely a concern for national security that galvanized official U.S. support for international family planning.

In the 1950s and 1960s, demographers, economists, and military leaders, responding to work that had been piloted by private foundations and nongovernmental organizations, began to draw attention to the challenges of global population growth. A series of high-level and popular reports warned that soaring growth rates would lead to environmental degradation, food and water shortages, stalled development, and political strife.

Among those most concerned was William H. Draper Jr., a retired Army general who led President Eisenhower's Committee to Study the U.S. Military Assistance Program. The Draper committee concluded in 1959 that economic development was essential to political stability and national security and that rapid population growth posed a significant obstacle to such development. "Unless population trends are reversed,"

---

*The terms *family planning* and *birth control*, like the terms *family planning assistance* and *population assistance*, are often used interchangeably, as they are in this chapter. Population size is a function of fertility, mortality, and migration. Population programs have historically focused on reducing fertility through voluntary family planning or birth control.

†It should be noted here that the origins of the family planning movement, both domestic and international, were not always altruistic. While women's rights and public health were key concerns among supporters, race and class-based eugenics were early factors as well.

the committee warned, "grave political disorganization must be antici-pated in many areas of the world."[7] It recommended that President Eisenhower boldly address the challenge of population growth (under the seemingly noncontroversial rubric of "maternal and child health assistance") by supporting birth control overseas.

But Washington was not yet ready to take on this issue. Eisenhower dismissed the Draper committee's report, saying of birth control, "I can-not imagine anything more emphatically a subject that is not a proper political or governmental activity or function or responsibility."[24]

Eisenhower's reticence stemmed, in part, from fear of offending the Catholic Church—particularly after the U.S. Conference of Catholic Bishops publicly attacked the Draper report. While Eisenhower would reverse his position after his presidency,* his early opposition, com-bined with President John F. Kennedy's careful separation of religion and politics, effectively shelved serious government consideration of the issue for several more years.[5, 20]

Despite the lack of government support, private foundations and nongovernmental organizations continued to sound the alarm about population growth and to generate support for birth control at home and abroad.† Those organizations' pioneering work was seen as vital by many, including then U.S. Representative to the United Nations George H. W. Bush. In the foreword to Phyllis Piotrow's seminal 1973 *World Population Crisis: The United States Response*, Bush gratefully acknowledged the important work these organizations had done "in assisting government policy makers and in mobilizing the United States response to the world population challenge."[20]

UNITED STATES LAUNCHES
INTERNATIONAL FAMILY PLANNING PROGRAMS

While early support came from outside the White House, Lyndon B. Johnson became the first American president to endorse U.S. popula-

---

*Eisenhower recorded a public service announcement for the Planned Par-enthood Federation of America in February 1968, in which he said, "The facts changed my mind. . . . I have come to believe that the population explosion is the world's most critical problem."[14]

†The Rockefeller and Ford foundations were major funders in these early days, with the Population Council, Planned Parenthood Federation of Amer-ica, and the Population Crisis Committee (now Population Action Interna-tional) among the first NGOs working in this area.

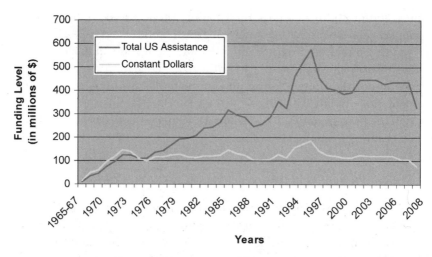

FIGURE 21.1 U.S. international family planning assistance, 1965–2008.
Note: 2008 funding is proposed.

tion assistance. In his 1965 State of the Union speech, Johnson made the first of several powerful statements about the challenges of rapid population growth—pledging U.S. funds to address "the explosion in world population and the growing scarcity in world resources."[12] With this presidential endorsement and a special allocation of $2 million, the U.S. Agency for International Development (USAID) initiated an official program of population research and technical assistance for family planning. (Figure 21.1 shows the history of funding levels for U.S. international family planning assistance.)

Concern about population growth went public in the late 1960s, fueled by a south Asian famine, Paul Ehrlich's best-selling *The Population Bomb*, a three-year-long series of congressional hearings, and President Johnson's impassioned, high-profile statements on the issue. In his 1967 State of the Union speech, Johnson argued that, "next to the pursuit of peace, the really greatest challenge to the human family is the race between food supply and population increase."[12] He implored Congress to provide funds to address this challenge. Shortly thereafter, with broad bipartisan support, Congress appropriated $35 million to USAID's fledgling program, increasing U.S. commitments to population assistance seventeenfold.[10] This funding would support technical assistance in family planning to developing country governments and nongovernmental organizations, training for health care workers, the provision of family planning information, services and contraceptive supplies, and research on population growth and related issues.

As this ambitious program got under way, Congress and the Johnson administration, along with a handful of allied nations, successfully encouraged the United Nations to join them in addressing the issue of population growth. The United Nations Fund for Population Activities (UNFPA*) was established in 1969, with 80 percent of initial contributions coming from the United States.[18]

PROGRAMS GROW, CONTROVERSY FLARES

Despite some election-year jitters around the issue in 1972, President Richard M. Nixon proved an ardent supporter of family planning programs, both at home and abroad, and Congress maintained broad bipartisan support for these programs throughout the 1970s. Nixon assigned a high-level committee to study the issue of unsustainable population growth[†] and encouraged a leadership role for the United States at the 1974 United Nations World Population Conference in Bucharest.[‡]

That same year, National Security Advisor Henry Kissinger signed a National Security Study Memorandum that urged "greatly intensified population programs" and stressed the need to ensure that family planning was never forced or coerced.[22] The Kissinger strategy was adopted a year later, when National Security Advisor Brent Scowcroft, on behalf of President Gerald R. Ford, signed a National Security Decision Memorandum affirming that "United States leadership is essential to combat population growth . . . and to advance United States security and overseas interests."[21]

By mid-decade, the United States was contributing $110 million annually for USAID and UNFPA family planning programs, which were reaching women in need across the globe. Presidents Ford and Carter also pronounced their moral and financial support for these efforts, increasing their budget allocations and sprinkling speeches and policy statements with expressions of concern about population growth and its impact on the environment and national security.

*Now called the United Nations Population Fund.
†The Commission on Population Growth and the American Future, known as the Rockefeller Commission, for its chairman, John D. Rockefeller III.
‡Despite political resistance from some developing countries who pushed for a New International Economic Order at Bucharest, 135 countries adopted a World Population Plan of Action that affirmed the rights of couples and individuals to determine the number and spacing of their children.

Yet even during the heyday of bipartisan support for population programs, when high-level political commitment seemed secure and appropriations were on the rise, signs of contention began to appear. For example, President Nixon rejected his own population commission's findings, issued months before the 1972 election. Why? According to commission member Congressman James Scheuer,

> After *Roe v. Wade* legalized abortion in the United States, population policy was hopelessly ensnared in the volatile politics of abortion.

"the reasons were obvious—the fear of attacks from the far right and from the Roman Catholic Church because of our positions on family planning and abortion."[16]

So even before *Roe v. Wade* legalized abortion in the United States, population policy occasionally became entangled in the volatile politics of abortion. But after that 1973 Supreme Court decision, it was hopelessly ensnared. Only months after *Roe v. Wade* was decided, Senator Jesse Helms (R-NC) led a successful effort to prohibit U.S. foreign aid funds for abortion services. That law, the "Helms Amendment," formally severed any connections between family planning and abortion in U.S.-funded population programs. However, as we shall see, the connection would live on in the minds of conservative thinkers and lawmakers.

A decidedly different voice of criticism also arose in the 1970s. While endorsing the right to family planning, feminists at the First World Conference on Women, held in Mexico City in 1975, condemned top-down population control policies. They took issue with population programs' singular focus on fertility, calling instead for an agenda that considered women's health, rights, and equality. A broader agenda, they argued, not only would benefit women but would be more effective in reducing fertility rates as well. This new voice urged a greater focus on—and greater say for—the women in developing countries who were most directly affected by family planning programs. The feminist voice would gather strength over the years, eventually compelling a significant shift in global population policies.

THE UNITED STATES IN RETREAT

The election of Ronald Reagan augured a significant and long-lasting shift in the U.S. approach to international family planning. In 1981, the wall that had separated church from state around these issues for years began to crumble. Beginning with the election of Reagan, members of the "New Right"—an unlikely alliance of evangelical social-conservative Christians and free-market thinkers who shared an aversion to governmental involvement in individual decision-making—were allowed an unprecedented role in formulating U.S. policy around sexuality-related issues. The New Right perceived government-supported family planning as an unwelcome intrusion into family—and opposed it at home and abroad.[9]

Exploiting their newfound access to the White House, leaders of the New Right engineered a 35 percent cut in Ronald Reagan's population assistance budget request for fiscal year (FY) 1982. An early draft of the president's FY 1983 budget proposed eliminating population funds altogether.[10] In both years, Congress came to the rescue by appropriating more funds than the president requested. But even on Capitol Hill, where bipartisan support had been the norm, religious conservatives began to rally behind Christopher Smith (R-NJ), a devout Catholic who played upon the perceived connection between abortion and family planning to call for greater restrictions.

A major turning point came at the Mexico City International Population Conference in 1984. The Reagan administration—abetted by leaders of the Christian Coalition and other New Right groups—surprised the international community by announcing a stunning reversal of U.S. population policies. Rejecting a stance that had endured for nearly two decades, the U.S. delegation proclaimed that population was "a neutral phenomenon" in regard to economic development—thereby removing an important rationale for population policy and programs.

U.S. representatives then unveiled the "Mexico City policy,"* which denied funds to any foreign nongovernmental organization that provided abortion counseling or services, or lobbied for legalized abortion, even with its own funding.[25] Citing concerns about China's one-child policy, and again making a spurious connection between abortion and family planning in U.S. funding, the delegation further announced that it would require UNFPA to prove that it did not support abortion or coercive family planning programs in China.[25]

The Mexico City policy effectively slashed funding for the International Planned Parenthood Federation, a longtime recipient of U.S.

---

*This policy was later called the "Global Gag Rule" by its opponents.

support.* Soon thereafter, the Reagan administration used the newly passed "Kemp-Kasten" legislation† to effectively block funding for UNFPA. From 1985 to 1992, and again from 2002 to 2008, hundreds of millions of dollars appropriated by Congress to the United Nations Population Fund were withheld by Republican administrations citing the Kemp-Kasten amendment.

BUSH CONTINUES THE RETREAT

Much to the dismay of advocates who hoped he would stay true to the pro–family planning positions he had held as a congressman and as a UN ambassador, George H. W. Bush maintained Reagan's restrictive policies upon assuming the presidency in 1989. Reversing his earlier support for UNFPA, Bush vetoed the 1990 Foreign Operations bill because of an earmark for that agency—just one of several vetoes and veto threats during his administration over international family planning. Like Nixon's, Bush's about-face no doubt stemmed from a desire to maintain the support of the Religious Right, which President Reagan had so effectively cultivated and which had become a central and powerful constituency of the Republican Party.[6]

The 1980s and early 1990s saw increasing opposition to population programs, not only from the Religious Right and economic conservatives, but also from feminists. Feminists and women's health advocates from around the world became increasingly concerned that family planning programs were too focused on demographic goals, with inadequate regard for the health and rights of women.[1, 8, 11] Building on the alliances and networks forged through the UN Decade for Women (1976–1985), women's groups began to advocate more vociferously for family planning within a context of women's empowerment, reproduc-

*It is not known how many organizations lost USAID funding because of their refusal to sign the Mexico City policy. IPPF was the largest, and it would lose an estimated $11 million per year as a result. Its affiliates throughout the developing world would be forced to cut services to their clients.

†Initially known as the Kemp-Inouye amendment, this 1985 law prohibits U.S. funds to an organization or program that "supports or participates in the management of a program of coercive abortion or involuntary sterilization." Numerous reviews and investigations have concluded that UNFPA does not support coercive abortion or sterilization, but the Reagan and both Bush administrations claimed that by virtue of working in China, UNFPA was culpable.

tive health, and rights.[9] While these arguments gained no political traction under Presidents Reagan and Bush, they became sharper and more refined during this time.

A CHANGED PARADIGM

Women's groups had a sympathetic ear in the next president, Bill Clinton. Accordingly, in his first official act as president, Clinton overturned the Mexico City policy. He then resumed funding for UNFPA, citing the agency's critical and lifesaving work around the world. During his tenure, Clinton would restore the United States' leading position in the international population arena, but within a new paradigm.

That revived leadership was crucial to the success of the 1994 International Conference on Population and Development (ICPD), in Cairo, where the United States helped formulate a new framework for the population field. A decade after the Mexico City conference, the United States had come full circle. The U.S. delegation again affirmed the need to achieve a sustainable global population, but it also gave credence and support to those concerned with women's health and rights. In the end, the United States helped forge a consensus in Cairo that reflected the successful advocacy of women's groups from around the world. Stabilizing global population remained an essential goal, but women's health and rights were placed squarely at the center of global priorities.[1, 13] The Cairo consensus shifted the focus of population programs away from top-down, demographically driven, contraceptive-centered policy and toward more comprehensive approaches that prioritized the basic rights and needs of individuals.

After the ICPD, USAID continued its core areas of assistance, while undertaking new initiatives to support integrated reproductive health services, expand services for young people, treat the complications of unsafe abortion, address female genital mutilation and other gender issues, and develop innovative linkages between population and environmental programs.[4] Changes were made, not only in rhetoric, but in reality.

The largely positive role of the United States did not last long, however. While the 103rd Congress approved the highest level of U.S. funding ever for international family planning ($542 million for USAID and $35 million for UNFPA) in 1994, the Republicans who swept into office later that year had different priorities. With the Religious Right now steering the Republican agenda, Congress pressed persistently over the next six years for budget cuts and new restrictions on family planning assistance. The relentless debates between Congress and the Clinton White House contributed to several funding delays, metering of appropriations,* government shutdowns, vetoes, and eventually, significant funding cuts for international family planning.

Regardless of the enormous pressure he faced around these issues, President Clinton's support for international family planning held firm. At an event hosted by the White House on World Health Day in 2000, Clinton chided Congress for its intransigence, asserting that "America has a profound interest in safe, voluntary family planning, a moral interest in saving human lives, a practical interest in building a world of healthy children and strong societies."[3] Clinton remained an adamant supporter of international family planning for both moral and practical reasons, but Congress—controlled by the Republican Party and the Religious Right—had gained the upper hand.

THE RELIGIOUS RIGHT CONSOLIDATES ITS GAINS

As is now well documented, the Religious Right obtained unprecedented access and influence in the administration of George W. Bush, and issues related to sexuality—including both domestic and international family planning—were at the core of their agenda.†

The importance of these issues became evident as soon as Bush was sworn in. Just as Bill Clinton began his presidency by rescinding the Mexico City policy, George W. Bush reinstated the same policy on his first day in office. Bush claimed that the policy was necessary to prevent taxpayer funds from being "used to pay for abortions," even though public funding of abortion had already been prohibited for some twenty-eight years.

While Bush initially affirmed the importance of family planning and promised to maintain funding for international population assis-

---

*The FY 1998 Foreign Operations bill contained a complicated agreement under which international population assistance would be disbursed at a rate of 8.34 percent per month.

†Esther Kaplan, David Kuo, Chris Mooney, Michelle Goldberg, Deborah McFarlane, Dianne diMauro, and Carole Joffe, as well as Americans United for Separation of Church and State and People for the American Way, are among the many who have written of the ways in which the Religious Right guided—or misguided—the policies of the Republican Party and the Bush administration in recent years.

tance, that support was short-lived. After overlooking a U.S. contribution to UNFPA in its first year, for example, the Bush administration proceeded to withhold $235 million in congressionally appropriated funds to the organization. Population assistance through USAID was increasingly restricted, and if left to the White House, such programs would have received a sharp 25 percent decrease in 2009. Ironically, in its proposed budgets for both FY 2008 and FY 2009, the Bush administration justified this cut by arguing that forty years of family planning had been so successful that significant U.S. funding was no longer essential. These arguments came at a time when—as highlighted by the other chapters in this book—tremendous and even growing need existed, and still exists, around the world.

Funding aside, the Bush administration's policies showed an indisputable disdain for international family planning programs. The administration peopled delegations to international conferences with representatives of the Religious Right who sought to roll back long-standing agreements and language on reproductive health and who threatened to cut assistance to countries that refused to support the U.S. position.* It disparaged condoms, proffering misinformation about their effectiveness in preventing sexually transmitted diseases and pregnancy. It withdrew funding from groups providing refugees with basic, yet literally lifesaving, reproductive health care.† It insisted on the promotion of abstinence-only-until-marriage policies for young people in the developing world, disregarding conclusive evidence that those programs are ineffective and even harmful. (For example, abstinence programs do nothing to protect young married women in developing countries, who are now the most vulnerable population to HIV/AIDS infection.)

The list could go on. The common theme of these policy decisions is that they were guided by ideology, not evidence. As Esther Kaplan writes in *With God on Their Side*, Bush "happily ceded huge swaths of his domestic and international policy" to the Christian Right. These reli-

---

*Dramatic examples, such as the tactics of the U.S. delegations to the 2002 UN Children's Summit and Asia and Pacific Regional Conference on Population, are described well by Esther Kaplan.

†The Reproductive Health Response in Conflict (RHRC) Consortium lost $1 million in U.S. funding for its work to prevent gender-based violence and provide emergency obstetric care, HIV/AIDS prevention services, and emergency contraception in refugee settings.

gious traditionalists are not concerned with the facts of either reproductive health or population dynamics; they want American public policy to mirror their values. But their values do not embrace the rights of women, men, and youth to determine freely and responsibly the number and spacing of their children. When policies are shaped by those values, reproductive decision-making is restricted and evidence is denied.

> People throughout the developing world are the ultimate victims of the ideological hijacking of U.S. population policy.

People throughout the developing world are, sadly, the ultimate victims of the ideological hijacking of U.S population policy. As access to comprehensive reproductive health services has become increasingly restricted, the world's most vulnerable women, men, and youth have endured high rates of unintended pregnancies, unsafe abortions, maternal deaths and injury, and sexually transmitted infections, including HIV/AIDS.

CONCLUDING THOUGHTS

Broad agreements about the impacts of population growth on national security, the environment, and human well-being placed international family planning on the agenda of the U.S. government in the 1960s and kept it there for nearly two decades. Although religion was always part of the discussion—both on behalf of and in opposition to family planning—the constitutionally guaranteed separation between church and state worked during this time to prevent religion from obstructing effective policy.

Beginning in the 1980s, however, the church-state barrier began to fall. The Religious Right was granted an unprecedented level of influence, and the issue became politicized and partisan—as it has been ever since.

The long-term strategies of the Religious Right worked quite effectively in the family planning arena, both domestically and internationally. They used public policy to impose laws and standards defined by their own interpretation of Christian values. Their deliberate conflation of abortion with family planning, their persistent entreaties to Congress and the executive branch, and their ability to develop coordinated messages and drive a common agenda enabled them to chip away at political support for international family planning. Opposition to family planning assistance, or at least support for ever more restrictions on it, has become the norm among conservative lawmakers. These tactics have vastly diminished financial support for key international and nongovernmental organizations providing family planning services.

This book contains many indisputable facts about the effects of population dynamics, poor reproductive health, gender inequality, and more. But while we understand—as lawmakers did some forty years ago—that voluntary family planning can help address many of the challenges associated with these issues, we must also recognize that we cannot rely on evidence alone to free family planning assistance from the political morass in which it has become entangled. Any successful effort to re-fund the field and reestablish the position of the United States as a global leader in international family planning must work from this reality.

> We cannot rely on evidence alone to free family planning assistance from the political morass in which it has become entangled.

The recent election of a president and members of Congress who have already proven to be supportive of international family planning and reproductive health will certainly ease the transition back to U.S. leadership on these issues. Advocates must not fall into the trap, however, of believing that short-term changes by the new administration and Congress will reverse trends and perceptions that took decades to build. Instead, they must work to create a sustainable base of support for international family planning by, as President Obama suggested when he overturned the Mexico City policy on January 23, 2009, moving beyond stale and fruitless debates about the role of family planning and reproductive health in U.S. policy.[19] Let us heed President Obama's call for a "fresh conversation on family planning" and find common ground in addressing the needs of women and families around the world.

Finally, this chapter cannot close without a word about the impact of women's rights advocates on family planning assistance. While women's health and empowerment were nearly always elements of U.S. programs, feminist advocates were unquestionably effective in achieving a much-needed increased focus on these issues over the years.

Some have argued that the success of feminists in shifting government priorities toward women's rights (and, by implication, away from demographics and national security) has made it easier for governments to retreat from international family planning (see Sinding,[23] for one). A thorough reading of the history, however, shows that the shift originated not with feminist advocates in the 1980s and 1990s but with opponents of family planning in earlier years. The feminists were granted a seat at the table much later. And when they were, the world was given a solid platform on which to act toward both improved women's health and rights and a sustainable population, with family planning a key component of that platform.

The fact that 179 countries, as well as advocates from the fields of women's rights, public health, environment, and more, joined the consensus in Cairo bespeaks a remarkable unanimity that should not be discounted. In seeking to overcome the current political impasse around international family planning, we should not turn back the clock to the days before Cairo, when global concerns too often outweighed individual rights. That solution would break the stalemate, but only by splintering the coalitions that were so carefully crafted. Instead, we should hearken back to the Cairo consensus, which placed freedom and informed choice front and center, and where voluntary family planning was enveloped in a comprehensive range of social and sustainable development policies. From that basis, we should seek new strategies that can extricate critical family planning programs—and the millions of women around the world who rely on them—from the grip of a small minority that has overtaken sexuality politics in this country. Doing so is critical, for only when individuals are given the means with which to decide freely and responsibly the number, spacing, and timing of their children will we be able to achieve the outcomes we all seek: improved quality of life and a healthy planet that can be sustained for generations to come.

REFERENCES

1. Ashford, Lori S., 1995, *New Perspectives on Population: Lessons from Cairo,* Population Reference Bureau, Washington, DC.

2. Bush, George W., 2001, Restoration of the Mexico City Policy, Presidential Documents, 17303, volume 66 of *Federal Register,* Government Printing Office, online via GPO Access.

3. Clinton, William J., 2000, Remarks by the President at World Health Day, Office of International Information Programs, U.S. Department of State.

4. Crane, Barbara B., 2007, personal communication, August, Washington, DC.

5. Crane, Barbara B., and Jason L. Finkle, 1989, The United States, China, and the United Nations Population Fund: Dynamics of U.S. policymaking, *Population and Development Review* 15(1): 36.

6. di Mauro, Diane, and Carole Joffe, 2007, The religious right and the reshaping of sexual policy: An examination of reproductive rights and sexuality education, *Sexuality Research & Social Policy* 4(1): 67–92.

7. Donaldson, Peter J., 1990, On the origins of the United States Government's international population policy, *Population Studies* 44(3): 385–399.

8. Eager, Paige Whaley, 2004, *Global Population Policy: From Population Control to Reproductive Rights*, Global Health, Aldershot, Hants, England, Ashgate, Burlington, VT.

9. Finkle, Jason L., and Barbara B. Crane, 1985, Ideology and politics at Mexico City: The United States at the 1984 International Conference on Population, *Population and Development Review* 11(1): 1–28.

10. Green, Marshall, 1993, The Evolution of U.S. International Population Policy, 1965–92: A Chronological Account, *Population and Development Review* 1992: 303–321.

11. Hartmann, Betsy, 1987, *Reproductive Rights and Wrongs: The Global Politics of Population Control and Contraceptive Choice*, first edition, Harper & Row, New York.

12. Johnson, Lyndon, 2007, State of the Union Speeches, 1965–1968, *The American Presidency Project* (online), John Woolley and Gerhard Peters, eds., University of California (hosted), Santa Barbara, March 2, http://www.presidency.ucsb.edu/ws/?pid=2132.

13. Kantner, John F., and Andrew Kantner, 2006, *The Struggle for International Consensus on Population and Development*, first edition, Palgrave MacMillan, New York.

14. Lasher, Craig, 2001, *A History of U.S. International Population Assistance Programs*, Population Action International, Washington, DC.

15. Lasher, Craig, 2007, personal communication, Washington, DC, April.

16. Mumford, Stephen, 2007, NSSM 200, the Vatican, and the world population explosion, *The Journal of Social, Political and Economic Studies*, Council for Social and Economic Studies, July 26, http://www.population-security.org/journal-spes.htm.

17. Nixon, Richard, 2007, Statement on Signing the Family Planning Services and Population Research Act of 1970, *The American Presidency Project* (online), John Woolley and Gerhard Peters, eds., University of California, Santa Barbara, CA, 1970 database, April 1.

18. Nowels, Larry Q., 2003, Population assistance and family planning programs: Issues for Congress, CRS issue brief, Congressional Research Service, volume IB96026, Library of Congress, Washington, DC.

19. Obama, Barack, 2009, Statement of President Barack Obama on Rescinding the Mexico City Policy, the White House, Washington, DC, January 23.

20. Piotrow, Phyllis Tilson, 1973, *World Population Crisis: The United States Response*, Praeger, New York.

21. Scowcroft, Brent, 1975, National Security Decision Memorandum 314,

Box 1, National Security Decision Memoranda and Study Memoranda, Gerald R. Ford Library, http://www.ford.utexas.edu/library/DOCUMENT/NSDMNSSM/nsdm314a.htm, March 3, 2007.

22. Scowcroft, Brent, and Gerald Ford, 1982, United States international population policy, *Population and Development Review* 8(2): 11.

23. Sinding, Steven, 2006, Population and Sexual and Reproductive Health and Rights: State of the Field and Some Suggestions for Future Program Actions, Papers from the International Grantmaking Review, Los Altos, CA, Packard Foundation, p. 8.

24. If we ignore the plight, 1965, *Time Magazine,* July 2.

25. U.S. Department of State, 1984, U.S. Policy Statement for the International Conference on Population, *Population and Development Review* 10(3): 574–579.

CHAPTER 22

....

Going to Extremes

Population Politics and
Reproductive Rights in Peru

SUSANA CHÁVEZ ALVARADO, with
JACQUELINE NOLLEY ECHEGARAY

In recent decades, reproductive health policy in Peru has veered between extremes. At one extreme, misguided "population control" policies in the 1990s led to horrific abuses—notably, the forced sterilization of poor and indigenous women. At the other, the Religious Right—with help from conservatives in the United States—has fought since the late 1990s to deny access to surgical sterilization and other forms of contraception. Although they may seem to be in diametrical opposition, both extremes have one element in common: a profound disregard for women's rights.

> In recent decades, reproductive health policy in Peru has veered between extremes, [which] have one element in common: a profound disregard for women's rights.

The Peruvian case offers examples of well-intentioned population policies that went disastrously amiss, with tragic implications for

....
Susana Chávez Alvarado is the Director of the Center for the Promotion and Defense of Sexual and Reproductive Rights/Centro de Promoción y Defensa de los Derechos Sexuales y Reproductivos (PROMSEX) in Lima, Peru. Jacqueline Nolley Echegaray is Associate for International Programs at the Moriah Fund.
Translated from Spanish by Josie Ramos and Jacqueline Nolley Echegaray.

women and their families. As such, it offers powerful lessons for twenty-first-century population policy.

PERU IN THE 1990S AND THE
ORIGINS OF FORCED STERILIZATION POLICY ·

During the late 1990s, at a time when Peru was mired in a bloody internal conflict,[1] allegations surfaced that the government family planning program forcibly sterilized poor, indigenous women in rural areas. Just as with reports of forced disappearances, officials attempted to dismiss the accusations. The allegations were characterized as part of a conspiracy to discredit the government or, alternatively, as merely the actions of a few "overenthusiastic" health-care practitioners.

As the Peruvian government was eventually compelled to acknowledge, however, forced sterilizations and related human rights abuses were neither a conspiracy against the government nor the work of a few rogue health care workers. And for the women who were sterilized, as well as their families, the sterilizations were not easily forgotten.

Forced sterilization campaigns were in fact an integral part of the population policy designed and implemented by the government of Alberto Fujimori during the 1990s.[2] Elected president in 1990, Fujimori assumed power at a particularly calamitous time in the country's history. The economy was in shambles, with an annual inflation rate of 7,500 percent. In the countryside, brutal insurgent groups were growing in strength and numbers. Both insurgent groups and Peru's armed forces were perpetrating mass atrocities. Nearly 70 percent of Peru's population lived in poverty; 24 percent in extreme poverty—especially in the country's mountainous region, which was also the area that bore the brunt of the terrorist violence. Peru's indigenous peoples, who comprised the vast majority of the population outside of the capital, fared even worse: 79 percent lived in extreme poverty, and 77 percent were illiterate.

Peruvians in every part of the country were desperate for peace and economic stability. Fujimori, a political unknown whose only previous professional experience had been as a professor of agricultural engineering and dean of a university, was expected by many Peruvians to be the country's "savior." Fujimori zealously embraced the neoliberal economic model as a solution to Peru's troubles—with the encouragement of international financial institutions such as the World Bank and International Monetary Fund. Accordingly, his first act upon taking power was to apply drastic "structural adjustment" measures. Numerous public companies were privatized, private-sector price controls were relaxed, and government subsidies and employment were sharply reduced, as were restrictions on investment, imports, and the flow of capital.

Food subsidies were rescinded, resulting in a 1,000 percent increase in the cost of staple foods. Electricity costs quintupled, water prices rose eightfold, and gasoline prices rose 3,000 percent. The impacts of these policies were particularly severe for women and indigenous people.[3]

Fujimori also sought to reform the country's health sector. By 1991, government health spending had fallen by three-quarters from the amount spent in 1980.[4] Fujimori's health sector reforms—like his economic reforms—were characterized by drastic reductions in the role of the state, particularly as a regulator and service provider. The government distanced itself from its identity as "benefactor to all," instead adopting a "focus on the most poor." International donors and creditors, including USAID, the World Bank, and the Inter-American Development Bank, enthusiastically encouraged—and financed—these reforms.

At the same time, Fujimori's administration embraced the era's growing enthusiasm for women's rights and reproductive health; in fact, the government's discourse on the topic was more progressive than at any other time in Peruvian history.[5] In 1994, Fujimori sent a high-level delegation to the landmark United Nations International Conference on Population and Development in Cairo. And the following year, Fujimori himself traveled to the Fourth World Conference on Women in Beijing, where he defended the provision of contraceptives, gender equality, and reproductive rights.

Fujimori's family planning program was launched with initial support from women's rights groups and reproductive health advocates. Contraception was made available at no cost, and voluntary sterilization was added to the contraceptive methods paid for by the state. Reportedly, Fujimori even considered legalizing abortion but stopped short because of fear of rebuke from the Catholic Church. The family planning program took on such importance that its budget grew to exceed those of all other health programs combined. The program practically achieved the status of an independent ministry; its director was said to receive his orders directly from Fujimori himself.

But while the Fujimori government rhetorically embraced the reproductive health paradigm, the central goal of its family planning program was to lower birthrates. The government associated rapid population growth with poverty and sought to break the "vicious cycle" of poverty and unwanted children.[6] This narrow focus on fertility—rather than on meeting clients' needs—would shape the strategy and goals of Peru's family planning program, leading to horrific abuses of human rights.

APPLYING THE NEW FAMILY PLANNING POLICY: FORCED STERILIZATIONS IN PRACTICE

Ensuring that family planning services reached the most remote areas in the country was a top priority of the program. To that end, the program launched sterilization "fairs" in the poorest areas of the country. Providers traveled to remote communities where most women speak only the indigenous languages Quechua or Aymara. The predominantly Spanish-speaking providers relied on male partners to serve as interlocutors between providers and their female patients, because in many rural communities, only the men speak Spanish.

Although the government never admitted to setting specific goals for the number of sterilizations to be performed, Peruvian attorney Giulia Tamayo uncovered overwhelming evidence that such mandates did indeed exist.[7] When asked about the pressure to achieve specific goals, one obstetrician replied:

> At that time it was obligatory for all providers, obstetricians as well as other doctors, to get patients to undergo [tubal] ligations. Those who were hired were told that it was a way to demonstrate their productivity. We were also threatened with being fired and even potentially having our medical licenses revoked [if we did not comply] (obstetrician, thirty-nine years old).

One woman who lived in a community targeted by the program recalls, "It was frightening to go into a health center; no matter what you went in for, they asked you how many children you had and offered you a [tubal] ligation" (woman, thirty-seven years old). Another woman, who was sterilized, recalls:

> They would get all of us mothers together and they would give us a talk. Then one by one, they visited us at home to talk to us about having the operation. One morning a doctor came with a truck and took us all in to the health center, and in the afternoon, they brought us all back with our tubes tied. They brought us back as if we were cattle, lying down and in pain (woman, forty-two years old).

Once the accusations of forced sterilizations became public, the women's movement mobilized in response, and the *Defensoría del Pueblo* (the human rights ombudsman's office) issued a series of reports on the implementation of "voluntary" surgical contraception. The Defensoría catalogued a host of abuses in the family planning program: tubal ligations performed without informed consent, surgery performed in unfit locations, numerical targets for the use of particular contraceptive methods, and a lack of follow-up treatment for patients who were sterilized. Tragically, the reports documented seventeen cases in which women died as a result of sterilization procedures.

CONTINUING VIOLATION: THE IMPACT OF FORCED STERILIZATIONS ON PERUVIAN WOMEN

The legacy of forced sterilization endures. For many rural Peruvian women, sterilization is a profound break from the natural cycles of menstruation, pregnancy, and childbirth that are seen as central to internal equilibrium. One woman observes, "Since they opened me up [and gave me a tubal ligation], nothing has ever been the same. I get sick all of the time, I can't carry any weight, and my husband gets mad at me and shouts at me because I can't work like before. I've become useless."

Rightly or not, many sterilized women blame the procedure for their current ailments; according to one peasant leader, who is currently a member of congress, "many of my sisters who have been sterilized are now repudiated by their husbands and suffer other damages, such as depression or even cancer" (indigenous peasant leader, forty years old).

Only a tiny number of cases related to forced sterilizations have been brought to trial. The closest the Peruvian government has come to accepting responsibility is in the "friendly settlement" negotiated by the Inter-American Commission for Human Rights in the case of Mamérita Mestanza (see Chapter 27). As part of this settlement, the government agreed to implement measures to avoid forced or coercive sterilizations in the future, including a mandatory 72-hour waiting period and the establishment of an accreditation process for clinics performing the procedure. Nevertheless, Peruvian authorities have been extremely reluctant to admit responsibility for the human rights violations committed as a result of the campaigns.

CO-OPTING THE STERILIZATION TRAGEDY

Another legacy of the era of forced sterilization is an ongoing backlash from the right. Fundamentalist groups in Peru—including Catholic

Church officials, NGOs affiliated with the Church, and ultraconservative policymakers—exploited evidence of abuses to advance their own agenda of restricting access to reproductive health services in general.

Beginning in the late 1990s, these groups demanded that the national family planning program be shut down and that sterilization be removed from the list of state-approved contraceptive methods. Mobilizing around this agenda, several right-wing leaders in Peru began coordinating their anti–family planning activities with like-minded groups in the Unites States, such as the Population Research Institute (PRI) and Human Life International (HLI). These groups carried out aggressive advocacy campaigns against USAID's support for Peru's national family planning program, even though USAID was not supporting or encouraging coercive activities.

In early 1998, a PRI representative traveled to Peru to meet with the ultraconservative physician and future congressman, Dr. Héctor Chávez Chuchón. Upon returning to the United States, PRI contacted Representative Chris Smith, a conservative ally and member of the U.S. Congress, claiming that USAID had financed human rights violations in Peru and therefore its program should be shut down immediately. In response, Smith ordered a congressional hearing and an investigation of forced sterilization in Peru. The investigation—carried out by the U.S. government—did not find any evidence that USAID had financed the abuses committed by the Peruvian government.[8]

Nonetheless, pressure from fundamentalist groups undermined USAID's support for reproductive health programs in Peru and led to other limitations. In 1998, USAID/Peru ended its financial support for postabortion care and insisted that the ministry of health remove emergency contraception from the list methods offered by Peru's national family planning program. USAID likely anticipated that right-wing groups would ease their pressure if the agency demonstrated that its family planning program in Peru did not include these controversial elements; however, this preemption strategy backfired. Instead of stepping back, right-wing groups continued their attempts to restrict access to contraceptive methods and limit reproductive choice. In the ensuing years, U.S.-based antiabortion groups joined PRI in working with conservative Peruvian allies to discredit and harass USAID and other organizations working in support of reproductive health and rights in Peru.

Post-Fujimori governments in Peru have also used the abuses committed under Fujimori to justify a radical swing back to conservative reproductive health policy. Rather than identifying and punishing those responsible for the violations committed during the 1990s, the government has chosen to vilify sterilization itself. One congressional investigation into the abuses resulted in a single recommendation: the outright ban of sterilization procedures. Although this recommendation was not implemented, the attitude it expresses has had a "chilling effect" on the use of sterilization. As a result, there are regions in the country were tubal ligations are practically nonexistent because doctors fear being investigated and/or harassed by government officials opposed to family planning. Today, Peru has come full circle: Where once women were sterilized against their will, now women who want tubal ligations often cannot obtain them.

> Today, Peru has come full circle: Where once women were sterilized against their will, now women who want tubal ligations often cannot obtain them.

WHAT ARE THE LESSONS LEARNED?

Today, as climate change and other environmental crises are reviving calls for "population control," it is important to consider the lessons learned from the Peruvian experience.

First, human rights abuses are likely where reproductive health services are seen as a means to an end, rather than as an end in themselves. Spurred by social and economic crises, Fujimori's family planning program did not exist to meet the needs of individual men and women but was believed to serve the broader interests of the state. The results were coercion, disregard for clients' very humanity, and ultimately, the horrors of forced sterilization, including death. Furthermore, the Peruvian case makes clear that where democracy is fragile and racism and social exclusion are rampant, it is the poorest, most marginalized women who will bear the brunt of these abuses.

Second, coercion provokes backlash. In Peru and elsewhere, heavy-handed tactics tarnished public opinion about family planning, eroding gains in maternal health and slowing the transition to smaller families. Moreover, abuses were quickly seized upon by right-wing groups that oppose contraception, abortion, and women's rights. Those groups—many of which are based in the United States—are poised to replay the Peruvian scenario wherever abuses occur.

Population policies that do not respect human rights run the risk of terrible abuses and can engender a powerful backlash against family planning and women's rights. Neither of these outcomes has a place in a justice-based approach to the climate crisis. Women—and the environment—deserve better.

REFERENCES

1. An armed internal conflict between the Peruvian government and self-proclaimed communist revolutionaries raged in Peru from 1980 to 2000. According to Peru's Truth and Reconciliation Commission, nearly 70,000 people were killed or disappeared during the conflict. Civilians—particularly in the rural, mountainous region where the revolutionary groups were based, and in Lima, the capital—were targeted by all sides.

2. http://www.elpais.com/articulo/internacional/PERU/Fujimori/ordeno/esterilizacion/forzosa/200000/mujeres/indigenas/Peru/elpepipor/20020725elpepiint_4/Tes/.

3. Yamin, A., with R. Hurtado, 2002, Castillos de Arena in The Road to Modernity: A Human Rights Perspective on the Reform Process of the Health Sector in Peru (1990–2000) and its Implication on Maternal Mortality, Centro de la Mujer Peruana Flora Tristan, 2003.

4. Ewig, Christina, 2002, The Politics of Health Sector Reform in Peru, http://www.future.org/files/Ewig%20Christina.%20Washington%20Abril%202002.%20The%20Politics%20of%20Health%20Sector%20Reform%20in%20Peru.pdf.

5. Vasquez del Aguila, Ernesto, 2002, Invisible women: Forced sterilization, reproductive rights, and structural inequalities in Peru of Fujimori and Toldeo, *Estudos e Pesquisas em Psicologia, UERJ, RJ, Ano 6, 1 Semestre de 2002*, accessed December 2008.

6. Ibid.

7. Tamayo, G., 1999, Nothing Personal: Report on Human Rights on the Application of Surgical Contraception in Peru.

8. Chavez, Susana, and Anna-Britt Coe, 2007, Emergency contraception in Peru: Shifting government and donor policies and influences, *Reproductive Health Matters*, May, accessed via AccessMyLibrary on Febuary 3, 2009, http://www.accessmylibrary.com/coms2/summary_0286-31830652_ ITM ?email=luchadora20@yahoo.com&library=Aurora percent20Hills.

CHAPTER 23

···········

Mobilizing Constituencies

Achieving Sexual and Reproductive Health and Rights for All

ADRIENNE GERMAIN

The 1994 International Conference on Population and Development (ICPD) set forth a comprehensive approach to population policy centered on sexual and reproductive health and rights (SRHR). That approach provides a base for much wider political support than earlier population policies, which focused narrowly on family planning. Indeed, the ICPD agenda has already inspired support from a diverse range of constituencies—from Brazilian feminists to Nigerian religious leaders. But much more remains to be done: To fulfill the promise of ICPD, we must consolidate—and expand—those constituencies of support. This chapter begins with examples of diverse coalitions that have been built and mobilized in support of ICPD, then outlines five challenges we must meet in order to achieve the ICPD agenda.

> A comprehensive approach to population policy provides a base for much wider political support than earlier population policies, which focused narrowly on family planning.

Adrienne Germain is President of the International Women's Health Coalition (IWHC).

MOBILIZING SUPPORT FOR SRHR

The SRHR framework encompasses women's empowerment and human rights, embraces adolescent health and rights for the first time, and acknowledges the central importance of sexually transmitted infections (STIs), including HIV/AIDS. ICPD recognizes that reproductive health and rights are a vital aspect of the lives of all people, male and female, which opens the door for men's engagement—largely neglected in earlier programs and policies. The ICPD agenda thus speaks to the interests of numerous constituencies, including women, men, youth, and a wide range of vulnerable groups. In the examples below, those constituencies have joined forces to support SRHR at the regional, national, and international levels.

Bangladesh

From 1996 to 1998, a coalition of donors and civil society stakeholders partnered with the government of Bangladesh to create and implement a national health and population policy rooted in the ICPD agenda. This coalition negotiated to shift national policy from a narrow focus on family planning to more-comprehensive reproductive health services, including emergency obstetric care, initiation of a national HIV/AIDS program, and work with young couples. Health outcomes improved noticeably. For example, after the policy took effect, there were these outcomes:

- The proportion of women receiving prenatal care increased from 26 percent to 56 percent.
- The use of emergency obstetric care rose from 5 percent to 27 percent.
- Maternal mortality decreased from 440 to 322 deaths per 100,000 live births.
- The total fertility rate dropped from 3.3 to 3.0.
- Contraceptive prevalence rose from 50 percent to 58 percent.
- Infant mortality declined by 24 percent.[2]

Unfortunately, however, neither government nor donors continued to support stakeholder and broad civil society participation as planned. When a new government came to power, the program was politically vulnerable, implementation stalled, and the key ingredient—unification of health and family planning services—was reversed. The World Health

Organization's 2005 World Health Report concludes, "Experience from Bangladesh in the mid-1990s shows that time and money invested in mobilizing [civil society] constituencies is well worth it; failing to do so can have serious negative consequences."[7] Now, civil society is mobilizing again: In December 2006, advocates launched Health Watch, which is now flourishing. Modeled on influential efforts in the environment and education sectors, Health Watch has produced two annual reports, which it uses to draw press and public attention to the sector, convene meetings with senior government officials and donors, and recommend actions.

Brazil

Across the world, the Brazilian Initiative for the Right to Legal and Safe Abortion (*Jornadas Brasileiras pelo Direito ao Aborto Legal e Seguro*, or Jornadas) serves as a powerful example of mobilization for SRHR. Abortion is currently legal in Brazil only in cases of rape or life-threatening pregnancy, but an estimated 1.2 million Brazilian women have been hospitalized in the last five years with infections, vaginal bleeding, and other complications resulting from illegal abortions.[6] In addition, women eligible for abortion under the restrictive law are, more often than not, denied services. In response to this major public health concern, Brazilian feminists and nongovernmental organizations formed Jornadas to advocate for safe and legal abortion as a matter of public health, social justice, and human rights. Members also work to ensure that women eligible under the law can actually get services. They inform medical professionals about the law, work for protocols in hospitals, and provide information so that women themselves know their rights. Jornadas members know that legal reform is a long, hard road and is broadening the alliance to include key constituencies such as journalists, health professionals, and young people.

Nigeria

In one of the most conservative states of northern Nigeria, the Adolescent Health and Information Project (AHIP) has forged critical partnerships with religious, traditional, and government leaders to ensure that adolescents have the information they need to make a healthy transition to adulthood. AHIP provides information, education, and counseling on sexuality and reproductive health to thousands of young people each year.

As founding director Mairo Bello explains in Box 23.1 (page 305),

AHIP initially faced strenuous opposition from religious leaders who misunderstood the group's mission. Following intensive advocacy with the imams (Islamic male religious leaders), AHIP was able to enlist those leaders in educating their congregants about reproductive health and HIV/AIDS, safe motherhood, and youth rights and health. Since 2006, AHIP has trained over 1,000 imams across nineteen states in northern Nigeria. Each of these leaders reaches an average of 5,000 people through Friday congregational prayers and sermons every week.

United States

ICPD has also provided a platform for U.S.-based groups to organize around shared foreign policy goals. For example, in recent years, a coalition of environmental, faith-based, population, global health, and women's organizations worked together, through outreach to members of Congress and the public, to promote the Focus on Family Health Worldwide Act (HR 1225). This legislation would significantly increase global investments in reproductive health and family planning, but unfortunately Congress has not taken action on this bill (perhaps because the focus of health assistance has been on HIV/AIDS). An even larger coalition of the above constituencies regularly works to increase funding for reproductive health in the context of the appropriations process—together with service providers, professional organizations, domestically focused organizations, and others.

Global

At the global policy level, sexual and reproductive health and rights advocates have worked with young people's networks and UN organizations to defend young people's health and rights from the assaults of ideologues (including, but certainly not limited to, the Bush administration). At the 2002 UN General Assembly Special Session (UNGASS) on Children, for example, the United States, the Holy See, and some Islamic countries attempted to make abstinence-until-marriage the centerpiece of sex education and to secure language against making abortion available to young people.[4] Building on links established during and after the ICPD process, women's health groups, children's advocates, and young people themselves successfully mobilized to block these attempts.

Coalition building is also being used proactively to advance the ICPD agenda, not just defend it from assault. One such initiative is the group With Women Worldwide: A Compact to End HIV/AIDS. With

Women Worldwide represents a coalition of seven constituencies (those for women's rights, young people, faith-based groups, HIV/AIDS, human rights, SRHR, and people living with HIV/AIDS) mobilized around an agenda that includes three core action items: investment of HIV/AIDS funds in sexual and reproductive health services, including expedited access to and research on women-initiated protective methods; universal access to comprehensive sexuality education; and equitable access to treatment, care, and support. As a result of the coalition's work at the UN High-Level Meeting (HLM) on HIV/AIDS in June 2006, the outcome document contains several strong SRHR elements, including reaffirmation of the core ICPD goal.[5] These HLM commitments have since been reaffirmed by global officials at the highest levels, including the World Health Organization director-general and the executive director of UNAIDS.

TODAY'S SEXUAL AND REPRODUCTIVE HEALTH AND RIGHTS CHALLENGES

SRHR challenges vary throughout the world. In some places, access to family planning and contraception is lacking; in others, those services are well established, but women have little access to safe abortion or skilled help in childbirth. In parts of the world, rates of STIs, including HIV/AIDS, are currently very high among girls and women; in others, the rates are low but increasing. Given the diversity of needs, ensuring access requires that we meet five primary challenges: policy support for comprehensive services, increased funding, HIV/AIDS prevention approaches that empower women and girls through SRHR, information and services for the largest generation ever of young people, and successful opposition to the opposition.

Policy Support for Comprehensive Services

Advocacy work is critical at the local, national, and global levels to make sure strong policies are in place. Once sufficient funds are allocated (see below), continued policy support for such services is required at all levels to ensure that these funds are used effectively. For example, as health systems in many developing countries are decentralized and policy decisions are made at local rather than national levels, it is crucial that state or district policymakers are educated about and engaged in the importance of the ICPD agenda.

BOX 23.1

An interview with Mairo Bello, Executive Director, Adolescent Health and Information Project (AHIP)

Can you talk a little bit about how AHIP began and how it has evolved over the years?
It started with the women's movement. During the 1980s and 1990s in Nigeria, there was a lot of research being done, a lot of talk, and a lot of demonstrations in the streets, trying to better women's lives. But we felt like we were not making enough headway just working with women and men and thought maybe we should begin with the younger generations. And so AHIP came to life. At first, in 1989, it was just a club that organized activities for young people. But we discovered that young people need more than recreation. They need information about their health and rights, things that will enable them to become useful citizens and useful human beings. That was what gave me the vision to reorganize our little group. It was restructured in 1992 to become the Adolescent Health and Information Project.

We started by addressing adolescent reproductive health in northern Nigeria. At the time we started, no other organization would touch issues of sexuality education with a 10-foot pole. Today, AHIP has evolved from one organization with one staff person to three offices in three states with eighty staff. We have programs in twelve additional states. Before all this could happen, though, we ran into strong opposition, some brick walls, even some iron curtains.

(continued)

Increased Funding

Investing in comprehensive reproductive health services has the potential to simultaneously address multiple urgent public health burdens, including maternal mortality and morbidity, HIV/AIDS, and violence against women. Yet an analysis of the most recent available data on development assistance funds for health in low- and middle-income countries found an imbalance among the components in 2005: $2.6 billion on HIV/AIDS and STIs, $300 million for family planning, and $500 million for all other components of reproductive health care.[3] Particularly notable is the dramatic increase in HIV/AIDS funding (the largest portion for treatment, and little investment in ICPD-mandated

BOX 23.1 *(continued)*

Can you say more about the obstacles and brick walls that you have encountered?

Misconceptions, misunderstandings, ignorance, and religious interpretations. Religion itself is not the problem, but religion misinterpreted. We were accused of ridiculous things out of fear and misunderstanding, sometimes by competitors, sometimes religious bodies—and I'm talking about all religions. We were accused of encouraging girls and boys to have sex, even providing rooms for them to copulate. People said we wanted to stop child marriages so that girls could instead go prostitute themselves. We were accused of being an arm of an American organization that was trying to defeat Islam and promote Western values.

These accusations really came to a head when I was named as a proponent of sexuality education in a newspaper piece—one of two people named in all of northern Nigeria. The resulting tension and disapproval in the air around us was so strong that a lot of staff resigned. We didn't know what to do. The International Women's Health Coalition actually offered to help me travel to Ghana, where we thought I might be safe. But I believed that if I ran away, I might never come back. I believe strongly in what I'm doing. So the alternative was to go back to the drawing board and see what we could do to make it better.

So how did you address the ignorance and fear?

By engaging in a dialogue. We consulted with many groups in the community, including a group of Muslim religious leaders. Unfortunately, we were not allowed to explain our program or what we meant by certain concepts or things. We were just bombarded with all the accusations and advised to put them right, desist from being a Western propagation tool, and Islamatize AHIP programs. They went away as ignorant as they came into the meeting. I still shed a tear any day I talk about it.

But we decided to keep trying. We thought if we held two-day meetings with these leaders and gave them a chance in the first few hours to air their accusations, we could then try to respond. We could talk with them about the true objectives of AHIP, the present situation of sexual and reproductive health in the state, and what we all want to see in the future.

At one meeting, after all the accusations were aired, I spoke with them about HIV/AIDS. This time, they listened to what I was saying. After the session, they said they'd never known these facts about HIV/AIDS, how it happens and how it is spread. "My God. This is something that people must do something about." And I thought, okay, something is starting to happen here.

So how did you transition from these meetings to doing the trainings for imams?
Well, we had several of these meetings with both Christian and Muslim leaders, and each time we learned how to plan and strategize for the next meeting. Ultimately, there were a lot of recommendations that came out, which included working with local organizations and organizations with religious knowledge and connections. We were invited to collaborate with the Center for Islamic and Legal Studies (CILS) of the Ahmadu Bello University (ABU) to train forty imams. It's grown from there. Sometimes it takes six months to get them in the room. It demands a lot of patience, and we must be strong in our knowledge of the issues and in our own values.

Do you think you are making an impact?
Yes. As an example, a wife of one of the imams called me and asked, "What was it you did to our husbands at that training you organized?" I was speechless for some seconds, then I asked, "What do you mean?" She responded, "No, no, I did not mean that you did any harm. I was going to appeal to you that whatever it was, please invite them again for more." When I asked why, she said that she noticed that her husband started talking more positively and not shouting as much as usual. At first she did not understand where the change came from, until two women living near the mosque where he leads prayers came to ask her what training her husband was referring to that made him start talking of women as human beings. These women concluded by saying, "Whatever training it was, Allah will bless the trainers. They should do some more of those trainings for the imams."

Now, we are training a lot of leaders throughout the community. There's hardly any gathering today where you don't find at least a few of the people we have sensitized or trained. Also, since our work began so many years ago, there are many more organizations working on reproductive health. That's something that we should be proud of.

On the other hand, we still have not achieved what we can be really proud of. If women are still dying in childbirth, if women are still being denied their rights at different levels, we need to keep working.

SRHR prevention strategies), a significant decline in family planning funds, and some increase—from a low base—for other elements of reproductive health. While comprehensive data are not yet available, these trends seem to have continued in recent years. U.S. funding for global HIV/AIDS programs has greatly increased since 2003—an important and notable achievement. Unfortunately, however, U.S. investments in child health, maternal health, and family planning have remained stagnant at far lower amounts for nearly a decade.[1]

What level of funds is required? As a starting point, experts estimate that $3.9 billion annually would provide contraception to 200 million women who currently lack access, and $6.1 billion per year would provide essential maternal and child health care to 75 percent of women in seventy-five countries.[1]

HIV/AIDS Prevention Approaches
That Empower Women and Girls through SRHR

HIV/AIDS epidemics around the world are primarily driven by heterosexual sex. HIV/AIDS is thus essentially a sexual and reproductive health and rights issue, especially for girls and women. The realities of girls' and women's lives that violate their sexual and reproductive rights and undermine their health are the same realities that put them at serious risk of HIV/AIDS—and that impose disproportionate burdens on them once they are infected and/or caring for others who are sick.

> The realities of girls' and women's lives that violate their sexual and reproductive rights and undermine their health are the same realities that put them at serious risk of HIV/AIDS.

HIV/AIDS policies and budgets should invest in expanding access to sexual and reproductive health services. Reproductive health services have critical core capacities and are generally well accepted by families and communities. With increased investment, they can provide subsidized female and male condoms and, eventually, microbicides, as well as HIV/AIDS counseling, testing, and other services, to women and their partners.

Information and Services for the
Largest Generation Ever of Young People

The ICPD recognized that sexual and reproductive health and rights, including HIV/AIDS prevention, require not just access to health services but also new social norms. To that end, we must ensure universal access to comprehensive sexuality education that provides full and accurate information and helps young people build skills to establish equality within their relationships, respect the right to consent in sex and marriage, and end violence and sexual coercion. Young people must also be guaranteed access to health services, not necessarily through separate services but at least through appropriately trained staff, adjusted hours of service, and "youth-friendly" outreach. We must also create safe learning spaces for girls in and out of school and promote SRHR, including HIV/AIDS prevention, in political life and in national development.

Investing in youth has gained considerable momentum since ICPD, and major efforts have been made to develop programs for young people. Yet in many parts of the world, adolescence is still barely recognized as a significant period of life, and many programs are designed for young adults.

Successful Opposition to the Opposition

Before, during, and after the ICPD, conservatives attempted to undermine or destroy the ICPD agenda. Particularly contentious items have included access to safe abortion services and postabortion care, education and services for young people, and recognition of the sexual and reproductive rights of the individual and of various forms of family. Yet over time, the ICPD agenda has been protected and even strengthened by the broad array of groups with a vested interest in its commitments. This underscores the importance of maintaining ICPD as a bold and comprehensive agenda and not relinquishing those elements that may be challenged.

CONCLUSION

The ICPD agenda has already rallied diverse constituencies to the cause of SRHR. Now there are opportunities to expand that base of support. It is widely agreed that women's rights and reproductive health are central to the achievement of the Millennium Development Goals (MDGs). A good deal of the work and leadership for this consensus came from the constituencies described here, led by women. But even more work is needed to generate the policies and funding to fully implement these commitments. To fully realize the ICPD agenda, we must sustain and expand the constituencies already engaged, while we draw in those with whom partnerships could be much stronger—environmental, poverty, and development groups.

REFERENCES

1. Daulaire, Nils, 2007, Written Testimony to the Senate Committee on Appropriations, Subcommittee on State, Foreign Operations and Related Operations, April 18.

2. Demographic and Health Surveys, Bangladesh 1993/94–1999/2000, Infant mortality figures, available at http://www.measuredhs.com/.

3. Kates, Jennifer, and Eric Lief, 2007, *Donor Funding for Low- & Middle-Income Countries, 2001–2005*, The Henry J. Kaiser Family Foundation and the Center for Strategic and International Studies, Washington, DC, August.

4. Sengupta, Somini, 2002, Goals set by UN conference on children skirt abortion, *New York Times*, May 11.

5. UN General Assembly, 2006, Resolution Adopted by the General Assembly: 60/262. Political Declaration on HIV/AIDS, UN, New York, June 15.

6. Welsh, Andrea, 2007, Brazil's women at risk from unsafe abortions: Study, Reuters, May 30.

7. World Health Organization, 2005, *World Health Report 2005: Make Every Mother and Child Count*, World Health Organization, Geneva.

CHAPTER 24

....

Women at the Center

ELLEN CHESLER

Today's challenges in global health, development, and the environment
will not be solved by technological innovation alone. Before we can
hope to improve human well-being, we must address the gross violations of human rights that prevent individuals—and especially
women—from developing the will, claiming the authority, and taking
the necessary actions to improve their own lives.[1]

Women are central to environmental sustainability. To protect the
Earth's natural resources and reverse potentially catastrophic changes
in climate, we must address living patterns and substantially limit consumption in affluent countries, where women most often make major
decisions about the well-being of their families and are the world's
principal consumers. Meanwhile, in the developing world, women
carry primary responsibility for the production of food, the transport
of water, and the gathering of wood for fuel—all activities with important environmental consequences. These responsibilities also render
women especially vulnerable to the impacts of climate change (see
Chapter 7). But, too often, women's voices are not heard because they
have little power or opportunity to speak. To make environmental policies that are sustainable and just, women must be at the table.[2]

The world's population has grown to 6.8 billion. Our numbers have

Ellen Chesler is a Distinguished Lecturer and Director of the Eleanor Roosevelt Ini-
tiative on Women and Public Policy at Hunter College, City University of New York.

more than doubled in the past fifty years alone. Organized family planning efforts have dramatically increased the voluntary use of modern contraception and slowed the rate of growth substantially, but still our absolute numbers increase.[3]

Moreover, with more than half the world's people now dwelling ever precariously in sprawling metropolises, demand is growing exponentially for freshwater, food, and shelter—and most of all for energy to fuel the industrial, housing, and transportation requirements of modern urban life. Improvements in living standards are desperately needed by the half of humanity still living in poverty, but without concerted action those improvements will pose further environmental hazard, accelerating climate change, depleting natural resources, and straining fragile ecosystems, most seriously in the very places that most need stability and growth.[4]

Yet so deep are the scars from more than a century of conflict over flawed population and development policies and practices that many of us hesitate to even acknowledge these sober realities. How do we find our way forward through the intellectual conundrums and ideological divides that have for so long plagued this field to confront what is now widely accepted as an emerging environmental challenge with serious human consequences? And what can environmentalists learn from the battles over birth control that have for so long captivated our attention and polarized our thinking?[5]

In the 1970s and 1980s, high-profile abuses of human rights by numbers-driven population programs, especially in India and China, undermined a well-established consensus that family planning programs are an essential tool of sound public health and development practice. During the 1990s, an alternative discourse wed reproductive health to individual women's rights, but this framework has also engendered criticism. Reproductive rights are now regularly attacked by conservatives for fostering a corrosive individualism that threatens the presumed cohesion of traditional families and communities. At the same time, they are criticized across the political spectrum from right to left as a doctrine rooted in Western enlightenment values, denying cultural variation and rights of self-determination.[6]

Why risk getting involved with the subject of women's status in a world rife with political conflict around gender; a world where so many cultures are attempting to negotiate the assaults of modernization, secularization, and globalization; a world where fundamentalisms are resurgent?

The answer is simple: There is no greater dividend than the payoff from women's rights. For all the missteps of the past, for all our conflicts and disputes, there is one thing we know for certain. More than four decades of international development experience demonstrates that investments in women and gender equality yield immediate gains. Prioritizing women is not just a moral imperative but also a necessary condition of success if our aim is to meet the world's most critical challenges—if we hope to reduce poverty, grow economies, improve health and nutrition, stabilize fragile countries, secure good governance, and achieve political stability. We no longer need to rely on anecdotal evidence. We now have solid metrics from more than a hundred countries to demonstrate the effectiveness of aid that specifically disaggregates women, targets their needs, and makes them more powerful.[7]

> Why risk getting involved with the subject of women's status in a world rife with cultural and political conflict around gender? The answer is simple: There is no greater dividend than the payoff from women's rights.

Consider just a few powerful illustrations:

In sub-Saharan Africa, where agriculture remains a key economic activity, women who received the same levels of education, skill development, and access to farming assistance as men increased their agricultural yields by 22 percent, substantially improving food security and reducing hunger and malnutrition. Women constitute the majority of Africa's farmers but receive less than 10 percent of farm credit and own just 1 percent of all land. The World Bank estimates that economic growth in the region would have doubled since the 1960s had education and employment opportunities for women been equal to those of men.[8]

In South Africa, new research suggests that it may be possible to dramatically change the status of women in a short time, even in the most troubled and impoverished communities, by coupling microfinance programs for local craftswomen with life skills workshops. These successful "consciousness-raising" interventions provide open-ended discussions about women's rights, including the right to healthy and safe sexuality. By fostering domestic relationships based on mutual respect, they are credited with having led to diminished rates of violence and substantial declines in new infections of HIV and AIDS.[9]

In India, where economic growth rates vary tremendously by region, those states with the highest percentage of women in the formal economy grew the fastest and enjoyed the steepest reductions in poverty. Studies there also show that girls' education substantially lowers birthrates by delaying age of marriage and increasing effectiveness of voluntary family planning interventions. Increased political participation, as a result of recent laws reserving a third of local council membership to women, has also been linked to greater health and social infrastructure investments, which boost many basic indicators of well-being, including safe drinking water.[10]

In the Middle East, the UN correlates progress with the educational, economic, and civil status of women. For example, Tunisia has a long history of progressive measures banning polygamy and guaranteeing basic civil rights, education, and employment to women. Tunisia has a vastly superior record of economic growth and significantly lower birthrates than neighboring Morocco, which only recently introduced reforms to the highly restrictive personal status laws that were placed on women when the country gained independence from France more than fifty years ago.[11]

PUTTING WOMEN AT THE CENTER

What does it really mean to put women at the center of health, development, and environmental policies? The most successful programs have deep roots in the households and communities they serve. They provide access to quality services and encourage people to make free and informed decisions about their lives. In the words of the powerful alliance of women in the global south known as Development Alternatives with Women for a New Era (DAWN), a comprehensive framework for sustainable development must have a "double lens" of accountability: It must safeguard the rights and full citizenship of every man and woman *and* promote collective well-being.[12]

Effective models for responsible, women-centered development policies must also be guided by strong standards of performance accountability. A recent analysis of best practices by the International Center for Research on Women demonstrates how development agencies in Australia, Canada, and Great Britain have improved the execution and results of their foreign assistance programs by focusing on women. The top leaders of these programs are schooled in gender analysis and apply those skills in all aspects of their work. Technical specialists in women's rights and development are integrated into all departments, and all investments are assessed by tools of evaluation that take impacts on women into account. Mandates for expenditures on women are also in effect across all sectors of foreign investment from health and education to water supply and sanitation, where these kinds of considerations have never before existed.

By contrast, extremely weak guidelines for integrating women—dating back to the 1970s—still govern much of U.S. foreign assistance. The U.S. Millennium Challenge Corporation established in 2004 offers a single exception and serves as a model for reforms mandating that all new U.S. foreign assistance programs require recipient countries to disaggregate beneficiaries by sex.[13] These priorities are even more essential in light of current prospects for increased U.S. expenditure on reforestation and other investments to "green" the global economy.

There is no better example of how environmental and gender goals intersect than the Green Belt Movement founded by African Nobel Prize winner Wangari Maathai of Kenya. Under Maathai's inspired leadership, nearly 40 million trees have been planted on community lands in Kenya and neighboring countries since she introduced the idea thirty years ago.

Maathai takes great pride in restoring African landscapes, but she speaks with greatest conviction about the impact her movement has had in bringing opportunity and hope to ordinary women who comprise 90 percent of its membership. She is also the first to acknowledge that her earliest programs had to be restructured to address deep cultural barriers to women's advancement. In recent years she has incorporated education in basic life skills, or what she now calls "consciousness raising," into her job training for women. And she has also opened residential facilities for battered women who are victims of an epidemic of violence.[14]

EQUALITY: A PROGRESS REPORT

Women are central to progress on health, development, and environmental protection, but progress toward improving women's lives is mixed. We are now more than halfway to the 2015 deadline for achieving the UN's Millennium Development Goals, adopted with great fanfare in the year 2000. The MDGs explicitly iden-

> Women are central to progress on health, development, and environmental protection, but progress toward improving women's lives is mixed.

tify women's empowerment as a goal and require the integration of gender analysis into all other substantive priorities of the United Nations, including environmental sustainability.

Currently, the only numerical targets that measure women's progress in the MDGs are universal access to primary education and gender parity within secondary school. Primary school goals are said to be within reach by 2015, but more than half of children not in primary school are girls. More problematically, many girls are not receiving secondary education, which is known to have the greatest impact on social change. Only 23 percent of girls in sub-Saharan Africa attend secondary school; only 35 percent in south Asia.[15]

By other measures women's equality remains even more elusive. Higher education for women, which is essential for entry into government, business, and the professions, has seen negligible gains. Moreover, there are no enforceable UN targets for increasing women's opportunities in civic and political life, where their participation has been clearly and favorably linked by the World Bank and others to enhanced concern for human rights, social welfare, peace, and security, as well as to reduced corruption and greater effectiveness in government.[16]

Today only one in five members of parliament worldwide is a woman, and at the existing rate of increase of female parliamentary representation in the developing world, it will take at least forty years for women to reach parity in their national assemblies. Fixed quotas or other incentives to accelerate the political participation of women may therefore be advisable. Quotas have been successful in France and Rwanda, where women now hold half of legislative seats. And the new constitutions of Iraq and Afghanistan require that at least 25 percent of legislative seats be reserved for women. Ironically enough, this process was facilitated by the United States, which does not hold itself accountable to such practices and still has a dismal record of female representation (about 20 percent) in both chambers of Congress. As it happens, however, it was Afghani women who many years ago drafted language in CEDAW, the UN's visionary 1979 Convention to Eliminate All Forms of Discrimination Against Women, to support affirmative action for women.[17]

At the same time, women's share of nonagricultural employment, which gives greater control of family income and decision-making, is slowly increasing around the world. Women currently hold 39 percent of all nonagricultural jobs in the "formal" economy, but that figure is skewed by extreme regional disparities. In China, as in most developed countries, women's labor force participation is high. In the Middle East, north Africa, and south Asia, however, women hold fewer than one in five paid jobs, and the vast majority of these jobs are in low-paid positions, where workers are without protection and vulnerable to abuse.[18]

The MDGs have been roundly criticized for failing to address sexual and reproductive health and rights (SRHR). And, as Carmen Barroso and Steven W. Sinding explain in Chapter 19, the goal of universal access to SRHR remains unmet. With U.S. leadership restored under the Obama administration, that goal is likely to receive renewed support, but enforcement in the face of continued conservative opposition will require continued vigilance by coalitions of powerful interests.[19]

> Meaningful progress on women's rights will require . . . sustained advocacy, not just by women's rights organizers, but also by environmentalists and others who have a big stake in improving the status of women.

Meaningful progress on women's rights will require stronger mandates, clearer performance indicators, and better incentives overall. This, in turn, will require sustained advocacy, not just by women's rights organizers, but also by environmentalists and others who have a big stake in improving the status of women.

FORECAST: BETTER WEATHER

President Obama's appointment of Hillary Rodham Clinton as secretary of state brightens prospects for a more rigorous consideration of the intersections between advancing women's status and achieving the goals of U.S. foreign policy and foreign assistance, including global environmental sustainability.[20] The eyes of the world are now upon her to deliver on the promise of her landmark speech at the UN's Fourth World Conference on Women in Beijing in 1995, which forcefully identified the rights, health, and safety of women—their education and economic roles, their opportunities to participate in public life and to have the number of children they desire—as fundamental to human security and well-being. Clinton referenced that speech in her confirmation hearing and again pledged to remain vigilant in support of its ideals. She also noted that President Obama supports her commitment because he is, after all, the son of Ann Dunham, a pioneer in this work in Indonesia from the 1970s through the 1990s, whose critical illness prevented her from traveling to Beijing.[21]

Hearing Clinton's testimony, I was reminded of the unseasonable storms that created havoc during the Beijing conference, which I was fortunate to attend. In China, ancient lore extols a natural harmony between yin and yang, two opposing elements, one believed to be female and the other male. And some of our Chinese hosts sheepishly blamed the foul weather on an atmospheric imbalance brought on by the excess of women descending on their country.

At first I was dubious about this proposition, but then I decided to turn it on its head. Perhaps nature wasn't protesting the presence of too many women but was instead marking the end of their long subordination. After all, just before Clinton spoke, and just as agreement on important principles of gender equality was reached, the skies miraculously cleared.

Maybe it was just a coincidence, or maybe it was a sign of the symbiotic relationship between the natural environment and the well-being of the world's women. If so, let us all hope—and work—for better weather in the future.[22]

REFERENCES

1. This observation is also made in Epstein, Helen, and Julia Kim, 2007, AIDS and the power of women, *New York Review of Books,* February 15, http://www.nybookscom/articles/19874. Also see Farmer, Paul, 2003, *Pathologies of Power: Health, Human Rights, and the New War on the Poor,* University of California Press, San Francisco.

2. Goetz, Anne Marie, lead author, *Progress of the World's Women 2008/2009,* UN Development Fund for Women, New York, p. 132. One of many examples of environmental policy recommendations that ignore fundamental questions of human rights is Independent Task Force Report 61, *Confronting Climate Change: A Strategy for U.S. Foreign Policy,* Council on Foreign Relations, New York, 2008.

3. Cohen, Joel E., 1998, How many people can the Earth support? *New York Review of Books,* October 8, p. 29–31.

4. UN predicts urban population explosion, 2007, *New York Times,* 28 June, p. A6.

5. Contrasting viewpoints on battles over population and environmental sustainability are in Sen, Amartya, 1994, Population: Delusion and reality, *New York Review of Books,* September 22, p. 64–71, and Speidel, J. Joseph, et al., 2007, *Family Planning and Reproductive Health: The Link to Environmental Preservation,* http://crhrp.ucsf.edu/publications/files/Speidel_Family Planning_2007.pdf.

6. Chesler, Ellen, 2008, Paradigm and Perceptions: Exploring the Triggers of Major Socio-political Change, testimony at the Climate Change Solu-

tions Summit with Vice President Al Gore and the Alliance for Climate Protection, Nashville, Tennessee, May 15.

7. Selvaggio, Kathleen, Rekka Mehra, Ritu Sharma Fox, Geeta Rao Gupta, 2008, *Value Added: Women and U.S. Foreign Assistance for the 21st Century,* International Center for Research on Women, Washington, DC. Also see, Coleman, Isobel, 2004, The payoff from women's rights, *Foreign Affairs* 83(3): 80–95, May/June, and Sen, Amartya, 2001, The many faces of gender inequality, *The New Republic* 225(12): 35–40, September 17.

8. Food Policy Research Institute, 2000, Women: The Key to Food Security, June; Gender: Working towards greater equality, 2007, *Gender Equality as Smart Economics: A World Bank Group Action Plan,* World Bank, Washington, DC, www.worldbank.org/gender; Udry, Christopher, et al., Gender differentials in farm productivity: Implications for household efficiency and agricultural policy, *Food Policy* 20:407–443, all cited in Selvaggio et al., 2008, *Value Added,* p. 4–5.

9. Pronyk, P. M., et al., 2006, Effect of a structural intervention for the prevention of intimate partner violence and HIV in rural South Africa: Results of a cluster randomized trial, 1973–1983, *The Lancet* 368: 9551, December 2, also cited in Epstein, Helen, and Julia Kim, 2007, AIDS and the power of women. Also see *A Social Revolution: Report of the United Nations Secretary-General's Task Force on Women, Girls and HIV/AIDS in Southern Africa,* 2004, UN, New York, p. 3–6.

10. Besley, Timothy, et al., 2005, *Operationalising Pro-Poor Growth: India Case Study,* World Bank, Washington, DC, cited in Selvaggio et al., 2008, *Value Added,* p. 5; Goetz, Ann Marie, *Progress of the World's Women,* p. 29.

11. Coleman, Isobel, 2004, The payoff from women's rights, p. 87–89; Coleman, Isobel, and Tamara Cofman Wittes, "Economic and political development in the Middle East: Managing change, building a new kind of partnership," in Council on Foreign Relations and Saban Center for Middle East Policy at Brookings, 2008, *Restoring the Balance: A Middle East Strategy for the Next President,* Brookings Institution Press, Washington, DC, p. 162–165.

12. Petchesky, Rosalind P., and Karen Judd, eds., 2003, *Global Prescriptions: Gendering Reproductive Health and Rights,* Zed Books, London.

13. Rao Gupta, Geeta, 2009, Recommendations on How to Integrate Gender Successfully in U.S. Development Assistance Institutions, memorandum to the Presidential Transition Team for U.S. Foreign Assistance, International Center for Research on Women, Washington, DC, January 8.

14. Sears, Priscilla, Wangari Maathai: You Strike the Woman . . . , *In Context: A Quarterly of Humane Sustainable Culture,* http://ww.context.org/ICLIB/IC28/Sears.htm.

15. Goetz, Anne Marie, *Progress of the World's Women 2008/2009,* p. 120–121.

16. Ibid.

17. Chesler, Ellen, Introduction, Chesler and Chavkin, eds. p. 27.

18. Goetz, Anne Marie, *Progress of the World's Women 2008/2009,* p. 1–15. Excellent current data on women's status are also available in Seager, Joni, 2009, *The Penguin Atlas of WOMEN in the World,* Penguin Books, New York.

19. On Unmet Need for Family Planning, see The marathon's not over,

2008, *Economist,* July 19, http://www.economist.com/world/international/
PrinterFriendly.cfm?story_id+11708001.

20. Global environmental sustainability falls under the purview of the
State Department through its Bureau of Oceans and International Environmental and Scientific Affairs.

21. The Caucus, 2009, thecaucus.blogs.nytimes.com/2009/01/13/
live-blog-clintons-confirmation-hearing.

22. A version of this story is in Chesler, Ellen, and Joan Dunlop, 1995,
Consensus on women's rights cleared the skies in China, *The Christian Science
Monitor,* September 19.

CHAPTER 25

·············

Taking Stock

Linking Population, Health, and the Environment

ROGER-MARK DE SOUZA

"We have to take stock of our people and natural resources," says Felipe Hilan Nava, a physician and mayor of the municipality of Jordan in central Philippines. "If we don't strike a balance now, we lose eventually." Thanks to an integrated program that addresses population growth, public health, and environmental conservation, Nava's neighbors are taking stock—and working to strike a better balance.

Just a 15-minute boat ride off the coast of the Philippines province of Iloilo, Nava's town is located on the island of Guimaras. Tourist companies advertise the seas here as teeming with fish and tout the island's white-sand beaches and pristine waterfalls and springs. But the casual visitor smitten by the natural beauty might miss the hard realities of life on the island.

Despite its bountiful marine resources, Guimaras is among the twenty poorest provinces in the country. Many people in Guimaras live beyond the ambit of modern technology—without even electricity. These are men who toil on the land and seas in earnest; women with limited access, if any, to economic opportunities; and children who trek 2 miles of dirt road to school every day. The children are observant: They notice that their fathers have been catching fewer fish. The

Roger-Mark De Souza is the Director of Foundation and Corporate Relations at the Sierra Club.

town has also lost some of its mangroves and seagrasses, which provide breeding grounds for fish and economic opportunities for the residents. And every year, the island's resources must be shared among more people: Guimaras's population has doubled in the last thirty years, and a youthful population ensures continued growth.

In Guimaras—as in coastal areas around the world—population growth exerts unprecedented pressure on natural resources. Today, more than 3 billion people live along a coastline or within 125 miles of one, and the coastal population may double by 2025. Population growth and its effects degrade coastal ecosystems, yet maintaining healthy coastal habitats for marine organisms is critical because most of the world's fish catch produces its young inshore and feeds on organisms in coastal waters. Coastal fish stocks in some regions are down to 30 percent since 1980.

> Here in Guimaras, people are testing new approaches to staving off poverty, sustaining nature's bounty, and minimizing a growing population's demands on resources.

But here in Guimaras, a quiet revolution is taking place: People are taking matters into their own hands, testing new approaches to staving off poverty, sustaining nature's bounty, and minimizing a growing population's demands on resources.

In 2001, Save the Children launched an innovative "population-health-environment" (PHE) project in Guimaras and Iloilo that combined reproductive health service delivery and environmental management. Save the Children worked with local government officials and fishing families to provide assistance and training in coastal resource management and fish-catch monitoring. Activities included designating and monitoring marine protected areas, replanting mangroves, and delivering improved reproductive health services—efforts that are helping communities achieve their long-term goals of poverty reduction and economic well-being.[1]

The project has had an impact: "There is a certain awareness now in the *barangay* [village] about coastal resource management," says Barangay Captain Fernando Balidiong, of Alegria in Guimaras. "Residents take care of the mangroves and monitor them, and we have also become more alert in looking out for illegal fishing activities."

There is also a new awareness about the importance of family planning. Mayor Nava and Dr. Esteban Magalona, municipal health officer of Sibugnay, both believe that linking human population issues to the environment accounts for an increase in family planning usage in their community. Nava says the project "fine-tuned" the government's family planning program through regular information sessions and skills development training for family planning trainers.

The residents of Guimaras are not alone in their efforts to strike a better balance between people and resources. Since the late 1990s, a number of communities in developing countries have initiated PHE programs in a variety of settings, including environmentally fragile biodiversity "hot spots," cities, and coastal areas. The key objective of PHE efforts is to simultaneously help communities manage natural resources in ways that improve health and livelihoods, conserve the critical ecosystems on which they depend, and improve access to family planning and other health services. The list of issues addressed by these projects is long, including family planning, food security, basic health, nutrition, income generation, conservation, disease prevention, and access to safe water and sanitation.

> Population-health-environment programs help communities manage natural resources in ways that improve health and livelihoods, conserve the critical ecosystems on which they depend, and improve access to family planning and other health services.

In this chapter, I reflect on the successes—and challenges—of the first generation of PHE projects around the world, and I offer recommendations for future directions in this promising approach.

MEETING COMMUNITY NEEDS

The most successful PHE projects are explicitly designed to meet community needs. For example, in Kenya, the World Wildlife Fund (WWF) increased the impact of its conservation work in the Kiunga Marine National Reserve by addressing an unmet need for family planning. In the remote area near the reserve, WWF launched a mobile clinic that provides family planning information and services. The clinic's popularity has drawn new participants to WWF's conservation work.[2]

Community involvement is also essential; the best projects directly involve community members and leaders in project implementation and the celebration of success. In Tanzania, a project implemented by the Jane Goodall Institute served more than 200,000 people in Kigoma region, which borders Gombe National Park. To achieve its goal of biodiversity protection, the institute enlisted community leaders to identify projects that would help the community and promote sustainable resource management. These projects included construction of dispensaries, spring protection structures, gravity water schemes, classrooms, and ventilated improved pit latrines. The communities established their own priorities and provided one-fourth of the total cost and much of the labor required to complete each project. In return, they received clean, safe water and better schools and health care facilities.[3]

Similarly, an approach called the "champion community," piloted in Madagascar in the mid-1990s, helps achieve development goals by emphasizing local needs, community involvement, and public celebrations of success. In this approach, program planners and communities work together to select specific goals and supportive activities to be carried out in a defined time frame (usually six to nine months). Communities monitor their own progress toward achieving the goals and celebrate their successes in public celebrations where they are held up as champion communities.[4]

> An integrated approach yields a greater impact than single-sector approaches.

COST-EFFECTIVE INNOVATION

PHE programs can offer cost-effective approaches to meeting community needs. Many project managers report that integrated PHE approaches have lower operating costs and greater overall project efficiency (economies of scale in terms of staff time and effort) than single-sector approaches. Efficiency is also measured in terms of pooling expertise from different fields, leveraging efforts across programs, and merging funds from different funding streams.

Another PHE project in the Philippines, the Integrated Population and Coastal Resource Management (IPOPCORM), examined the project's central hypothesis that integrated or cross-sectoral approaches yield a bigger payoff than single-sector strategies.[5] The results show that IPOPCORM's integrated approach has had a significantly more positive impact on reproductive health, coastal resource management, and food security than stand-alone programs. These results strongly suggest that an integrated approach yields a greater impact than single-sector approaches.

There are other benefits of the PHE approach, as well, such as finding an innovative way to introduce family planning despite objections from the Catholic Church, encouraging NGOs and local government to address community needs from different viewpoints, and facilitating community involvement in program implementation.

ADVOCATING THROUGH PHE

Whether in the halls of the national congresses or at the local government level, within the NGO community or among the public at large, PHE advocacy takes many forms. In the congress in the Philippines, for example, local activists have helped develop legislative proposals on population and human development and have tracked progress on congressional bills.

PHE advocacy is important for reproductive rights activists, particularly in a country like the Philippines that has no family planning laws. Activists there are currently fighting for legislation that will guarantee universal access to safe, affordable, and quality reproductive health care services, including family planning.[6] The legislation has been vigorously opposed by the Philippine Catholic Church, which defeated a similar bill in 2005.

In this context, PHE offers a potential for advocacy that transcends any single issue. Those advocating solely for reproductive health and rights face the wrath and lobbying efforts of the Catholic Church. But the PHE approach allows advocates to increase their impact by presenting reproductive health and rights among a portfolio of issues, such as the right to a clean environment and meaningful livelihood opportunities.

For those who seek to advance natural resource management issues, particularly at the community level, the PHE approach offers an entrée into local communities—driven by communities' requests for reproductive health, public health, and livelihood services. The PHE approach is also tied to poverty alleviation, disaster mitigation, and food security, allowing activists to develop multipronged approaches and build on grassroots movements, joint advocacy campaigns, and relationships of trust.

PHE CHAMPIONS: MIDWIVES TO POLITICIANS

At the forefront of sharing the PHE message are a variety of champions—from midwives to local politicians. In the Philippines, an unusual champion from one project is an "environmental midwife," Susan Dignadice. Like other Filipina rural midwives, she manages a village health center as well as a few smaller centers. In 2003, Susan was recognized as one of the country's outstanding midwives by the Philippines Integrated Midwives Association and Johnson and Johnson for her efforts to link family planning with environmental conservation.[7]

She does this in two ways. The first is her community-based nutrition and "clean-green" projects that involve backyard gardening; she works hard to motivate people to plant root crops and vegetables. The second is the support she and her network of health workers have given to mangrove reforestation projects in the local village. "As a family planning project linked to the environment," she says, "we were part of the entire process, from the moment the barangay made an assessment of its coastal resources to the time when the residents planted mangroves. It is important to be part of the process so that the people will believe you. Leadership by example is key."

Other champions are helping to spread the message, including the popular mayor of the Philippine town of Concepción, Dr. Raul Banias. "We grow by three babies a day," Banias says. "The town of Concepción has a population growth rate of 2.8 percent, higher than the national average of 2.4 percent. . . . It stretches our resources, it stretches our services, and if you factor the vulnerability of the ecosystem, especially in the islands, that will be a very big social problem if we do not address it now."

PROMOTING BEHAVIOR CHANGE
THROUGH POPULAR AVENUES

The word is spread in other ways as well. For instance, the teenagers of the Barangay Hoskyn Theater Arts Group in the Philippines include messages about adolescent sexual and reproductive health in their plays. They have even produced a play that brings out the relationships between people and their natural resources.

And a coastal management partnership in Tanzania used theater to highlight the impact of gender discrimination on the community and the environment.[8] As a result, the common practice of marrying off young girls to older men has become increasingly stigmatized, and parents now often reject "short-term marriages" between visiting fishermen and local girls.

Other projects have developed catchy slogans that are printed on

T-shirts, posted at health clinics, and used in public celebrations. In the Philippines one such slogan, roughly translated from Tagalog, reads, "With family planning, your health is ensured, your environment is saved." Such messages appear on boats at annual water parades and on T-shirts of contestants at local beauty pageants.

FUTURE CHALLENGES OF PHE INTEGRATION

Despite PHE success stories in countries such as Madagascar and the Philippines, significant challenges remain. First, there is the challenge of "scaling up" these promising pilot projects. This effort has been hindered by a shortage of models that can be followed and replicated easily and by a lack of solid, scientifically based evidence to prove that the PHE approach is a viable method of achieving development goals. Now, a new PHE operational manual and operations research provide preliminary examples and guidelines.[9] These models need to be more widely disseminated and applied.

There is also a strong need to evaluate what works. Evaluation efforts would ideally start with a detailed assessment of the program and policy environment and data available to illuminate problems and solutions, then move into the design and implementation of PHE activities. The development of indicators and benchmarks could help future interventions identify where to start and how to gauge their accomplishments. In the real world, many confounding factors in the political environment trump efforts to promote change, but a careful documentation of that environment would nevertheless be valuable and instructive for future efforts.

Insufficient funding is another constraint. The field continues to depend on outside donor support, even as local governments in the Philippines and Madagascar are increasing their contributions. And there is a great need to build local capacity to make policy decisions, support expert partnership, and increase understanding of PHE linkages, particularly with regard to livelihood/income-generation efforts.

TAKING STOCK: A STEP IN THE RIGHT DIRECTION

Today there is a growing awareness of interactions among population dynamics, economics, environment, culture, and justice—and a new imperative to develop operational models to address these complexities. As the PHE approach is refined and tested, it can offer an effective long-term strategy to alleviate poverty, improve health, and manage natural resources.

PHE offers a step in the right direction at a pivotal moment—a flexible, innovative way for policies and programs to keep pace with today's rapidly changing world. It can also help empower our children to manage these changes for generations to come.

REFERENCES

1. De Souza, Roger-Mark, 2008, *Scaling Up Integrated Population, Health and Environment Approaches in the Philippines: A review of early experiences*, World Wildlife Fund and the Population Reference Bureau, Washington, DC.

2. U.S. Agency for International Development, 2008, Fact Sheet: Balancing People and the Environment to Promote Resilient Communities, USAID, Washington, DC.

3. Thaxton, Melissa, 2007, *Integrating Population, Health, and Environment in Tanzania*, Population Reference Bureau, Washington, DC.

4. Mogelgaard, Kathleen, and Kristen P. Patterson, 2006, *Linking Population, Health, and Environment in Fianarantsoa Province, Madagascar*, Population Reference Bureau, Washington, DC.

5. Castro, Joan, and Leona D'Agnes, 2008, Fishing for families: Reproductive health and integrated coastal management in the Philippines, *Focus*, issue 15, April, Environmental Change and Security Program, Washington, DC.

6. The Reproductive Health and Population Development Act of 2008.

7. De Souza, Roger-Mark, 2004, Harmonizing population and coastal resources in the Philippines, *WorldWatch* 17(5): 38–40, September/October, WorldWatch Institute, Washington, DC.

8. The project was called SUCCESS—Sustainable Coastal Communities and Ecosystems, and it worked in eight villages (total population 13,000) bordering on or surrounded by the Saadani National Park, about 200 kilometers north of Dar es Salaam on the Indian Ocean (Thaxton, Melissa, 2007, *Integrating Population, Health, and Environment in Tanzania*).

9. Castro, Joan, and Leona D'Agnes, 2008, Fishing for families; Oglethorpe, J., C. Honzak, and C. Margoluis, 2008, *Healthy People, Healthy Ecosystems: A Manual for Integrating Health and Family Planning into Conservation Projects*, World Wildlife Fund, Washington, DC.

CHAPTER 26

············

From Crisis to Sustainability

JAMES GUSTAVE SPETH

Given all the impacts of the current global economic crisis, it is easy to miss one: There's a lot less garbage. In early 2009, as the stock market tanked and growth went negative, the amount of trash sent to American landfills declined by as much as 30 percent.[1] Americans are also using less energy and thus are generating less of the emissions that cause global warming.

There is a paradox at the heart of our economic system: The imperatives of today's economy are sharply at odds with those of the environment. Yet the environment is the very basis of the economy and of human well-being.

Long before the global economy started crashing, it was crashing against the Earth. The damage has been enormous. For all the material blessings economic progress has provided, for all the disease and destitution avoided, and for all the glories that shine in the best of our civilization, the costs to the natural world—the costs to the glories of nature—have been huge and must be counted in the balance as tragic loss.

····················

James Gustave Speth is a Professor at the Yale School of Forestry & Environmental Studies and author, most recently, of The Bridge at the Edge of the World: Capitalism, the Environment, and Crossing from Crisis to Sustainability *(Yale University Press, 2008), from which this chapter is adapted.*

Moreover, despite the staggering economic growth of the last half century, half of humanity still lives in abject poverty, and inequality continues to grow in the United States and many other countries.

It is no accident that environmental crisis is gathering momentum as social injustice is deepening and inequality is impairing democratic institutions. Each is the result of a system of political economy—today's capitalism—that is profoundly committed to profits and growth and profoundly indifferent to nature and society. Left uncorrected, it is an inherently ruthless, rapacious system, and it is up to citizens, acting mainly through government, to inject human and natural values into that system. The best hope for change is a fusion of those concerned about the environment, social justice, and strong democracy into one powerful progressive force. That fusion must occur before it is too late.

Time is short. Half the world's tropical and temperate forests are now gone.[2] About half the wetlands and a third of the mangroves are gone.[3] An estimated 90 percent of the large predator fish are gone, and 75 percent of marine fisheries are now overfished or fished to capacity.[4] Twenty percent of the corals are gone, and another 20 percent are severely threatened.[5] Species are disappearing at rates that are about a thousand times faster than normal.[6] Over half the agricultural land in drier regions suffers from some degree of deterioration and desertification.[7] Persistent toxic chemicals can now be found by the dozens in essentially each and every one of us.[8]

Human impacts are now large, relative to natural systems. Human activities have pushed atmospheric carbon dioxide up by more than a third and have started in earnest the dangerous process of warming the planet and disrupting climate. Everywhere Earth's ice fields are melting.[9] Industrial processes are fixing nitrogen, making it biologically active, at a rate equal to nature's; one result is the development of more than 200 dead zones in the oceans due to overfertilization.[10] Human actions already consume or destroy each year about 40 percent of nature's photosynthetic output, leaving too little for other species.[11] Freshwater withdrawals doubled globally between 1960 and 2000 and are now over half of accessible runoff.[12] The following rivers no longer reach the oceans in the dry season: the Colorado, Yellow, Ganges, and Nile, among others.[13]

All we have to do to destroy the planet's climate and biota and leave a ruined world to our children and grandchildren is to keep doing exactly what we are doing, with no growth in the human population or the world economy. Just continue to release greenhouse gases at current rates, impoverish ecosystems and release toxic chemicals at current rates, and the world in the latter part of this century won't be fit to live in. But the scale of the human enterprise is not holding at current levels—it is accelerating dramatically.

> All we have to do to leave a ruined world to our children and grandchildren is to keep doing exactly what we are doing, with no growth in the human population or the world economy.

As noted elsewhere in this volume, world population is likely to grow by 30 percent—and could grow by more than half—by 2050. The world economy has more than quadrupled since 1960 and is projected to quadruple again by mid-century. At recent rates of growth, it will double in less than twenty years. It took all of human history to grow the $7 trillion world economy of 1950. We now grow by that amount in a decade. Societies face the prospect of enormous environmental deterioration just when they need to be moving strongly in the opposite direction.

The escalating processes of climate disruption, biotic impoverishment, and toxification that continue despite decades of warnings and earnest effort constitute a severe indictment, but an indictment of what, exactly? If we want to reverse today's destructive trends, forestall further and greater losses, and leave a bountiful world for our children and grandchildren, we must return to fundamentals and seek to understand both the forces driving such destructive trends and the economic and political system that gives these forces free rein. Then we can ask what can be done to change the system.

The underlying drivers of today's environmental deterioration have been clearly identified. They range from immediate forces like the enormous growth in human population and the dominant technologies deployed in the economy to deeper ones like the values that shape our behavior and determine what we consider important in life. Most basically, we know that environmental deterioration is driven by the economic activity of human beings. The struggle of the poor to survive creates a range of environmental impacts in which the poor themselves are often the primary victims—for example, the deterioration of arid and semiarid lands due to the press of increasing numbers of people who have no other option.

But the much larger and more threatening impacts stem from the economic activity of those of us participating in the modern, increasingly prosperous world economy. This activity is consuming vast quantities of resources from the environment and returning to the environment vast quantities of waste products. The damages are already huge and are on a path to be ruinous in the future. So, a fundamental question facing societies today—perhaps the fundamental question—is how can the operating instructions for the modern world economy be changed so that economic activity both protects and restores the natural world?

> How can the operating instructions for the modern world economy be changed so that economic activity both protects and restores the natural world?

With increasingly few exceptions, modern capitalism is the operating system of the world economy. I use *modern capitalism* here in a broad sense, meaning an actual, existing system of political economy, not an idealized model. Capitalism as we know it today encompasses the core economic concept of private employers hiring workers to produce products and services that the employers own and then sell with the intention of making a profit. But it also includes competitive markets, the price mechanism, the modern corporation as its principal institution, the consumer society and the materialistic values that sustain it, and the administrative state actively promoting economic strength and growth for a variety of reasons.

Inherent in the dynamics of capitalism is a powerful drive to earn profits, invest them, innovate, and thus grow the economy, typically at exponential rates, with the result that the capitalist era has in fact been characterized by a remarkable expansion of the world economy. The capitalist operating system, whatever its shortcomings, is very good at generating growth.

These features of capitalism, as they are constituted today, work together to produce an economic and political reality that is highly destructive of the environment. An unquestioning society-wide commitment to economic growth at almost any cost; enormous investment in technologies designed with little regard for the environment; powerful corporate interests whose overriding objective is to grow by generating profit, including profit from avoiding the environmental costs they create; markets that systematically fail to recognize environmental costs unless corrected by government; government that is subservient to corporate interests and the growth imperative; rampant consumerism spurred by a worshipping of novelty and by sophisticated advertising; economic activity so large in scale that its impacts alter the fundamental biophysical operations of the planet—all combine to deliver an ever-growing world economy that is undermining the planet's ability to sustain life.

In many ways, this system has also failed to deliver on its promise of human well-being. If we look at per capita income—an admittedly crude measure of human welfare—we see that the benefits of capitalism have been distributed very unevenly. Despite a century of unprecedented growth, half of humanity lives on less than two dollars a day. Currently, the richest 1 percent of the world's people receives as much income as the bottom 57 percent.[14] Even in industrialized countries, inequality has grown: The gap between rich and poor has increased in more than three-quarters of OECD countries over the past two decades. The United States has the highest inequality and poverty among leading industrialized nations, and the gap has widened dramatically since 2000.[15]

Even for the "haves" of the global economy, growth beyond a certain level does not produce commensurate increases in welfare. The never-ending drive to grow the economy undermines families, jobs, communities, the environment, a sense of place and continuity, even national security. In America, especially, we have not applied our growth dividend to meeting social and environmental needs. And there is good evidence that increased incomes do not lead to greater satisfaction with life. In affluent countries we have what might be called "uneconomic growth," to borrow Herman Daly's phrase, in which, if one could total them up, all the costs of growth would outweigh the benefits.

AVERTING THE COLLISION

The fundamental questions, then, are about transforming capitalism as we know it. Can it be done? If so, how? And if not, what then? The good news is that there is more than one prescription to take the economy and the environment off a collision course and to transform economic activity into something benign and restorative. The most important of these prescriptions range far beyond the traditional environmental agenda.

Market failure can be corrected by government, perverse subsidies can be eliminated, and environmentally honest prices can be forged. The laws, incentives, and governance structures under which corporations operate can be transformed to move from shareholder primacy to stakeholder primacy. But even more vital is the need to challenge economic growth and the consumerism it depends on. This challenge is as relevant to addressing social problems as environmental ones.

Overriding commitment to economic growth—mere GDP growth—is consuming environmental and social capital, both in short supply. Affluent countries must become "postgrowth" societies where jobs and work life, the environment, communities, and the public sector are no longer sacrificed to push up GDP.

There are many steps to slow growth while improving social and environmental well-being, such as shorter workweeks and longer vacations; greater labor protections, job security, and benefits; restrictions on advertising; a new design for the twenty-first-century corporation; strong social and environmental provisions in trade agreements; rigorous environmental and consumer protection, including full-cost pricing; greater economic and social equality, with progressive taxation of the rich and greater income support for the poor; heavy spending on public services and environmental amenities; and a huge investment in education, skills, and new technology.

> Instead of merely pursuing GDP growth, we need policies that address social needs directly.

Instead of merely pursuing GDP growth, we need policies that address social needs directly—that strengthen families and communities and address the breakdown of social connectedness and the erosion of social capital; that provide for universal education and health care; that ensure care and companionship for the chronically ill and incapacitated. Such measures, wise in their own right, should be seen as environmental measures too: central parts of the alternative to the destructive path we are on.

But even if the affluent countries make the transition to postgrowth societies, the developing countries must continue to grow their economies in order to lift half of humanity from poverty and achieve acceptable levels of economic well-being. And it is in those countries that the lion's share of future population growth will occur. For these reasons, the prospect of a sustainable future for the human community depends as much on the developing countries as on the industrial ones. Yet what they do will be influenced, perhaps decisively, by what we in the industrialized world do (or fail to do) in forging an equitable, workable, and cooperative set of economic arrangements with them. It is therefore essential that industrialized countries meet the commitments they have made through a series of UN agreements—including the Millennium Development Goals and the International Conference on Population and Development.

BEYOND CAPITALISM

Taken together, these prescriptions, if implemented, would take us beyond capitalism as we know it today. The question of whether we would then have an operating system other than capitalism or a reinvented capitalism is largely definitional. In the end, the answer is probably not important. I myself have no interest in socialism or centralized economic planning or other paradigms of the past. The question for the future, on the economic side, is how do we harness economic forces for sustainability and sufficiency?

The creativity, innovation, and entrepreneurship of businesses operating in a vibrant private sector are essential to designing and building the future. We will not meet our environmental and social challenges without them. Growth and investment are needed across a wide front: growth in the developing world—sustainable, people-centered growth; growth in the incomes of those in America who have far too little; growth in human well-being along many dimensions; growth in new solution-oriented industries, products, and processes; growth in meaningful, well-paying jobs, including green-collar ones; growth in natural resource and energy productivity and in investment in the regeneration of natural assets; growth in social and public services and in investment in public infrastructures, to mention a few. These are the things we should be growing, and it makes good sense to harness market forces to such ends. Even in a postgrowth society, many things still need to grow.

These prescriptions for change in the fundamental arrangements of capitalism are difficult, to put it mildly. What circumstances might make deep change plausible? A mounting sense of imminent crisis, wise leadership—both will help. Most of all, we need a new politics and a new social movement powerful enough to drive change.

American environmentalists must join social progressives to address the crisis of inequality in the United States and around the world. We must also join those seeking to reform politics and strengthen democracy. America's gaping social and economic inequality poses a grave threat to democracy. We are seeing the emergence of a vicious circle: Income disparities shift

> The creativity, innovation, and entrepreneurship of businesses operating in a vibrant private sector are essential to designing and building the future.

political access and influence to wealthy constituencies and large businesses, which further imperils the potential of the democratic process to correct economic disparities. Corporations have been the principal economic actors for a long time; now they are the principal political actors as well. Neither environment nor society fares well under corptocracy. Environmentalists need to embrace public financing of elections, lobbying regulation, nonpartisan congressional redistricting, and other reforms as a core of their agenda. Politics as usual will never deliver environmental sustainability.

> Today's environmental reality is linked powerfully with other realities, including growing social inequality and the erosion of democratic governance and popular control.

Today's environmental reality is linked powerfully with other realities, including growing social inequality and the erosion of democratic governance and popular control. We as citizens must mobilize our spiritual and political resources for transformative change on many fronts. Our best hope for change is a fusion of those concerned about environmental sustainability, social justice, and political democracy into one progressive force.

One area where fusion is beginning is the conversation between environmental and social justice activists on solutions to the climate change threat. That's encouraging, but it's a small part of what's needed. Mostly, our movements remain in separate silos. A sustained dialogue is urgently needed to build a common agenda for action and a shared commitment to build a new social movement for change. We are all communities of a shared fate. We will rise or fall together.

Humanity is fast approaching a fork in the road. Beyond the fork, down either path, is the end of the world as we have known it. One path beyond the fork continues us on our current trajectory. Presidential science adviser John Gibbons used to say with a wry smile that if we don't change direction, we'll end up where we're headed. And right now we're headed toward a ruined planet. That is one way the world as we know it could end, down that path and into the abyss.

But there is another path, and it leads to a bridge across the abyss. Of course, where the path forks will be the site of a mighty struggle, a struggle that must be won even though we cannot see clearly what lies beyond the bridge. Yet in that struggle and in the crossing that will follow, we are carried forward by hope, a radical hope, that a better world is possible and that we can build it. "Another world is not only possible. She is on her way," says Arundhati Roy. "On a quiet day, I can hear her breathing."[16]

REFERENCES

1. Schulte, Brigid, 2009, A trashed economy foretold: Intake at landfills has been falling, *The Washington Post*, March 14.

2. Millennium Ecosystem Assessment (MA), 2005, *Ecosystems and Human Well-being: Synthesis*, Island Press, Washington, DC, p. 31–32.

3. MA, 2005, *Ecosystems and Human Well-being: Synthesis*, p. 2; MA, *Ecosystems and Human Well-being*, volume 1, *Current State and Trends*, Island Press, Washington, DC, p. 14–15. See also Duke, N. C., et al., 2007, A world without mangroves? *Science* 317(41). And see Revenga, Carmen, et al., 2000, *Pilot Analysis of Global Ecosystems: Freshwater Systems*, WRI, Washington, DC, p. 3, 21–22; World Resources Institute et al., 2000, *World Resources, 2000–2001*, WRI, Washington, DC, p. 72, 107; and Burke, Lauretta, et al., 2001, *Pilot Analysis of Global Ecosystems: Coastal Ecosystems*, WRI, Washington, DC, p. 19.

4. Food and Agriculture Organization, 2006, *World Review of Fisheries and Aquaculture*, FAO, Rome, p. 29; Myers, Ransom A., and Boris Worm, 2003, Rapid world-wide depletion of predatory fish communities, *Nature* 423: 280. See also Pearce, Fred, 2003, Oceans raped of their former riches, *New Scientist*, August 2, p. 4.

5. MA, 2005, *Ecosystems and Human Well-being: Synthesis*, p. 2.

6. MA, 2005, *Ecosystems and Human Well-being: Synthesis*, p. 5, 36.

7. UN Environment Programme, 2002, *Global Environment Outlook 3*, Earth-scan, London, p. 64–65. Drylands cover about 40 percent of the Earth's land surface, and an estimated 10 to 20 percent suffer from "severe" degradation. Reynolds, James F., et al., 2007, Global desertification: Building a science for dryland development, *Science* 316: 847. See also Key facts about desertification, 2006, Reuters/Planet Ark, June 6, summarizing UN estimates.

8. Pearce, Fred, 1997, Northern exposure, *New Scientist*, May 31, p. 25; Enserink, Martin, 2003, For precarious populations, pollutants present new perils, *Science* 299: 1642. See also the data reported in Thornton, Joe, 2000, *Pandora's Poison*, MIT Press, Cambridge, MA, p. 1–55.

9. UN Environment Programme, 2007, Global Outlook for Ice and Snow, June 4. See also http://www.geo.unizh.ch/wgms. See generally Collins, William, et al., 2007, The physical science behind climate change, *Scientific American*, August, p. 64.

10. UN reports increasing "dead zones" in oceans, 2006, Associated Press, October 20. See generally Shrope, Mark, 2006, The dead zones, *New Scientist*, December 9, p. 38; and Mee, Laurence, 2006, Reviving dead zones, *Scientific American*, November, p. 79. On nitrogen pollution, see Driscoll, Charles, et al., 2003, Nitrogen pollution, *Environment* 45(7): 8.

11. Vitousek, Peter M., et al., 1986, Human appropriation of the products

of photo-synthesis, *Bioscience* 36(6): 368; Rojstaczer, S., et al., 2001, Human appropriation of photosynthesis products, *Science* 294: 2549. See also Haberl, Helmut, et al., 2007, Quantifying and mapping the human appropriation of net primary production in Earth's terrestrial ecosystems, *Proceedings of the National Academy of Sciences.*

12. UN Environment Programme, *At a Glance: The World's Water Crisis*, and MA, 2005, *Ecosystem and Human Well-being: Synthesis*, p. 32.

13. MA, 2005, *Ecosystem and Human Well-being: Synthesis*, p. 12.

14. University of California Atlas of Global Inequality, http://ucatlas.ucsc .edu/income.php.

15. Organization for Economic Cooperation and Development, 2008, *Growing Unequal? Income Distribution and Poverty in OECD Countries*, OECD Publishing.

16. Roy, Arundhati, 2004, Come September, in *The Impossible Will Take a Little While: A Citizen's Guide to Hope in a Time of Fear*, Paul Rogat Loeb, ed., Basic Books, New York.

PART IV

Thoughts for the Journey

CHAPTER 27

....

Reproductive Rights Are Human Rights

JACQUELINE NOLLEY ECHEGARAY and SHIRA SAPERSTEIN

Sexual and reproductive rights . . . emerge from the recognition
that equality in general, gender equality in particular, and the
emancipation of women and girls are essential to society.
Protecting sexual and reproductive rights is a direct path to
promoting the dignity of all human beings, and a step forward
in humanity's advancement towards social justice.[1]

—*Constitutional Court of Colombia*

FIGHTING GLOBAL CLIMATE CHANGE— BY ANY MEANS NECESSARY?

As alarms are sounded about climate change, concerns about what con-
stitutes a sustainable number of people on the planet are also on the
rise. In some quarters, fears of an impending "population bomb"—like
that predicted by Paul R. Ehrlich during the late 1960s and 1970s—are
reemerging, only this time the catastrophe predicted to follow the
"bomb" is devastating global climate change.[2] As in the past, such fears
are provoking calls for "population control": A letter published in the
Medical Journal of Australia in December 2007, for example, proposed

*Jacqueline Nolley Echegaray is Associate for International Programs at the Moriah
Fund. Shira Saperstein is the Moriah Fund's Deputy Director and Program Director
for Women's Rights and Reproductive Health.*

applying a "baby levy" in the form of a carbon tax on families with more than two children.[3]

All too often, efforts to limit population growth have resulted in the use of coercive, draconian methods that infringe directly on human rights, and particularly on the rights of women. Examples abound, but one has only to look at China's one-child policy, and the forced abortions that accompany it, or India's forced sterilization of millions of poor men and women during the 1970s[4] to see the human rights violations that can result from population control programs.[5]

> If climate change and population policies are to pass muster vis-à-vis human rights standards, they must also meet sexual and reproductive rights standards.

As the world searches for solutions to the climate crisis, there is cause for concern that policymakers may adopt a "by any means necessary" approach to forestalling environmental catastrophe. Such an approach poses great threat to sexual, reproductive, and other human rights. Sexual and reproductive rights (SRR) are integral human rights, yet they are commonly misunderstood and ignored;[6] as a result, even policymakers conscious of their state's obligations under international human rights law may neglect to consider a climate policy's SRR dimensions. This is unacceptable. If climate change and population policies are to pass muster vis-à-vis human rights standards, they must also, necessarily, meet sexual and reproductive rights standards.

In this chapter, we trace the evolution of SRR in international human rights instruments and explore what sexual and reproductive rights guarantees—or the lack thereof—look like in reality. To conclude, we urge decision makers—including those grappling with climate change—to secure and advance the progress that has been made toward the universal realization of SRR.

REPRODUCTIVE RIGHTS IN INTERNATIONAL LAW: A BRIEF HISTORY

Sexual and reproductive rights are rooted in the Universal Declaration of Human Rights (UDHR),[7] the founding document of the modern human rights era and a landmark in the global women's rights movement. The UDHR established for the first time that "discrimination against women is an appropriate matter for international concern"[8] and that the oppression of women—whether based in law or in custom, and whether occurring behind closed doors or in public—should no longer enjoy legal sanction. The UDHR also established several rights that are relevant to population policy, including the right to life, liberty, and security; the right to privacy; and the right to medical care and necessary social services.[9]

The International Covenants on Civil and Political Rights (ICCPR) and on Economic, Social and Cultural Rights (ICESCR), both of which entered into force in 1976, gave legal teeth to the UDHR.[10] The ICCPR and the ICESCR require states to report to the UN regarding the implementation of each treaty. In addition, an optional protocol to the ICCPR established a mechanism through which individuals may file complaints against member states. Women have since utilized this mechanism to lodge complaints against policies that discriminate based on gender, a violation of the ICCPR.[11]

In 1979, the UN adopted the landmark Convention on the Elimination of All Forms of Discrimination against Women (CEDAW).[12] CEDAW is rightfully known as the "international bill of rights for women" because of its comprehensive approach to guaranteeing women's rights in all arenas—including within what were long considered "private" spheres, such as marriage and the family, as well as in "public" spaces such as the workplace. CEDAW affirms women's rights to property, inheritance, access to health care, suffrage, political representation, and education, among others.[13]

CEDAW was the first—and is still the only—human rights treaty that explicitly affirms the *reproductive* rights of women,[14] and its definition of *discrimination* extends states' responsibility into the "private" spheres of family and community life.[15] It affirms the centrality of SRR to women's human rights, including access to family planning and maternal health services and information and, most notably, the right to "decide freely and responsibly the number and spacing of children and to have access to the information, education and means"[16] to exercise these rights.[17] Upon signing and ratifying CEDAW, signatories commit to undertaking all measures necessary to guarantee its provisions.[18]

Following world conferences on women's rights during the 1970s and 1980s; the UN Decade for Women from 1975 to 1985; and years of organizing, agitating, and behind-the-scenes efforts by thousands of women's rights activists, the Vienna World Conference on Human Rights in 1993 produced another watershed moment in the history of women's rights. The meeting's final declaration and program of action—adopted by 171 countries—established unequivocally that "the human rights of women are an inalienable, integral and indivisible part of universal human rights."[19]

The following year, at the 1994 International Conference on Population and Development in Cairo (ICPD), 179 states adopted the Cairo Program of Action (PoA) which included—and expanded—language from CEDAW. The Cairo PoA defined reproductive rights for the first time in an international policy document:

> Reproductive rights embrace certain human rights that are already recognized in national laws, international human rights documents and other consensus documents. The rights rest on the recognition of the basic rights of all couples and individuals to decide freely and responsibly the number, spacing and timing of their children and to have the information and means to do so, and the right to attain the highest standard of sexual and reproductive health. It also includes their right to make decisions concerning reproduction free of discrimination, coercion and violence, as expressed in human rights documents.[20]

The ICPD definition of *reproductive rights* is now widely considered the definitive definition of the term; it has been adopted as the basis of public policy, countless advocacy efforts, and numerous high-impact legal cases in countries around the world.

The ICPD marked the first time an international consensus document addressing population growth was grounded in a reproductive rights approach.[21] The PoA explicitly recognizes the central role of human rights in the context of population policies; its preamble "affirms the application of universally recognized human rights standards to all aspects of population programs."[22] Subsequent international meetings on women's rights, including the five-year follow-up to the ICPD, the UN Fourth World Conference on Women, in Beijing (1995), and the five-year follow-up to the Beijing Conference, all further developed the concept of a rights-based approach to population and reproductive health.[23]

In addition to these global efforts, important progress in establishing reproductive rights has been made regionally. Signatories to the Protocol to the African Charter on Human and Peoples' Rights on the Rights of Women in Africa,[24] for example, agree to ensure the right of women to control their own reproduction and to choose any method of contraception, and they commit to "protect[ing] the reproductive rights of women by authorizing medical abortion in cases of sexual assault, rape, incest, and where the continued pregnancy endangers the mental and physical health of the mother or the life of the mother or the fetus."[25] In the Americas, the Inter-American Convention for the Prevention, Punishment and Eradication of Violence Against Women (Convention of Belém do Pará)[26] is the first international treaty to specifically address the issue of violence against women. The convention's definition of *violence against women* is broad and includes sexual violence such as rape, sexual assault, forced prostitution, trafficking, and sexual harassment.[27]

REPRODUCTIVE RIGHTS TODAY: NOT YET A REALITY

Reproductive rights are rooted within a constellation of human rights, including

- the right to life, liberty and security;
- the right not to be subjected to torture or other cruel, inhuman, or degrading treatment or punishment;
- the right to be free from gender discrimination;
- the right to modify customs that discriminate against women;
- the right to health, reproductive health, and family planning;
- the right to privacy;
- the right to marry and found a family;
- the right to decide the number and spacing of children;
- the right to education;
- the right to be free from sexual assault and exploitation; and
- the right to enjoy scientific progress and to consent to experimentation.[28]

Although, as outlined above, much progress has been made in establishing an international legal framework for these rights, the realization of reproductive rights remains elusive. Consider the following examples.

Inherent in *the right to life, liberty, and security* is the right to survive pregnancy and childbirth. Yet, unconscionably, complications during pregnancy are the leading cause of death globally for young women aged fifteen to nineteen,[29] and women die of pregnancy-related causes at a rate of about one per minute, or approximately 536,000 maternal deaths worldwide in 2005.[30] (See Chapter 19.) The shameful fact is that maternal mortality rates have held steady at unacceptably high levels for the past fifteen years when, according to *The Lancet*, "simple, cheap and effective [reproductive health] interventions have existed for more than 50 years . . . but . . . are not available in many parts of the world."[31] Clearly, despite their international commitments, many countries choose to neglect the rights of women, literally to the point of death.

The right to decide the number and spacing of children is not only fundamental to securing the right to life but is also a right in itself, as asserted in the Cairo PoA. Nevertheless, globally 200 million women lack access to contraceptives.[32] The UNFPA estimates that universal access to contraception would save the lives of one in three women who die of causes related to pregnancy and childbirth, or roughly 160,800 women per year.

The right not to be subjected to torture or other cruel, inhuman, or degrading treatment or punishment is enshrined in the UDHR, yet this right is violated where women are forcibly sterilized. As Susana Chávez Alvarado recounts in Chapter 22, during the 1990s the Peruvian government undertook a "massive, compulsory, and systematic"[33] sterilization campaign targeting the country's most marginalized population: the indigenous and poor. María Mamérita Mestanza, a thirty-three-year-old mother of seven, was coerced by government officials into undergoing a tubal ligation; she died as a result of complications stemming from the surgery. In 2003, the Peruvian government acknowledged that it had violated, among other rights, Ms. Mestanza's right to humane treatment as protected by the Inter-American Convention on Human Rights.[34]

Bans on abortion violate not only the *right to health* but rights to *life, liberty, bodily integrity, privacy, nondiscrimination,* and *equality,* as well as the *right to choose the number and spacing of children.*[35] Abortion bans do not, however, prevent women from terminating pregnancies; in fact, abortion remains one of the most common medical procedures in the world, regardless of whether it is legal.[36] When the state fails to ensure access to safe abortion, the costs to women's health and lives are high: Almost 70,000 women per year die from unsafe abortion, and at least 5 million others suffer serious injury. These deaths and injuries are almost entirely preventable.[37]

Events in Nicaragua following a 2006 decision to criminalize abortion under all circumstances—even when necessary to save a woman's life—demonstrate the consequences when these rights are denied. A Human Rights Watch investigation carried out shortly after the ban was implemented found that, despite protocols requiring doctors to treat certain obstetric emergencies, a pervasive fear of prosecution among health care providers resulted in the denial or delay of care.[38] Women who were refused treatment suffered sterility, maiming, and in at least eighty cases, death.[39] Responding to the ban, the Inter-American Commission on Human Rights asserted that "therapeutic abortion has been internationally recognized as a specialized and necessary health service for women," and that "denial of this service endangers women's lives as well as their physical and psychological integrity."[40]

In the United States, the Supreme Court established a right to legal abortion in 1973; however, this right is effectively denied to poor women today as a result of subsequent rulings and policies. Since 1976, the Hyde Amendment has limited public funding for abortion, leaving approximately 12 million women of reproductive age enrolled in Medicaid without coverage for this procedure. Almost all of these women are poor, and a majority are women of color. As a result of the Hyde Amendment and other funding bans, women struggle to find the funds they need for abortion care; they forgo food, shelter, and other necessities and often delay treatment until their second trimester, thereby increasing the health risks.[41]

HOPE ON THE HORIZON— MAKING REPRODUCTIVE RIGHTS A REALITY

For too many women, reproductive rights are more rhetoric than reality. But reproductive rights activists around the world are working to close the gap.

For example, a young Chilean woman who was expelled from school because she became pregnant successfully challenged her expulsion, based on international rights that guarantee privacy and equal protection under the law.[42] In Peru, when a young woman was drugged, raped by a doctor at a state medical facility, and denied justice by local courts, a lawsuit filed before an international human rights tribunal resulted not only in justice for the victim but also in improvements related to the treatment of rape victims by state authorities.[43] Activists in Africa successfully invoked the international rights to nondiscrimination, health, and bodily integrity to achieve the criminalization of female genital mutilation in eighteen countries.[44] And because international human rights law takes precedence over domestic law in Colombia, in 2006 the Colombian Constitutional Court overturned the country's long-standing absolute ban on abortion, thanks to a legal challenge from a reproductive rights activist.[45]

Reproductive rights are indivisible from human rights, and states committed to implementing civil, political, and other rights are similarly bound to implement reproductive rights.[46] Rather than violating these rights, policymakers concerned about humanity's growing numbers would find their interests best served by guaranteeing women everywhere the full spectrum of human rights. For example, fertility rates drop when women and girls are able to realize their right to education; in fact, according to the World Bank, it is estimated that one year of female schooling reduces fertility by 10 percent. Women with formal education are much more likely to use reliable family planning methods, delay marriage and childbearing, and have fewer and healthier babies than women with no formal education.[47]

> Policymakers concerned about humanity's growing numbers would find their interests best served by guaranteeing women everywhere the full spectrum of human rights.

Simply put, rather than attempting to control population growth, policymakers would be wise to promote the full realization of the widest possible array of reproductive rights and guarantee that control over reproduction is squarely in the hands of the world's women. The Cairo Program of Action—the international community's clearest articulation of SRR standards—emphasizes that the best way to slow population growth is by advancing human rights: by ensuring that all people have the means and power to make their own decisions about childbearing and by eradicating discrimination and injustice that deny men and women reproductive self-determination and dignity. Reproductive rights are good not only for women but for the planet; it is time to make the Cairo Program of Action, and all other reproductive rights instruments, a reality.

REFERENCES

1. Women's Link Worldwide, 2007, *C-355/2006: Excerpts of the Constitutional Court's Ruling That Liberalized Abortion in Colombia*, http://www.womenslink worldwide.org/pdf_pubs/pub_c3552006.pdf.

2. Population growth is identified as "the single world issue that binds all the others together" and "the root of many problems affecting the tree of humanity" in Redding, Terry M., 2007, *The Population Challenge: Key to Global Survival*, http://www.populationinstitute.org/cms/modules/Publications/front/lib/pdf.php?id=38.

3. Walters, Barry N. J., 2007, *Personal carbon trading: A potential "stealth intervention" for obesity reduction?* http://www.mja.com.au/public/issues/187_11 _031207/wal10921_fm.html, October. For other examples, see the *Optimum Population Trust's Green Planet Petition,* http://www.optimumpopulation.org/ opt.petition.html; McKie, Robin, 2007, *Science chief: Cut birthrate to save the earth,* http://www.guardian.co.uk/science/2007/jul/22/climatechange .climatechange, July; and Rapley, Chris, 2006, *Earth Is Too Crowded for Utopia,* http://news.bbc.co.uk/2/hi/science/nature/4584572.stm, January.

4. Chandiramani, R., 2005, Sexual health and rights in India, p. 127–153, *Where Human Rights Begin: Health, Sexuality, and Women in the New Millennium,* Wendy Chavkin and Ellen Chesler, eds., Rutgers.

5. Pronatalist policies can also lead to human rights violations. In Romania, for example, based on findings of a census conducted in 1966 that showed fertility rates falling below replacement levels, the government abruptly reversed its previously liberal abortion policy and instituted a number of measures to spur population growth. As a result, maternal mortality increased dramatically, and thousands of children were abandoned by their parents to the care of orphanages.

6. Shaw, Dorothy, 2006, *Sexual and reproductive health: Rights and responsibilities,* http://www.thelancet.com/journals/lancet/article/PIIS01406736066 94877/fulltext, December.

7. Adopted by the General Assembly of the United Nations on December 10, 1948.

8. Chesler, E., 2005, Introduction, p. 1–34, *Where Human Rights Begin: Health, Sexuality, and Women in the New Millennium,* Wendy Chavkin and Ellen Chesler, eds., Rutgers.

9. *Universal Declaration of Human Rights,* 1948, http://www.un.org/ Overview/rights.html (December).

10. Together with the UDHR, these treaties are commonly referred to as the "International Bill of Rights." Gaer, Felice D., 1998, And never the twain shall meet? The struggle to establish women's rights as international human rights, p. 4–89, *International Human Rights of Women: Instruments of Change,* Carol Elizabeth Lockwood et al., eds., American Bar Association.

11. *Examples of Cases Where Women Have Used the First Optional Protocol to the ICCPR to Challenge Sex Discrimination,* http://www.un.org/womenwatch/daw/ cedaw/protocol/cases.htm.

12. Currently 185 countries—although notably not the United States—are party to the convention.

13. Chavkin, Wendy, and Ellen Chesler, eds., *Where Human Rights Begin: Health, Sexuality, and Women in the New Millennium,* Rutgers.

14. UN Department of Economic and Social Affairs Division for the Advancement of Women, *Introduction to the Convention on the Elimation of All Forms of Discrimination against Women,* http://www.un.org/womenwatch/daw/ cedaw/cedaw.htm.

15. Lockwood, Carol Elizabeth, et al., eds., *International Human Rights of Women: Instruments of Change,* American Bar Association.

16. UN General Assembly, 1979, *Convention on the Elimination of All Forms of*

Discrimination against Women, http://www.un.org/womenwatch/daw/cedaw/text/econvention.htm#article16 (December).

17. As with the ICCPR and the ICESCR, to fulfill CEDAW obligations, States Parties must submit regular reports to a standing UN committee that semiannually reviews progress toward implementation of the treaty. In December 2000, the UN General Assembly adopted the Optional Protocol to CEDAW, which allows women to submit claims of violations of rights to the UN and authorizes inquiries into situations of grave or systematic violations of women's rights. Bora Laskin Law Library, *The Optional Protocol to the Women's Convention*, http://www.law-lib.utoronto.ca/diana/whrr/cfsearch_display_details.cfm?ID=2438&sister=utl&subjectid=1&type=documents&searchstring=1 (January 2009).

18. The United Nations Association of the United States of America, 2007, *UNA-USA Fact Sheet: Convention on the Elimination of All Forms of Discrimination Against Women*, http://www.unausa.org/site/pp.asp?c=fvKRI8MPJpF&b=337333 (April).

19. Vienna Declaration and Programme of Action, 1993, http://www.unhchr.ch/huridocda/huridoca.nsf/(Symbol)/A.CONF.157.23.En.

20. Programme of Action of the International Conference on Population and Development, 1998, Paragraph 7.3, p. 402–439, *International Human Rights of Women: Instruments of Change*, Carol Elizabeth Lockwood et al., eds., American Bar Association.

21. Center for Reproductive Rights, 2003, *Rethinking Population Policies: A Reproductive Rights Framework*, http://reproductiverights.org/sites/default/files/documents/pub_bP_rethinkingpop.pdf (February).

22. Ibid.

23. Ibid.

24. Adopted by the African Union in July 2003.

25. African Commission on Human and Peoples' Rights, 2003, *Protocol to the African Charter on Human and Peoples' Rights on the Rights of Women in Africa*, http://www.achpr.org/english/_info/women_en.html (July).

26. Adopted by the General Assembly of the Organization of American States in June 1994.

27. *Inter-American Convention on the Prevention, Punishment and Eradication of Violence against Women*, 1994, http://www.oas.org/cim/english/convention%20violence%20against%20women.htm (June). In addition, as of this writing, there is a regional campaign underway, led by feminist and reproductive rights organizations in countries across the continent, calling for an Inter-American Convention on Sexual and Reproductive Rights.

28. Center for Reproductive Rights, 2003, *Rethinking Population Policies: A Reproductive Rights Framework*, http://reproductiverights.org/sites/default/files/documents/pub_bP_rethinkingpop.pdf (February).

29. UNICEF, 2008, *Young People and Family Planning: Teenage Pregnancy*, http://www.unicef.org/malaysua/Teenage_Pregnancies_-_Overview.pdf (July).

30. UNFPA, 2007, *Maternal Mortality Figures Show Limited Progress in Making Motherhood Safer*, http://www.unfpa.org/mothers/statistics.htm (October).

31. Glasier, Anna, and A. Metin Gülmezoglu, 2006, Putting sexual and reproductive health on the agenda, *The Lancet,* http://www.thelancet.com/journals/lancet/article/PIIS0140673606694853/fulltext (November).

32. Doña, Rena, 2007, *Safe Motherhood Begins with Access to Family Planning,* http://www.unfpa.org.ph/speeches/safe-motherhood-begins-access-family-planning (August).

33. Inter-American Commission on Human Rights, 2003, Report No. 71/03 Petition 12.191, *Friendly Settlement in the Case of María Mamérita Mestanza Chávez v. Perú,* http://www.cidh.org/annualrep/2003eng/Peru.12191.htm (October).

34. Ibid.

35. Human Rights Watch, 2005, *Decisions Denied: Women's Access to Contraceptives and Abortion in Argentina,* http://www.hrw.org/en/node/11694/section/7 (June).

36. Guttmacher Institute, 2006, *Abortion in Women's Lives,* http://www.guttmacher.org/pubs/2006/05/04/AiWL.pdf (May).

37. Grimes, David A., 2006, *Unsafe Abortion: The Preventable Pandemic,* http://www.thelancet.com/journals/lancet/article/PIIS0140-6736(06)69481-6/fulltext#article_upsell (November).

38. The Pan American Health Organization estimates that approximately 730 women per year require emergency obstetric care in Nicaragua because of miscarriages, ectopic and molar pregnancies, and cancer-related complications. Human Rights Watch, 2007, *Over Their Dead Bodies: Denial of Access to Emergency Obstetric Care and Therapeutic Abortion in Nicaragua,* http://www.hrw.org/sites/default/files/reports/nicaragua1007webwcover.pdf (October).

39. Nicaragua abortion ban killing women-rights group, 2007, Thomson Reuters, http://uk.reuters.com/article/latestCrisis/idUKN0240981920071002 (October).

40. Sanahuja, Juan C., 2007, *Nicaragua: Presiones abortistas internacionales,* http://www.noticiasglobales.org/comunicacionDetalle.asp?Id=981, January.

41. National Network of Abortion Funds, 2005, *Abortion Funding: A Matter of Justice,* http://www.nnaf.org/pdf/NNAF%20Policy%20Report.pdf.

42. Inter-American Commission on Human Rights, 2002, Report No. 33/02 Petition 12.046, *Friendly Settlement in the Case of Mónica Carabantes Galleguillos v. Chile,* http://www.cidh.org/annualrep/2002eng/Chile12046.htm (March).

43. Center for Reproductive Rights, 2008, *MM v. Peru,* http://www.reproductiverights.org/pr_99_1022peruvic.html (December).

44. Center for Reproductive Rights, 2008, *Female Genital Mutilation (FGM): Legal Prohibitions Worldwide,* http://reproductiverights.org/en/document/female-genital-mutilation-fgm-legal-prohibitions-worldwide (December).

45. Women's Link Worldwide, 2006, Colombia's highest court rules in favor of easing one of the world's most restrictive abortion laws, http://www.womenslinkworldwide.org/pdf_press/press_release_2006510_col.pdf (May).

46. The responsibility to promote the full realization of reproductive rights extends to international development assistance. Bueno de Mesquita, Judith, and Paul Hunt, 2008, *International Assistance and Cooperation in Sexual*

and Reproductive Health: A Human Rights Responsibility for Donors, http://www2
.essex.ac.uk/human_rights_centre/rth/docs/Final%20pdf%20for%20
website.pdf (December).

47. World Bank, *Girls' Education,* http://web.worldbank.org/WBSITE/
EXTERNAL/TOPICS/EXTEDUCATION/0,,contentMDK:20298916~menu
PK:617572~pagePK:148956~piPK:216618~theSitePK:282386,00.html#why.

CHAPTER 28

....

Over-Breeders and the Population Bomb

The Reemergence of Nativism and Population Control in Anti-Immigration Policies

PRISCILLA HUANG

At the start of 2008, news of a "baby boomlet" made headlines.[1] For the first time in thirty-five years, the U.S. fertility rate, or average number of children born to each woman, reached 2.1—the number statisticians say is needed for a population to replace itself.[2] Demographers pointed to the growing immigrant population as a reason for the higher birthrate.[3]

Many economists welcomed the surge in population growth as a sign of the country's likely future prosperity. While most industrialized nations struggle with shrinking populations due to low birthrates, the United States can look forward to a stable tax base and a steady workforce.[4] However, not everyone greeted the news with enthusiasm. Many interpreted the increased birthrate as an indication of the country's failed immigration laws and turned a hostile eye toward immigrant women.[5]

News of the rising growth rate also prompted anti-immigration groups to claim that immigrants damage the environment. For example, the Center for Immigration Studies released a report in August 2008 that links immigration with soaring greenhouse gas emissions.[6] At the same time, a coalition of anti-immigration groups took out a

....

Priscilla Huang is Policy and Programs Director at the National Asian Pacific American Women's Forum.

series of print and Internet ads that blamed immigrants for a host of environmental problems, including deforestation and suburban sprawl.[7] By linking immigration to environmental degradation, these groups have revived an issue that has long divided the U.S. environmental movement.

Interestingly, some of those who decry high immigrant birthrates are simultaneously working to limit *all* women's access to contraception and abortion. The position is a peculiar one: Anyone who is concerned about the environmental effects of immigration and population growth would seem unlikely to support measures that are guaranteed to boost birthrates. This apparent contradiction can only be understood by examining the tangled history of immigration, population control, and conservation in U.S. politics—a history that still reverberates today.

> Some who decry high immigrant birthrates are simultaneously working to limit *all* women's access to contraception and abortion.

"LOVE YOUR MOTHER—DON'T BECOME ONE"[8]— IMMIGRANT WOMEN AND POPULATION POLITICS

Throughout U.S. history, oppressive tactics have been used to limit the reproductive capacities of women who were seen as socially "undesirable." Beginning in the late 1800s, the infamous eugenics movement sought to curb pregnancies by women deemed unfit for motherhood, namely, immigrant, poor, nonwhite, or socially deviant women. Eugenicists promoted coercive policies, such as compulsory sterilization, as necessary tactics to "prevent the American people from being replaced by alien or negro stock, whether it be by immigration or by overly high birth rates among others in this country."[9] Supporters advocated other "negative" eugenics methods, including involuntary confinement and immigration restrictions, to prevent "undesirable" people from reproducing.[10]

At the same time, a corollary movement encouraged large families among "racially superior" middle- and upper-class Anglo-Saxon women using "positive" eugenics methods, such as tax incentives and awareness campaigns.[11] Both the negative and positive eugenics movements sought to regulate women's fertility, and neither trusted women to make their own reproductive decisions. Both sought the advancement of one ultimate goal: to protect the political and economic interests of the "old stock" elite.[12]

THE RISE OF ENVIRONMENTAL NATIVISM

The U.S. conservation movement emerged in the late nineteenth century, and with it came new theories that perpetuated anti-immigrant sentiments. Some early conservationists coupled their campaigns for wildlife protection with anti-immigrant rhetoric, believing that immigrants "threatened 'native' values and wildlife."[13] As the conservation movement gained in popularity, nativists also adopted environmental rhetoric to argue for slowing population growth, on the grounds that increasing human numbers threatened to upset the delicate balance of the natural environment.[14] Reducing immigration, they reasoned, was the solution.

Concern about population growth resurfaced with the birth of the modern environmental movement in the mid-twentieth century. In 1968, the theory of overpopulation went mainstream when the Sierra Club published Paul Ehrlich's *The Population Bomb,* which warned against rapid population growth in a resource-limited world.[15] Ehrlich followed in the intellectual footsteps of Thomas Malthus, the eighteenth-century British economist who theorized that population growth would outstrip increases in food production.[16] The book's popularity prompted the creation of several national organizations to address overpopulation.[17]

In 1978, the immigration control and conservation movements joined forces to form the Federation for American Immigration Reform ("FAIR").[18] Known today as the largest anti-immigrant policy organization in the United States, FAIR was founded by several Sierra Club members and John Tanton, then president of Zero Population Growth, a group that advocated birth control and tax incentives as a strategy for limiting population growth.[19] Paul Ehrlich endorsed the organization in his anti-immigrant publication *The Golden Door.* FAIR came under public scrutiny when it became known that between 1985 and 1994, the organization accepted $1.2 million from the Pioneer Fund, a foundation that supports research in eugenics and "race science."[20]

Over the years, FAIR worked to pass anti-immigrant bills at the federal and state levels. One of its lobbyists coauthored Proposition 187, the 1994 California ballot initiative that imposed harsh and punitive measures against undocumented immigrants, such as barring undocumented children from attending public schools.[21] Men and women with ties to white supremacist groups continue to occupy some of FAIR's top positions, and the organization's influence extends into Congress.[22] For example, Representative Brian Bilbray, the current chair of the House Immigration Reform Caucus, is a former FAIR lobbyist.[23]

THE GREENING OF HATE

Anti-immigrant activists assert that immigration-fueled population growth is a major cause of environmental degradation in the United States. FAIR and other immigration restrictionists promote three main arguments when they make this link. First, they claim that there is an ecological equilibrium between population and the environment that is being pushed to its limits by immigrants and their offspring.[24] Although "equilibrium theory" was the dominant school of thought in American ecological sciences at the beginning of the twentieth century, ecologists challenged its underlying assumptions in the 1970s, and it soon fell out of favor.[25] The dominant theory today is that the environment never reaches a stable point.[26] Rather, it is fluid and dynamic, without boundaries, and constantly in flux.[27]

Second, immigration control proponents assert that Americans consume more per capita than people in other countries, and when immigrants enter the United States, they adopt similar consumption patterns.[28] As one such writer explained, "grain and legume eaters become meat eaters, walkers or bus riders become car drivers, and users of one gallon of water daily consume fifty here."[29] Proponents of this theory go on to blame immigrants for causing other environmental consequences of overconsumption, including urban sprawl and traffic jams.[30] Yet, this line of reasoning erroneously assumes that immigrants, who are disproportionately poor and without access to private transportation, have the same consumption patterns and environmental impact as their wealthier U.S. citizen counterparts.[31] Moreover, it shifts focus away from Americans' ecologically damaging systems of production and consumption. These arguments simply use immigrants as convenient scapegoats for unsustainable systems they did not create.

Population control advocates also contend that immigrants create a drain on public resources, degrading the "social environment."[32] This argument is based on two long-standing myths: first, that immigrants, particularly undocumented immigrants, do not pay taxes, and second, that immigrants overuse social service programs.[33] Both myths support the anti-immigrant movement's economic narrative: that immigrants burden U.S. citizen taxpayers, who must "pick up the tab" and spend their hard-earned money on undeserving "foreigners."

However, numerous studies and social science research analyzing the economic contributions of immigrants show quite the opposite. Immigrants, both documented and undocumented, represent an important base of taxpayers at the federal and state levels.[34] In fact, between one-half and three-quarters of the undocumented immigrant population pays federal and state income taxes, Social Security taxes, and Medicare taxes.[35] The office of the inspector general of the Social Security Administration has even noted that undocumented immigrants "account for a major portion" of the billions of dollars paid into the Social Security system, which are benefits those immigrants cannot collect while undocumented.[36]

At the same time, a host of legislative and administrative changes have rendered most immigrant women ineligible for public benefits.[37] Notably, the Personal Responsibility and Work Opportunity Reconciliation Act of 1996 (PRWORA) barred the use of federally funded Medicaid for resident immigrants who have lived in the United States for less than five years. Medicaid coverage has declined by half among all immigrant women during the years that this policy has been in effect, even among longtime resident immigrant women who should not have been affected by the change—largely because of displaced fears of deportation.[38] This policy deters immigrant women's access to vital preventative and primary reproductive health services such as prenatal care.

Despite evidence to the contrary, the anti-immigration movement today continues to espouse pseudoscientific theories about the "over-breeding" of immigrants and their impact on the environment. Fortunately, many mainstream environmental groups that supported immigration control efforts in the past have rejected this approach. Of particular note is the dramatic face-off that occurred between anti-immigrant population control activists and the mainstream environmental movement during a 1998 referendum vote among Sierra Club members. Members were asked to vote on a mail-in ballot initiative to reverse a 1996 board decision to "take no position on immigration levels or on policies governing immigration into the United States."[39] Those who pressed for reversal argued that the Sierra Club needed a "comprehensive population policy for the United States that continues to advocate an end to U.S. population growth . . . through reduction in net immigration."[40]

Proponents of the referendum joined forces with FAIR and other anti-immigrant groups such as Negative Population Growth and

Population Environment Balance.[41] Critics of the initiative sought to affirm the board position and urged the Sierra Club to "address the root causes of global population problems through its existing comprehensive approach."[42] Those who opposed the referendum also recognized the inherent racism of immigrant population control arguments. As Sierra Club Executive Director Carl Pope explained, the referendum "is offensive to people of color. . . . It is seen by people in the immigrant communities as saying: You are a form of pollution."[43]

The controversial referendum was defeated with 60 percent of the vote, but the debate revealed a deep schism within the environmental movement.[44] Immigrant scapegoating has only intensified in recent years, and supporters of immigration control continue to use environmental messages to further their agenda. Said John Tanton, founder of FAIR, "The Sierra Club may not want to touch the immigration issue, but the immigration issue is going to touch the Sierra Club."[45]

THE DRIVING FORCE OF NATIVISM IN THE ANTI-IMMIGRANT AND ANTICHOICE MOVEMENTS

In recent years, anti-immigrant policymakers introduced a new framework encapsulated by a signboard recently on display outside the New England Baptist Church: "If we didn't abort our children the United States wouldn't need to hire illegals to work."[46]

Given the long history of population control activism within the anti-immigration movement, blaming the nation's abortion laws appears to be a radical departure from the movement's traditional support for policies that help limit family size. However, the seemingly newfound alliance between two apparently antithetical movements is not so surprising once one recognizes their common goal: to maintain the population status quo and preserve the white American majority. In fact, an examination of the voting records of anti-immigrant lawmakers on issues relating to a woman's right to choose reveals that these lawmakers have long opposed pro-choice measures.[47]

Some politicians have become quite vocal about the supposed link between legal abortion and "illegal" immigration. For example, former Republican House Majority Leader Tom DeLay told attendees at a 2007 College Republicans gathering, "I contend [abortion] affects you in immigration. If we had those 40 million children that were killed over the last 30 years, we wouldn't need the illegal immigrants to fill the jobs that they are doing today."[48] Months earlier, former Democratic Senator turned Republican Zell Miller had also advanced this theory at an anti-choice fundraiser. He claimed that 45 million babies had been "killed" since *Roe v. Wade*, and "if those 45 million children had lived, today they would be defending our country, they would be filling our jobs, they would be paying into Social Security."[49]

It is clear from these statements that immigrants are not a desired demographic, even though, contrary to Senator Miller's claims, immigrants already serve in the military, work for U.S. businesses, and contribute to Social Security. Rather, this anti-immigrant, antichoice rationale refuses to acknowledge the critical contributions of immigrants and strips them of their American identity.

A DOUBLE STANDARD: THE ANTI-IMMIGRANT MOVEMENT AND THE PRONATALIST MOVEMENT

Concerns about unsustainable population growth cast an unfavorable light on Asian and Latina immigrant women, who have higher-than-average birthrates. Yet little is mentioned of the high birthrates of certain predominantly white religious groups, such as Mormons and evangelical Protestants. The Mormon birthrate is significantly higher than the national average; in 2006, Utah, which has a population that is over 70 percent Mormon and only 8 percent foreign born, reported the highest general birthrate in the country.[50]

Likewise, in recent years, white fundamentalist Protestants have quietly boosted their birthrates as part of a pronatalist effort called the "Quiverfull" movement, named after Psalm 127: "Like arrows in the hands of a warrior are sons born in one's youth. Blessed is the man whose quiver is full of them." Adherents eschew birth control and instead trust that God will bestow as many children as He chooses.[51] Quiverfull families often have six or more children, concentrate in counties that are nearly 100 percent white, and view parenthood as a calling.[52] They also believe they are fulfilling God's command to "be fruitful and multiply" and see their childbearing practices as part of the missionary effort to build an army of Christians for God.[53]

U.S. religious conservatives also see the Quiverfull movement as a solution to their concerns about immigration and "race suicide," and as a strategy for building up their political power. Many U.S. religious conservatives are acutely aware of the declining birthrates among Americans of European ancestry and warn of a coming "demographic winter."[54] When Dr. John Wilke, founder of the National Right to Life

Committee, testified as a medical witness for the 2005 Report of the South Dakota Taskforce to Study Abortion, he stated the following:

> Muslim countries forbid abortion. Furthermore they have large families. Germany's birth rate is 1.2. . . . That is the Aryan Germans. What is happening? They're importing Turkish workers who do all of the more menial labor and right now there are over 1500 mosques in Germany. The Muslim people in Germany have an average of four children. The Germans are having one. So it's only a question of so many years and what do you think Germany is going to be? It's going to be a Muslim country.[55]

Although Dr. Wilke's reference to Germany's rising Muslim population may seem out of context, his message is clear: The United States must restrict immigration and abortion or risk losing its white Christian majority. Other immigration opponents have proposed more blatantly xenophobic measures. In response to a *Washington Post* article about the projected population growth of Latinos in the United States, John Gibson, cohost of a nightly Fox News show, said, "By far, the greatest number [of children under five] are Hispanic. You know what that means? Twenty-five years, and the majority of the population is Hispanic."[56] Gibson, a Caucasian male, went on to say, "We need more babies."[57] It's clear from his use of *we* that he was speaking to nonimmigrant audience members of European ancestry, similar to himself.[58]

The Quiverfull movement is as much an electoral strategy as it is a religious one; there is a correlation between communities with high white fertility rates and those with conservative voters. In 2004, George Bush carried the nineteen states with the highest birthrates, while John Kerry took the sixteen states with the lowest rates.[59]

CONCLUSION

The population-environment connection is not a new one. However, it has reemerged in recent years as demographers continue to report shifts in the racial composition of the U.S. population, mainly due to immigration. Accordingly, anti-immigrant strategists are exploiting anxieties about the country's demographic future and using them as entry points for promoting their nativist messages to a broad audience of welfare restrictionists, environmentalists, and more recently, the faith-based pronatalist movement.

Thus, immigrant women's bodies have become the economic, demographic, and political battleground for America's future. Immigrant women are accused of overusing public benefits and burdening the environment with their larger family sizes and assumed adoption of American consumption patterns. Such population alarmism drives policymaking that seeks to reduce the birthrates of immigrant women of color and increase the birthrates of "native" white women and, in effect, restricts the reproductive rights of all women in the United States.

> We must increase access to quality reproductive health services for all women regardless of immigration status, and commit to hold all Americans—immigrants, corporations, and our government—accountable for reducing consumption levels.

A possible alternative to the current anti-immigrant rhetoric is one that many in the environmental movement, including the Sierra Club and Population Connection, have already embraced: We must engage in a unified strategy to increase access to quality reproductive health services for all women regardless of immigration status, and we must simultaneously commit to hold all Americans—immigrants, corporations, and our government—accountable for reducing consumption levels.[60] Environmental sustainability does not require a reduction in immigrant births or a rollback of public benefits. Rather, the United States must support strategies that promote the basic humanitarian goals of livable wages and access to affordable health care—including the full spectrum of reproductive health services such as prenatal health care and contraceptive equity. If we do not pursue this path, then all women—citizen and immigrant—will once again lack the ability to make autonomous life choices about their reproductive fates.

REFERENCES

1. Stobbe, Mike, 2008, Against the trend, U.S. births way up, Associated Press, Jan. 16, available at http://www.boston.com/news/nation/articles/2008/01/16/against_the_trend_us_births_way_up/.

2. Ibid.

3. Ibid. Experts also attributed the birthrate increase to declines in contraceptive use and access to abortion.

4. Ibid.

5. In 2006, Latinas had the highest fertility rate, at 2.96 per woman, followed by 2.11 for black women and 1.86 for whites. Leland, John, 2008, From the housing market to the maternity ward, *New York Times*, Feb. 1, p. A17. Asian and Pacific Islander (API) women also have higher than average birthrates, and the Asian population is expected to grow at almost the same rate as the Latino population. Passel, Jeffrey S., and D'Vera Cohn, 2008, Pew

Research Ctr., U.S. Population Projections: 2005–2050, p. 22. Although not all Latinos and APIs are noncitizens, 68 percent of the Latino population and 88 percent of the API population in 2005 were either foreign-born or U.S.-born children of immigrant parents. Passel, Jeffrey S., and D'Vera Cohn, 2008, p. 15, 17.

6. Kolankiewicz, Leon, and Steven A. Camarota, 2008, *Immigration to the United States and World-Wide Greenhouse Gas Emissions*, Center for Immigration Studies, Washington, DC.

7. The advertisement can be viewed online at http://www.splcenter.org/ images/dynamic/intel/fair_traffic062708.jpg.

8. Earth First! slogan used in the 1980s. Earth First! is a radical environmental organization formed in 1979 that promotes direct action tactics to prevent environmental destruction. See Earth First! home page, http://www .earthfirst.org/.

9. See Ross, Loretta J., 1993, African American women and abortion: 1800–1970, p. 141, 149, *Theorizing Black Feminisms: The Visionary Pragmatism of Black Women*, Stanlie M. James and Abena P. A. Busia, eds., (quoting Betsy Hartmann, 1987, *Reproductive Rights and Wrongs: The Global Politics of Population Control and Contraceptive Choice*, Harper & Row, New York).

10. Ibid.

11. See Silber, Rachel, 1997, Eugenics, family and immigration law in the 1920s, *Georgetown Immigration Law Journal* 11: 859.

12. See Silber, Rachel, 1997, Eugenics at p. 869–870.

13. Reich, Peter L., 1995, Environmental metaphor in the alien benefits debate, *UCLA Law Review* 42: 1577, 1579.

14. Reich, Peter L., 1995, Environmental metaphor at p. 1580–1581.

15. Ehrlich, Paul R, 1968, *The Population Bomb*.

16. Abrams, Paula, 1997, Population control and sustainability: It's the same old song but with a different meaning, *Environmental Law* 27: 1111, 1116.

17. Reich, Peter L., 1995, Environmental metaphor at p. 1579.

18. Ibid.

19. In 1994, the organization abandoned its Malthusian approach to immigration as a strategy for achieving environmental preservation. Instead, it seeks to address the "push" factors of migration and "call[s]on the United States to focus its foreign aid on population, environmental, social, education, and sustainable development programs." In 2002, Zero Population Growth changed its name to Population Connection. See Population Connection, Statement of Policy, http://www.populationconnection.org/index .php? (follow "About Us" hyperlink; then follow "Complete Statement of Policy" hyperlink).

20. Ibid.

21. See The Building Democracy Initiative, Center for New Community, Nativism in the House: A Report on the House Immigration Reform Caucus 2, 2007. A federal district court struck down Proposition 187's most punitive provisions because they violated Fourteenth Amendment protections, and Proposition 187 was later preempted by federal law after passage of the 1996 welfare and immigration reform laws.

22. See Beirich, Heidi, 2007, The Teflon nativists, *The Intelligence Report* (Southern Poverty Law Center, Montgomery, AL), winter, available at http://www.splcenter.org/intel/intelreport/article.jsp?aid=846.

23. The Building Democracy Initiative, Center for New Community, Nativism in the House: A Report on the House Immigration Reform Caucus 2, 2007, at p. 6.

24. Reich, Peter L., 1995, Environmental metaphor at p. 1580.

25. See Reich, Peter L., 1995, Environmental metaphor at p. 1584–1588.

26. See Reich, Peter L., 1995, Environmental metaphor at p. 1586; see also Slifer, Diane L., 2000, Growing environmental concerns: Is population control the answer? 11 *Villanova Environmental Law Journal* 111, 115, 151–152.

27. Reich, Peter L., 1995, Environmental metaphor at p. 1586.

28. Reich, Peter L., 1995, Environmental metaphor at p. 1581.

29. Reich, Peter L., 1995, Environmental metaphor at p. 1581, quoting David Simcox, 1992, The environmental risks of mass immigration, *Immigration Review* 1: 1.

30. See Slifer, p. 135.

31. See Federation for American Immigration Reform, 2002, *How Immigration Hastens Destruction of the Environment*, http://www.fairus.org/site/PageServer?pagename=iic_immigrationissuecentersa45e (stating that immigrants "do not maintain the old lifestyle of their home country" and therefore "become greater consumers and damagers of natural resources").

32. Czerwonka, Scott, 1992, Population-Environment Balance: Why Excess Immigration Damages the Environment, available at http://www.balance.org/articles/whyexcess.html.

33. See Reich, Peter L., 1995, Environmental metaphor at p. 1581.

34. See National Immigration Forum, 2003, *Top Ten Immigration Facts and Myths*, available at http://www.immigrationforum.org/documents/The Journey/MythsandFacts.pdf. See also Porter, Eduardo, 2005, Illegal immigrants are bolstering Social Security with billions, *New York Times*, April 5, p. A1.

35. Immigration Policy Center, 2007, *Undocumented Immigrants as Taxpayers*, available at http://www.immigrationpolicy.org (follow "Research and Publications" hyperlink; then follow "Immigration Fact Check" hyperlink; then follow "View Older Items" hyperlink; then follow "Undocumented Immigrants as Taxpayers" hyperlink).

36. Immigration Policy Center, 2007, quoting Office of the Inspector General, Social Security Administration, Rep. No. A-08-99-41004, Obstacles to Reducing Social Security Number Misuse in the Agriculture Industry 12 (2001).

37. Costich, Julia Field, 2002, Legislating a public health nightmare: The anti-immigrant provisions of the "contract with America" Congress, 90 *Kentucky Law Journal* 1043, 1047–1048, p. 1063.

38. Gold, Rachel Benson, 2003, Immigrants and Medicaid after welfare reform, *The Guttmacher Report on Public Policy* 6: 6–9, 7 (May), available at http://www.guttmacher.org/pubs/tgr/06/2/gr060206.html (last viewed June 17, 2008).

39. Crass, Chris, 2004, Controlling Gendered Immigrants and Racialized Populations: Overpopulation, Immigration and Environmental Sustainability, Alternative Media Project, Dec. 24, http://www.infoshop.org/texts/immigration.html.

40. Ibid.

41. Ibid.

42. Ibid.

43. Ibid.

44. Ibid.

45. Ibid.

46. Picture accompanying Eleanor J. Bader, Male Victims of Abortion, *The Public Eye*, Fall 2007, p. 1.

47. See The Building Democracy Initiative, Center for New Community, Nativism in the House: A Report on the House Immigration Reform Caucus 2, 2007, p. 9 (comparing the immigration and pro-choice voting records of members of the House Immigration Reform Caucus).

48. Roston, Michael, 2007, Tom DeLay tells college republicans that abortion, illegal immigration are linked, *The Raw Story* (July 17), available at http://rawstory.com/news/2007/Tom_DeLay_tells_College_Republicans_that_0718.html.

49. Ibid.

50. National Center for Health Statistics, 2007, *National Vital Statistics Reports* 56(7), Table 6; Migration Policy Institute, 2006, American Community Survey and Census Data on the Foreign Born by State, available at http://www.migrationinformation.org/datahub/acscensus.cfm#.

51. See Joyce, Kathryn, 2006, Arrows for the war, *The Nation*, November 27.

52. See Brooks, David, 2004, Op-Ed, The new red-diaper babies, *New York Times*, Dec. 7, p. A27.

53. Genesis 1:22, 9:7; see Joyce, Kathryn, 2006, Arrows for the war, *The Nation*, November 27, p. 126.

54. See Joyce, Kathryn, 2008, Missing: The "right" babies, *The Nation*, Mar. 3.

55. E-mail from Lynn Paltrow, Executive Director, National Advocates for Pregnant Women, December 15, 2006 (on file with author).

56. The Big Story with John Gibson (Fox News television broadcast, May 11, 2006).

57. Ibid.

58. See Ibid.

59. Sailer, Steve, 2004, Baby gap: How birthrates color the electoral map, *The American Conservative*, December 20; see also Joyce, Kathryn, 2008, Missing: The "right" babies, *The Nation*, Mar. 3 (describing the political impact Quiverfull advocates hope to have, including attaining "both houses of Congress and the majority of state governors' mansions filled by Christians").

60. See Beck, Roy, and Leon Kolankiewicz, 2000, The environmental movement's retreat from advocating U.S. population stabilization (1970–1998): A first draft of history, *Journal of Policy History* 12(1), available at http://www.numbersusa.com/about/bk_retreat.html.

CHAPTER 29

....

Christian Perspectives on Population Issues

REV. DR. JAMES B. MARTIN-SCHRAMM

The Program of Action adopted at the 1994 United Nations International Conference on Population and Development in Cairo prompted debates in Christian communities about many issues related to global population growth. Some have debated the relationship of reproductive rights to responsibilities for the common good. Others have focused on the place of abortion services in family planning and reproductive health care, the role of sexuality within and outside the bonds of marriage, definitions of the family, and the rights of adolescents. This brief essay summarizes several Christian theological perspectives that provide a common foundation for ecumenical reflection about these controversial issues.[*]

A THEOLOGY OF LIFE

From the creation accounts in Genesis 1–2 to the "river of life" in Revelation 22, one of the most fundamental themes in the Bible is the af-

---

[*]This essay is a revised and updated version of Chapter 7 in my book *Population Perils and the Churches' Response* (Geneva: World Council of Churches, 1997).

Rev. Dr. James B. Martin-Schramm is Professor of Religion at Luther College.

firmation that God is a God of life. After the flood, the Creator makes a covenant not only with Noah and his descendants, but also with "every living creature that is with you . . . as many as come out of the ark" (Gen. 9:10b). In the Gospel of John, Jesus declares, "I came that they may have life, and have it abundantly" (Jn. 10:10). God's Spirit infuses all of creation so that even the rocks and trees proclaim God's praise in the psalms.

These texts and others emphasize that human beings are members of a larger household of life that God has created and continues to create and value. The redemption that Christians celebrate through the work of Jesus Christ includes not only human beings but also all of creation. God's scope of moral concern extends beyond humanity to all that God has made. Earth is the "habitat of the Holy Spirit."*

Blessed with the gift and awesome responsibility to serve as good stewards of God's creation, Christians must examine not only the impact of their numbers but also the impact of their lifestyles on the household of God. Confronted with God's challenge to "choose life" (Dt. 30:19), we must reflect on how human patterns of production, consumption, and reproduction affect not only the quality of human life but the welfare of all on the planet. If Christians are to be good stewards of God's creation, we must support population and development policies that are just for each and sustainable for all.

DOCTRINES OF CREATION AND REDEMPTION

As Christians contemplate the stewardship of human reproduction, the doctrines of creation and redemption remain crucial resources. In particular, we must grapple with the injunction in the first creation account to "be fruitful and multiply, fill the earth and subdue it" (Gen. 1:28). This pronatalist directive made sense in a creation myth generated within the context of the Babylonian exile and the decimation of the Jewish people. Unrestrained human reproduction today, however, threatens God's creation and the quality of life of all who live within it. It is more accurate to view Genesis 1:28 as a blessing rather than a commandment. God asks human beings to reach all the corners of the earth so that they can take up their task of responsible dominion and care for it. The injunction to "be fruitful and multiply" was never

---

*James Nash coins this phrase in *Loving Nature: Ecological Integrity and Christian Responsibility* (Nashville: Abingdon Press, 1991), p. 233, note 69.

intended to be followed blindly and without concern for its impact on the common good.

A related matter concerns the purposes of human sexuality. Just as most Christians no longer follow the ancient practice of polygamy, so too should we reject ancient but now incorrect assumptions about the process of procreation and the purpose of human sexuality. Many traditional patriarchal cultures believed that women were

> The injunction to "be fruitful and multiply" should be seen as a blessing rather than . . . a commandment.

the fertile soil within which men planted the seed of life. The ancient Hebrews, for example, believed that any discharge of sperm that could not lead to procreation was an abomination. Today, we know that the egg of a woman plays an active role in the process of conception. Moreover, most people of faith now believe that God intended sexuality not only for procreation but also for the experience of pleasure and spiritual and physical union. The gift of human sexuality and the capacities for human reproduction—like all of the other gifts of God— call for wise and careful stewardship.

Finally, another area that requires theological attention is the relationship of salvation to reproduction. Much of the drama surrounding the stories of the ancient Hebrew ancestors and the leaders that followed them revolved around whether there would be an heir to inherit the covenant God made with Abraham and Sarah (Gen. 13–17). For Christians, the relationship of human reproduction to earthly redemption is fundamentally changed by the covenant God establishes in Christ: Association with God's covenant comes through neither birth nor the mark of circumcision, but rather through baptism into Christ. As a result, there is little emphasis in the New Testament on the virtues of childbearing. In fact, infused with an apocalyptic expectation of Christ's imminent return, the apostle Paul commends a life of celibacy to those who feel they can manage it (I Cor. 7). While the value of human life and the gift of children are affirmed throughout the New Testament, it is clear that the link between salvation and reproduction has been broken.

IMAGO DEI, DIGNITY, AND JUSTICE

One of the richest but frequently misunderstood claims of the Judeo-Christian tradition is that human beings are made *imago dei*, in the

BOX 29.1

Faith, Feminism, and the Next Generation of Justice Movements

Kate Ott

Young people are overwhelmingly present in the reproductive justice and environmental justice movements. When I show up in either group, I do so because my faith informs and obliges me to work toward a common good for others and the Earth.

The next generation of young spiritual and religious progressives has embraced the work of liberation and feminist theology. We refuse to "check our religion at the door" in grassroots movements. And we embody a desire to broaden single-issue movements to include all forms of oppression and adapt a more inclusive framework for justice.

For generations, people in all faith traditions have committed themselves to justice work. It is within these faith traditions—not in spite of them—that many young people of faith work for justice, some seeking more liberating interpretations of doctrines and scriptures, some working in grassroots movements, and still others refusing to let religion be a scapegoat or divisive wedge in secular justice movements.

Many young people of faith resist the trend to "modify justice"—to divide social justice movements into components seeking economic, gender, reproductive, and environmental justice. In the reality of people's lives, these injustices are intimately linked. For example, those who suffer most from environmental degradation are also subject to racial, economic, and gender oppression. Poor women suffer the most when contraception, emergency contraception, and abortion services are not readily available. Reproductive and environmental justice are also economic and racial justice issues.

An "intersectional" approach, which mirrors the experiences of the current generation of young people, brings questions of access, distribution, and self-determination to bear on the interrelated issues of environmental, economic, racial, and sexual oppression, to name a few. As we move forward in our justice-seeking movements, we must continue to forge a common vision of justice that reflects and addresses the interrelated nature of oppressions. If not, we will continue to work for pieces of justice, and forgo the realization of a just society in all its wholeness.

....

Kate Ott is Associate Director of the Religious Institute on Sexual Morality, Justice and Healing.

image of God (Gen. 1:26). Throughout much of the history of the West, this text has been used to elevate the status of humanity and set it above the status of other creatures. When joined with a logic of domination, the concept of *imago dei* has served to undermine genuine stewardship and care for creation. Understood correctly, creation in the image of God confers to each human being a profound and inalienable dignity that sets humans apart for loving service to the world. Created in the image of God, each human being deserves to be born into a world that cherishes, defends, and preserves his or her human dignity.

> Population policies that violate human dignity—by treating women as objects rather than subjects or by employing coercion of any kind—should be opposed.

In the arena of population policy, this important doctrine should help Christians recognize that the impetus and result of these policies should be to bolster and defend human dignity. Population policies that violate human dignity—by, for example, treating women as objects rather than subjects or by employing coercion of any kind—should be opposed. Instead, population policies that respect human dignity will seek to empower women and men to exercise voluntary and responsible stewardship of their reproductive lives.

Imago dei underlies the fundamental Christian concern for justice, which has important ramifications for contemporary discussions of population policy. As we have seen, population policy since 1994 has emphasized gender equity and equality. The equality of women is an intrinsic moral and theological good. In addition, the empowerment of women and the removal of the impediments to their full participation in society serve to lower the rate of population growth and improve the quality of their lives and those of their families. Christians concerned about justice must defend the rights of women, including the developing concept of reproductive rights and universal access to reproductive and primary health care services.

In addition, as fellow members of the body of Christ, men and women are created for mutual service to each other and the rest of humanity. In response, Christians should strive to promote gender roles that contribute to the common good and allow women and men to make full use of the gifts God has given them. Men, as well as women, must share in responsibility for the stewardship of reproduction, the care of children, and the maintenance of a loving home.

THE GIFT OF CHILDREN

Even a cursory reading of the Hebrew Bible and the New Testament reveals that children play a central role in these religious narratives. For example, the baby Moses is placed in the reeds by his young mother and is then cared for by Pharaoh's daughter (Ex. 2:1–10). In Matthew's gospel, Mary and Joseph bundle up their newborn son, Jesus, and flee to Egypt to avoid the repression of Herod (Mt. 2:13). The fate of the people of God hinges on the welfare of these children.

Not surprisingly, the care of children serves as a litmus test for justice in the Bible. The ethical focus of the Torah is consistently upon the welfare of the powerless: children, the poor, the orphan, and the widow (Dt. 24:14–22). Jesus reflects this concern with the special attention he shows to children. In response to the disciple's question, "Who is the greatest in the kingdom of heaven," Jesus calls a child and commends the humility and trust of children as a model of Christian faith; Jesus says, "Whoever welcomes one such child in my name welcomes me" (Mt. 18:5).

Tragically, Christian communities must confess that they have not consistently come to the defense of children generally, and girls in particular. It is a travesty of justice that children suffer from disease, hunger, abuse, illiteracy, and discrimination. Girls, especially, are subject to malnutrition, ill health, sexual assault, and social neglect around the world. This dereliction of moral obligation is deeply rooted in Christianity's failure to promote and defend the dignity and equality of women in words, deeds, and teachings. The forces of patriarchy and male domination have been and remain powerful forces in Christian communities.

Christians wishing to remedy these injustices should wholeheartedly support all efforts to close the gender gap and provide primary education for all children by 2015. This goal is not only a moral duty but also wise public policy. The education of girls demonstrably lowers fertility, increases economic productivity, and promotes the participation of women in all aspects of society. Christians should also support all efforts to protect children from physical and sexual abuse, including the exceedingly dangerous and painful practice of female genital mutilation. Children also have a right to understand the reproductive and sexual dimensions of their lives, and Christian communities have a responsibility to support this right through educational efforts in schools and churches. Every effort must be made to protect girls from all forms of discrimination and to provide all children with adequate health, nutrition, and education.

In a world driven by the values of wealth and power, the moral challenges posed by global population growth call Christians—and other people of faith—to honor their relationship with God by demonstrating a better stewardship of the Earth and compassion for their fellow human beings. Actions taken over the next decade will shape profoundly the size and welfare of not only the human population but also the planet as a whole. As Christians respond to these challenges and work together with others in an attitude of mutual respect and trust, they should be inspired by the vision of the prophet Isaiah:

> No more shall there be in [the world] an infant that lives but a few days, or an old person who does not live out a lifetime. . . . They shall build houses and inhabit them; they shall plant vineyards and eat their fruit. They shall not build and another inhabit; they shall not plant and another eat; for like the days of a tree shall the days of my people be, and my chosen shall long enjoy the work of their hands. They shall not labor in vain or bear children in calamity; for they shall be offspring blessed by the LORD— and their descendants as well. (Is. 65:20–23).

In a world driven by the values of wealth and power, the moral challenges posed by global population growth call Christians—and other people of faith— to honor their relationship with God by demonstrating a better stewardship of the Earth and compassion for their fellow human beings.

CHAPTER 30

··········

Ecomorality

Toward an Ethic of Sustainability

URSULA GOODENOUGH

MANY NARRATIVES

It is a privilege to contribute to this excellent book. The authors offer various, and often variant, perspectives on our current environmental dilemma, in general, and the influence of human population pressures, in particular. Each perspective, by definition, reflects the nationality, education, profession, experience, and ethical framework of each author, providing a buffet of rich insights from which to choose. The advantage of a buffet is that diners are able to select dishes that they like and pass over dishes that they dislike, choices also reflecting nationality, education, experience, and so on. So the question arises: What is needed if humanity is to share a common meal, share a common set of rich insights on how best to move forward and then act on them?

This question, needless to say, is anything but rhetorical. No matter which scenarios for a better planetary future are being considered, all are going to require collective engagement, motivation, and sacrifice on a global scale. So, restating the question, What kinds of human institutions are up to the task of framing such collective commitment?

Looking to human history for possible answers, religion emerges as

Ursula Goodenough is a Professor of Biology at Washington University in St. Louis.

an obvious candidate. To be sure, religions are as vulnerable to schisms and power moves as other institutions (political parties, trade alliances, unions). But they are far more likely to reconstitute than to disintegrate, and some have survived for millennia, and all have had palpable success in orienting their adherents toward common perspectives. Why is this?

A common denominator of religious traditions is that each articulates a core narrative, a mythos, which sets forth How Things Are and Which Things Matter.[1] To share a religion is to share common premises and values, and even after these undergo the occasional revision or reconfiguration, persons in a tradition derive meaning and purpose from their narrative core. Indeed, by this criterion, the founding documents of the United States can also be said to qualify as religious statements, articulating large Truths about humankind and Rights that follow therefrom. Religion is about educating human emotions to generate both personal wholeness and social coherence.[2]

There are heartening signs that some of the existing religious denominations are moving in an "ecomoral" direction, finding passages in their texts and traditions that proclaim the imperative of earthly stewardship. But there are inherent constraints on this trajectory. First, the narratives of most of the traditions, even after lifting up said passages, remain deeply anthropocentric, whereas a global perspective insists on a far broader horizon of engagement. Moreover, the narratives were crafted prior to contemporary understandings of How Things Are derived via scientific inquiry, generating all-too-familiar conflicts about which accounts are "true." And finally, even though some traditions have succeeded in spreading geographically well beyond their regions of origin, they generally retain ethnic and historical signatures that confound commonality, not to mention the many deep interreligious animosities that frustrate efforts at reconciliation. Given these constraints, it could be argued that the existing religious traditions are in fact poorly equipped to navigate global ecomoral consensus.

EVERYBODY'S STORY

I am optimistic that the existing traditions can and will make important contributions along this axis. That said, another path beckons, namely, to articulate a new religious tradition that is forged in our current context. Let me be clear at the outset: I am in no way suggesting that this new tradition would compete with, let alone "bury," the existing traditions. Rather, it would coexist with them, informing and guiding our global concerns while fully respecting, indeed celebrating and often appropriating, the many deep truths to be found in traditional texts and practice.

In recent times, a new core narrative has been crafted that tells of How Things Are for all planetary beings—a narrative, I will later argue, that is also rich in resources for articulating Which Things Matter on a planetary scale. It's a narrative that most persons either understand poorly or misunderstand or actively reject—daunting odds indeed. Yet I would submit that it is the lone candidate on offer for what could be called Everybody's Story:[3] It derives from scientific inquiry undertaken and verified from all corners of the globe, and it applies to every being on the planet. It has been called various names: the Universe Story,[4] the New Story,[5] the Epic of Evolution.[3, 6, 7] Here we can call it the History of Nature.

In brief, the History of Nature tells us that our observable universe came into being as a singularity some 13.7 billion years ago, generating hydrogen and helium atoms, and has expanded since then to an unimaginable size; that it contains hundreds of billions of galaxies, each with hundreds of billions of stars; that most kinds of atoms—like carbon, nitrogen, and oxygen—were forged in the depths of stars; that our star and its planets came into being 4.567 billion years ago in our Milky Way galaxy; that life on Earth originated from nonlife perhaps more than 3.5 billion years ago; that the common ancestor to all modern creatures lived perhaps 3 billion years ago; that life then radiated into three major domains—the archaea, the bacteria, and the eukaryotes; that the first animals (members of the Eukaryota) appeared about 1 billion years ago; that the hominid line diverged from other apes about 6 million years ago; and that modern humans showed up about 150,000 years ago. The evolution of life occurred on a planet whose continents are in continuous motion and in the context of major fluctuations in climate and atmospheric composition.

The biological chapters of this history are illustrated in Figures 30.1 through 30.3. Figure 30.1 shows the three great radiations, the ten major groupings of eukaryotes, and some of the numerous, possibly countless, lineages of bacteria and archaea. Ready for the *Where's Wallace* game? Go to Figure 30.1 and find our lineage. Got it? Yes, there in the Opisthokonts at about 8 o'clock. Animals. Right in there with the fungi.

And what is the arrow pointing to? The common ancestor to all modern life. How do we know that such an ancestor existed? Because all the creatures on this planet share striking commonalities even as

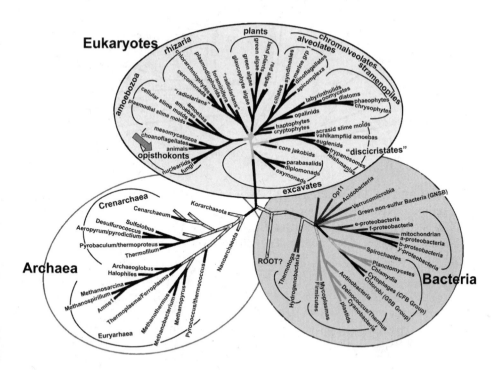

FIGURE 30.1 The three great radiations, the 10 major groupings of eukaryotes, and some of the innumerable lineages of bacteria and archaea. *Courtesy of Sandra Baldauf.*

they have diverged for over 3 billion years: All have genomes inscribed in DNA that use the same triplet genetic code; all share metabolic pathways and patterns of protein synthesis; all use many of the same protein building blocks; the list is very long. Hence, at some point some 3 billion years ago, a group of single-celled creatures must have come up with a core set of arrangements for generating cells that was then utilized by all subsequent evolutionary lines as they moved into every corner and crevice of the planet, novelty upon novelty, extinction upon extinction, generating the web we now behold and within which we are embedded.

Figure 30.2 illustrates the relationships between some of the major groups of animals. Note the major bifurcation between invertebrates and vertebrates, the first mammals at approximately 300 million years ago, and then, virtually in the present given the figure's timescale, the divergence of chimpanzees and humans.

Expanding this timescale in Figure 30.3, we see our immediate

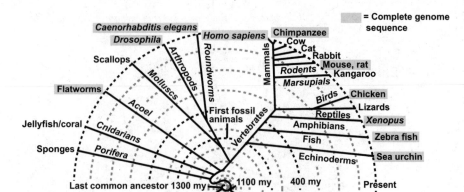

FIGURE 30.2 The relationships between some of the major groups of animals. Note: my = million years ago.

Source: Thomas D. Pollard, William C. Earnshaw, and Jennifer Lippincott-Schwartz, 2007, *Cell Biology*, Saunders, Philadelphia PA.

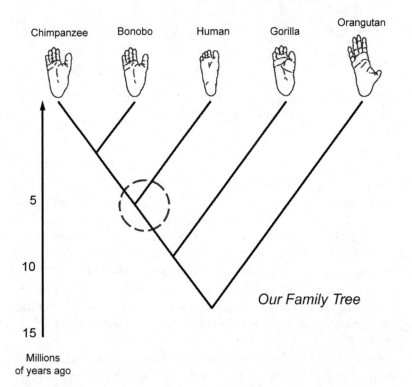

FIGURE 30.3 The human family tree, showing our immediate ancestors and relatives.

Source: Dale Peterson and Richard Wrangham, 1997, *Demonic Males: Apes and the Origins of Human Violence*, Mariner Books, New York.

family tree, with the common ancestor to humans and chimpanzees at 6 million years ago (dashed circle) and a more recent chimpanzee-bonobo divergence 2 million years ago.

The hominids originated in Africa (we are all Africans), and at least one species (*Homo erectus*) in our lineage migrated extensively about a million years ago before going extinct. Some 120,000 years ago, a second wave of migration resulted in modern *Homo sapiens* settling throughout the planet. By 10,000 years ago, agriculture had been invented, then towns—and now, there are 6.8 billion of us. Along the way, we came up with our particular form of symbolic language-based culture and our attendant ability to discover, and continue to explore, the History of Nature.

So the History of Nature tells us How Things Came to Be and How Things Are, offering a mind-boggling narrative that dwarfs any religious mythos in sweep and in time. Here are some obvious take-homes:

- We belong in the universe. Our very atoms are recycled star stuff.
- We earthly creatures are all connected. Not just in food chains and ecosystems. We share a common ancestor. We are connected all the way down and all the way back.
- All those ancestors, those intermediary forms, those forks in the road that no longer exist, all of them made our current existence possible. In a very real sense, they died for us.
- Our animal "twig" (Figure 30.1) is but one of countless others, and while we consider ourselves to be dominant, the microbes—most of the other twigs—really run the show. Take away the bacteria, for example, and the biosphere would shut down in a few months. Take away the marine phytoplankton, and we lose 80 percent of our oxygen supply.
- Our hominid twig ventured forth but a moment ago, the human twig less than that. We now occupy not only our ecosystem of origin (the African savanna) but virtually every other planetary ecosystem, usually with profound consequences for the ecosystems that we "develop" to support us.

THE DYNAMICS OF EVOLUTION

A story always includes a chronology—what happened next?—but a good story also weaves in a dynamic: What is driving the plotline? Why did these things happen? In theistic narratives, this part is straightforward: Deities are causal; their motives are integral to outcomes.

Although the History of Nature doesn't rule out deities, the story can be told without invoking them, a major plus for its "universality," given the lack of any consensus, despite millennia of debate, as to both the existence and the nature of deities.

That said, what *is* driving the plotline?

For the life chapters, Darwin's core insight—descent with modification and natural selection—continues to hold, amplified and deepened by 150 years of research. Yet from the outset, and continuing to this day, Darwinian dynamics tend to leave most people cold, eliciting such responses as that they are too mechanical, materialistic, reductionistic, depressing, meaningless, even nihilistic—not promising responses for a unifying narrative. Is there a way around this?

There is and, although anything like a full explanation would require a chapter of its own,[8] I can offer a three-paragraph outline of two key concepts.

The first is called *emergence*, defined as the generation of something-more from nothing-but as a consequence of relationships. Emergent properties arise in both nonlife and life. A classic example in nonlife is the surface tension of liquid water. Surface tension is not a property of a single water molecule. It arises ("pops through") as the consequence of water molecules in relationship with one another; water molecules (nothing-buts) generate surface tension (something more) when they interact under particular conditions. Other nonlife something-mores— the malleability of metals, the hardness of minerals—are also the outcome of relationships between nothing-buts (particular sets of atoms). In life, the nothing-buts are cellular components—DNA, proteins, lipids, and so on—which are, in turn, nothing but atoms in relationship. Relationships between cellular components in turn generate the something-mores called biological traits, like motility, metabolism, photosynthesis, vision, life cycles, and minds. Hence emergence is a layered concept: A something-more on one scale (the protein insulin, emergent from atoms) can serve as a nothing-but for a something-more at a different scale (the trait we call blood-glucose regulation).

A key insight from our understandings of molecular evolution is that novel biological traits emerge from previously novel traits, with slight variations in relationships generating new fodder for natural selection to scrutinize. The fins of a fish, that generate the swimming trait, evolved into the limbs of amphibians, that generate the walking trait. The fact that humans and mice share virtually all their genes is bewildering until it is realized that they often express these genes at different times and in different contexts, permitting new kinds of protein relationships and hence new kinds of traits. The brain of the common ancestor to chimps and humans was generated by one set of instructions; subsequent variation in the expression patterns of the same sets of genes generated chimp-style brains in one lineage and human-style brains in the other, each adapted to relevant salient circumstances.

The second key concept is called *teleodynamics.* Emergent properties in nonlife—surface tension, malleability—arise as the consequence of thermodynamics (energy considerations) and morphodynamics (shape considerations). These dynamics underlie biological traits as well, but traits also have a function, a purpose (telos)—they are *for* something, they operate *so that* some outcome will result. Motility and vision are *for* acquiring food and avoiding predators, and all forms of life are intensely aware of features of their ecosystems that are meaningful to them. To summarize teleodynamics with a quip—at the onset of life, matter acquired the property of mattering.

> At the onset of life, matter acquired the property of mattering.

Again we can list some take-homes:

- Nature is inherently creative. New relationships generate novel outcomes, and when these relationships are instructed and remembered by genomes, they generate biological traits. The History of Nature is the history of relationship and of creativity.

- Traits vary not only in degree (black moths versus white moths, narrow beaks versus stout beaks). Traits also serve as the nothing-buts for the emergence of the next suite of traits. Tweak the timing of gene expression in a previous kind of embryo, and you get a new kind of embryo with the potential for new adaptive possibilities.

- The teleodynamics of the evolutionary story introduce concepts like purpose and meaning, concepts infused with warmth and excitement. The plot becomes rich, inventive, and complex.

THE EMERGENT HUMAN

All evolutionary transitions entail emergence, and the hominid-to-human transition is no exception. The two salient human traits are our distinctive form of language[9] and our language-based cultures. Language, in turn, requires brains capable of language learning. Hence human evolution has entailed a three-way coevolution of language, brains, and culture, with culture serving as the repository of wisdom (and foolishness) acquired by language-based minds. Like all traits, these have emerged from previous traits—witness the examples of cultural transmission of skills in nonhuman primates[10] and the capacity of bonobos to learn human language.[11] That said, our lineage obviously took these traits and ran with them.

And at some point along the way, something else popped through: our narrative selves, the "I" that wakes up in the morning and falls asleep at night and has a rich autobiography. While it may be the case that other animals have versions of such selves, they clearly are front and center in human versions of consciousness.

So here we are, new kids on the block, with traits that emerged from preexisting traits and yet seem so discontinuous, so unique, that most people still struggle to acknowledge our relationship to other apes, let alone to all the creatures shown in Figure 30.1. I admit full-throatedly that I am thrilled and profoundly grateful to be a human. I am continuously astonished that I experience human-style awareness and can revel in the arts, the sciences, and the wisdom traditions transmitted via culture. Fully as thrilling, though, is the sense of my immersion in the totality of it all, generating a deep commitment to the proposition that this vibrancy continue. Which takes us to Which Things Matter.

WHICH THINGS MATTER

Some say that the History of Nature is devoid of resources as to Which Things Matter, that it tells us facts but is devoid of values, that you can't get an "ought" from an "is." Religious narratives, it is correctly noted, weave the values right into the plotline, spelling out the thou-shalts and thou-shalt-nots as they go; moreover, the human is the focus of these accounts, not some almost-an-afterthought in the grand scheme of things. How could the History of Nature begin to ground a common set of human insights on how best to move forward?

From my perspective, this is a no-brainer. Once the story is grasped and taken into one's mind and heart, once one absorbs the iconic satellite images of Earth and experiences both its wonder and its fragility, there follows an outpouring of reverence and care, foundational components of a moral sensibility. Which things matter? The whole thing matters! When I give lectures to diverse groups and project Figures 30.1 to 30.3 on the screen, there are two common responses: First, incredulity—"You mean we're just that one twig?" and then, deep testimonials to affiliation—"I get it. We aren't 'other' after all. We're part of an enormous family."

A dominant trait of the primate lineage—monkeys and apes—is its deep sociality. Relationships—friendships, hierarchies, families, troops—are critical to who we primates are. Humans canonize important ancestors, identify with our ethnic roots, and take DNA tests to ascertain our peoples of origin. We may experience our lives through our narrative selves, but we are fully dependent on and immersed in our social contexts.

> Which things matter? The whole thing matters! We aren't "other" after all. We're part of an enormous family.

And now we are called to recognize our deep affiliation with, as well as dependence on, all the creatures and habitats of the planet. This realization has been translated as an invocation to be *stewards*, to take responsibility for reversing the many alarming trajectories the planet is taking. But "stewardship" puts us back into the mind frame of humans-as-other, hence I better resonate with the invocation that we be *participants* in establishing right relations with our fellow beings and our earthly habitat.

Participation is at the core of the religious impulse. Those who share a story, share a common understanding of How Things Are, are moved to take part in its preservation, its continuation, and yes, its flourishing. They are motivated to put aside their own interests in the name of sustaining something that is much larger.

A word widely used in religious contexts is *faith*. Faith has everything to do with the conviction that one's story is true and that one's moral tenets are sound. At such time that the History of Nature becomes known, and celebrated, as the core of everybody's faith, then ecomorality, and its attendant commitment to achieve sustainability, will emerge as a guiding planetary principle.

REFERENCES

1. Rue, Loyal D., 1989, *Amythia: Crisis in the Natural History of Western Culture*, University of Alabama Press, Tuscaloosa.

2. Rue, Loyal D., 2006, *Religion Is Not About God: How Spiritual Traditions Nurture Our Biological Nature and What to Expect When They Fail*, Rutgers University Press, Rutgers, NJ.

3. Rue, Loyal D., 1999, *Everybody's Story: Wising Up to the Epic of Evolution*, SUNY Press, Albany, NY.

4. Swimme, Brian, 1994, *The Universe Story: From the Primordial Flaring Forth to the Ecozoic Era—A Celebration of the Unfolding of the Cosmos*, HarperOne, New York.

5. Berry, Thomas, 2006, *The Dream of the Earth*, second edition, Sierra Club Books, San Francisco, CA.

6. Wilson, Edward O., 1978, *On Human Nature*, Harvard University Press, Cambridge, MA.

7. Goodenough, Ursula, 1998, *The Sacred Depths of Nature*, Oxford University Press, New York.

8. Goodenough, Ursula, and Terrence W. Deacon, 2007, The Sacred Emergence of Nature, p. 854–871, *Oxford Handbook of Religion and Science*, Philip Clayton, ed., Oxford University Press, New York.

9. Deacon, Terrence W., 1998, *The Symbolic Species: The Co-Evolution of Language and the Brain*, W. W. Norton, New York.

10. De Waal, Frans, 2006, *Our Inner Ape: A Leading Primatologist Explains Why We Are Who We Are*, Riverhead Books, New York.

11. Savage-Rumbaugh, Sue, Stuart G. Shanker, and J. Taylor Talbot, 2001, *Apes, Language, and the Human Mind*, Oxford University Press, New York.

CHAPTER 31

············

Reconciling Differences

Population, Reproductive Rights, and the Environment

FRANCES KISSLING

In the wake of significant population growth following World War II, the modern population movement was born. From the beginning it was mired in controversy. Demographers and economists debated the effects of population size and growth on public health, economies, national and international security, and the environment. Would rapid population growth outrun the supply of food and other resources, causing massive human suffering and political unrest? Or would human creativity and scientific progress transcend the perceived limits of growth? Are smaller families a necessary precondition—or the inevitable result—of economic growth? And as modern contraception enabled couples to control their fertility, new questions arose about their effect on family life and religious and public values. Notwithstanding the complexity of the issues, an international consensus in favor of population assistance emerged, although it has wavered many times over the last fifty years in the face of political, cultural, and scientific change and criticism.

The complexity and controversy of population issues should not be surprising. After all, how we think and talk about population size and growth is related to fertility and therefore to sexuality and reproduc-

············

Frances Kissling is a Visiting Scholar at the Center for Bioethics, University of Pennsylvania.

tion, so public policies designed to influence or control fertility seek to affect the most intimate, complex, and value-laden aspects of human relationships. Our beliefs about the personal meaning of bringing new life into the world are challenged by the political understanding of the public consequences of procreation. Moreover, the right to decide the number and spacing of one's children has only recently been accepted as a fundamental human right. Even more revolutionary is the idea that women, separate from spouses or partners, might have some right to decide about sexual intimacy, reproduction, and childbearing and rearing. Given the sensitivity of these issues, and the fragility of these hard-won rights, family planning providers and state actors must proceed with caution in attempts to influence sexual and reproductive behavior, no matter how positive they believe the public consequences may be.

> Family planning providers and state actors must proceed with caution in attempts to influence sexual and reproductive behavior, no matter how positive they believe the public consequences may be.

This is truly a new restriction on state power. Throughout most of history, male, state, or religious control of reproduction was taken for granted. Even the Roman Catholic Church, which severely limits the methods couples may use to control fertility, has a long history of seeking to influence demographic trends. For example, after Prussia defeated the French in 1871, Catholic moralists blamed the defeat on waning fertility and predicted the decline and fall of an unpopulated France. "You have rejected God and God has struck you," a Swiss cardinal told the French on Bastille Day in 1872.[1] The more recent case of Iran is instructive. During the Iran-Iraq War in the 1980s, the Iranian government discouraged family planning and rewarded birth. Following the war and in the face of economic pressure, not only were family planning clinics expanded but abortion was legalized. Current concern about a "birth dearth" in Europe has led to pronatalist tax policies in many countries.

Taking all this into account, the patriarchal and paternalistic thrust of the population movement of the 1950s and 1960s can be understood as a product of its time. Questions of power and privilege, race, class, and gender were just beginning to have an impact on social thought and public policy. The men who led the population movement in the United States saw themselves as forward thinking and focused on the common good. Both Nehru and John D. Rockefeller III were advocates of population control, and neither thought of themselves as other than humanitarians. These leaders wanted to alleviate poverty, accelerate economic development, and advance their enlightened self-interest, which included protecting national security. While their definition of the common good was inadequate, reducing population size and slowing population growth rates through controlling women's fertility was, they believed, a means of achieving a good end for the many, including individual women and families.

In casting these leaders' motives in the most positive terms, readers should not assume that I am unaware or approving of the conscious and unconscious at-times racist, classist, and sexist aspects of their mindset. Nor do I dismiss the negative effects—particularly on women's lives—that resulted from their actions. It was possible even in that time period to understand justice and rights in a way that was radically different from the way mainstream leaders of the population movement understood them. However, for many of those leaders, the ethical questions that are a part of late-twentieth- and early-twenty-first-century thought were simply not on the radar screen.

As the use of birth control in Europe and the United States became almost universal and an unquestioned good, and as information about hunger and poverty in the developing world became more accessible, support for international family planning assistance increased. In the 1970s and 1980s, the population/family planning movement gathered momentum with the growth of the modern environmental movement and its concerns about deforestation, overcrowding, and the depletion of scarce resources.

A MEANS TO AN END, OR AN END IN ITSELF?

Meanwhile, new criticism of population policy was brewing, although it rarely reached the general public. An unlikely alliance of conservative religionists, Marxist-Leninists, and feminists argued that family planning and population programs violated the core ethical principle of *respect for persons*. In general terms, respect for persons requires that human beings must not be treated primarily as means to serve the ends of others. Specifically, they argued, women and some men in the developing world were being used to advance the demographic goals of the first world. For these critics, *intent* was key: Programs that were intended to serve objectives other than the needs of their clients were open to coercion and other abuses. The members of this alliance pointed to specific instances of abuse—from forced abortions and sterilizations to uncontrolled pilot studies of new contraceptives—as evidence of the harm that can result from misplaced intent.

For many, concerns about means and ends were sharpened by the growing involvement of environmentalists in the population field. In the 1980s and early 1990s, environmentalism was probably the largest and most powerful social movement in the United States. Both population groups and donors hoped the interest in the environment would rally support for international population stabilization programs. They actively courted these groups and touted population control as a solution to environmental problems. Those who were already concerned with ethical flaws in both national and international population control and family planning programs were alarmed by the rhetoric of the environmental and population axis. Poor people and women's fertility, they claimed, were being blamed for the sins of corporations who exploited third world resources and for the excesses of an untrammeled consumer culture in the global north. Words mattered, and population-environment discourse was believed by some to contribute to draconian efforts to curb population growth in China, India, Indonesia, Peru, and other countries.

> Population-environment discourse was believed to contribute to draconian efforts to curb population growth in China, India, Indonesia, Peru, and other countries.

These tensions erupted at the 1992 UN Earth Summit in Rio de Janeiro, where feminists and the Vatican took a seemingly similar and somewhat successful position against linking environmental degradation with population size and growth. Their arguments catalyzed a reassessment of how to think about population control and family planning.

That reassessment culminated at the 1994 UN Conference on Population and Development, where a new ethical paradigm was launched. The Cairo Program of Action holds that population policies should strive to meet the needs of individual women and men rather than demographic objectives, that human rights must underlie population program goals and objectives, and that services must be delivered in a comprehensive reproductive health model in which women's full humanity and well-being are central. In Cairo, the world's nations agreed that family planning and other reproductive health services are worthy ends in themselves. These programs should be part of national health systems and overseas development aid—not to protect the environment or promote national security and development, but because women want and need these services to be autonomous participants in society and to reach their own life goals. At the same time, the new paradigm recognizes that family planning and reproductive health services are necessary but insufficient to ensure women's reproductive autonomy. The Cairo agreement urged greater attention to girls' education, economic development, and other measures that would give women the power—as well as the means—to make their own reproductive decisions.

This was indeed a lofty values statement undergirded by practical experience. Many studies showed that the mere provision of family planning was insufficient to achieve population stabilization. Improvements in women's status and education are central to reducing the number of children women want. So, giving women what they want has the most salutary effect on population growth. This could be seen as a win-win situation for both those concerned about population and those concerned about women's rights and well-being.

For environmentalists, the shift was a mind bender. They had entered the field of population out of concern for the effect of population size and growth on the environment. If addressing this relationship was now considered unethical, was there any reason for them to stay in the field? After all, their core concern is for environmental quality, not women's rights or sexual and reproductive health. And many had taken a beating from the Religious Right and their own members for entering the population and family planning field, where the role of abortion in reducing population size and growth was a hot potato. Within a few years after Cairo, most environmental groups bowed out of population work.

The decade following the Cairo conference saw a substantial shift in discourse and in programs. Organizations that had routinely talked about "population control" found those words were now politically incorrect. There was considerable stumbling over the term in speeches, and it pretty much disappeared from written materials. It was replaced by the rather awkward and certainly less-well-understood mouthful "sexual and reproductive health and rights." Even talking about "family planning" was considered inadequate.

Linguistic challenges aside, the new language undergirded substantial changes in overseas development goals as well as goals in national programs. Program targets at the local level (goals for a set number of "contraceptive acceptors," incentives to workers who met their targets, and similar benchmarks) were abolished in many places and replaced by an emphasis on offering women a range of contraceptive choices along with comprehensive reproductive health care. The new approach assumes an "unmet need" for contraception: Women do not need to be coerced or offered incentives to reduce fertility; they simply need to be provided with a choice of methods and treated as persons rather than uteri. There is no doubt that better quality programs have developed under the new paradigm.

But just as the new paradigm took hold, public attention to sexual and reproductive health and rights (SRHR) declined. As funding stalled and media coverage waned, SRHR professionals debated the cause. Demographers and other members of the pre-Cairo population establishment, who felt de-legitimized and silenced by the new paradigm and its explicit rejection of demography as a frame for the discourse, blamed the shift away from population growth. Rather than rejecting the new paradigm, however, they called for a middle-ground approach in which *both* numbers and women matter. Without the larger geopolitical frame of national security, poverty alleviation, and environmental preservation, they argued, funding for SRHR and women's empowerment would continue to decline. This reflects a sad truth: Although we have come a long way toward recognizing that "women hold up half the sky," few countries put women's health, education, and advancement high on the finance agenda.

Most recently, as this book amply demonstrates, some policy analysts and scientists have called for renewed attention to the links between population growth and environmental crises, especially climate change. Some SRHR advocates believe it will be easier to secure public funding for that work if they can talk about why slowing population growth is important for reducing global warming. Some feminist leaders, though concerned that women's reproductive health and rights will again become a means to other good ends rather than an end in itself, are open to exploring the links between these issues. They believe that the Cairo paradigm has been sufficiently accepted by governments and NGOs to prevent a reemergence of the women-as-instrument approach to population and environmental stabilization.

In fairness to the full spectrum of views, another group of feminists believes that the reemergence of these links embodies the same dangers in the same degree as in the past. Once again, overstatement of the effect of population on the environment will justify putting the imperatives of developed countries above the rights and needs of women in the developing world. Moreover, focusing on population growth could offer an illusory quick fix to our environmental problems: Telling poor women what to do is so much easier than reducing our consumption or making financial investments in green technology or holding corporations accountable for their greed and ravaging of resources. Also, they ask, if we accept the conclusion that women's reproductive health will never be fully funded for its own sake, will we set the cause of women's rights back a decade or two?

> For groups working to achieve women's sexual and reproductive health and rights, there is no reason to include environmentalism or population stabilization advocacy in their agenda. . . . But those groups need not discourage *all* organizations from linking population, environment, development, and reproductive health.

A lot is at stake in determining the answer to this question, especially for SRHR groups and others who have embraced the Cairo paradigm. I would suggest that for one set of organizations, whose central goal is achieving women's sexual and reproductive health and rights, there is no reason to include environmentalism or population stabilization advocacy in their agenda. In fact, there are good reasons to avoid these issues. The social transformation needed for women's reproductive rights to be fully accepted as fundamental human rights is in process, but it is not complete. Some groups must continue to work single-mindedly for that transformation in culture and politics by insisting that women's rights are an end in themselves and not a means to a better life for children, men, and society at large. Groups working on population issues—from migration to aging and fertility—and those working on the environment face a different ethical challenge. Can they enter the policy discourse on the link between population and environment without taking a position on sexual and reproductive health and rights, including the right to abortion? Will they stand next to and with the advocates of SRHR, or will they ignore these issues—as they have in the past? It seems to me reasonable that the SRHR community make that ethical demand of their colleagues.

At the same time, there is no need for SRHR groups to attempt to prohibit *all* organizations from making links between population, environment, development, and reproductive health or to offer blanket public criticism of such efforts as unethical or unfounded. We have become extremely sensitive to the efforts by the Right to ignore or subvert evidence and science in service of ideology. We would fall prey to the same dishonesty were we to insist that these links cannot be explored. And to claim that they do not exist at all would be intellectually dishonest. More valuable would be a set of guidelines that would ensure that the ethics and values that inform SRHR are honored in population and environmental advocacy. Consider the two examples below:

Good Facts Make Good Ethics

It is essential to fairly and accurately represent population-environment linkages. The current literature, especially that produced by those rooted in the population movement, is too often reminiscent of the 1960s absolutist and excessive claims of the dire effects population growth and/or immigration would have on "the planet." It is in strong contrast to scientific articles by those in other fields who offer a far more nuanced connection between demographic change and environmental quality. The latter posit a moderate role for demographic phenomena and a far greater role for consumption and industrial causes.

Human Rights Are Fundamental

Violations of human rights in reproductive health programs must be immediately and vociferously condemned by all civil society organizations engaged in population and reproductive health advocacy. While these abuses may be few, we must not make excuses for them. The SRHR community lacks clear written guidelines on what practices constitute coercion, violation of informed consent, the use of incentives and disincentives, protocols for participation in drug trials, and confidentiality, among other issues. We have been reluctant to clean up our own act. China's population policies are a clear example of an inadequate response to violations of human rights. While the Chinese program has improved over the last decade, abuses continue. Yet China is a member in good standing of the International Planned Parenthood Federation and few organizations, including feminist groups active on SRHR, have publicly spoken out against violations reported in the press. The desire for "quiet diplomacy" as a way of improving the Chinese program is not sufficient reason to mute our voices. It is critically important that environmental groups that make the link between climate change and population size and growth rates do not fall into this trap. They should always stress the importance of adherence to human rights principles and voluntary, comprehensive services in reproductive health programs.

Personal autonomy is a core bioethical value and cannot be violated in the delivery of reproductive health services or in policy. Women and girls must be respected as competent moral agents capable of making good decisions about when and whether to bear children. It is not unreasonable for states to educate and inform men and women about the role that individual decisions about reproduction will have on the community; in fact, it is responsible for the state to bring these issues to people's attention. What is unacceptable is to assume that women will not make good decisions and to use coercion, including incentives and disincentives, that would cause people to make decisions they would not otherwise make or to deny choice in methods of fertility control. Draconian measures such as forced abortion, sterilization, and penalties for "excess births" such as loss of educational opportunities or employment are profound violations of human rights that should find their way to human rights commissions for adjudication.

Adherence to these ethical principles, among others, in discourse, communications, advocacy, and programs will make it possible to link population, reproduction, and the environment. It will also link the population movement's early and legitimate concern for the eradication of human suffering and poverty with the feminist movement's commitment to women's human rights. And it will contribute to a just and balanced approach that respects women's rights, protects the environment, and stabilizes population growth.

REFERENCES

1. Hume, Maggie, 1991, *Contraception in Catholic Doctrine: The Evolution of an Earthly Code,* Catholics for a Free Choice, Washington, DC.

AFTERWORD

···········

Work for Justice!

LAURIE MAZUR and SHIRA SAPERSTEIN

There's a bumper sticker often seen around the town where we live; it says, "If you want peace, work for justice." Substitute *sustainability* for *peace*, and you have the central message of this book.

We began kicking around the idea for this book a few years ago. Originally, it was conceived as an update of *Beyond the Numbers: A Reader on Population, Consumption and the Environment*, a contributed volume Laurie edited in 1994. That volume, published on the eve of the historic United Nations International Conference on Population and Development, in Cairo, brought together many architects of the Cairo Consensus—a pathbreaking plan to address population growth through comprehensive support for reproductive health services, gender equity, and human rights.

But *Beyond the Numbers* had fallen out of date; fifteen years later, for example, world population has grown by a billion. So we saw a need to update the data—and the thinking—on the relationship between demographic and environmental change. We recognized that the specter of climate change posed new challenges for environmental activists and that it might even offer an opportunity to rejuvenate support for the long-neglected Cairo consensus on reproductive health and rights. And we thought that policymakers, academics, the media, and activists would be open to exploring these issues.

Little did we know. In the process of pulling together this book, we found that the data on the relationship between population and environment are far more complex than even we imagined; that the range of issues that shape that relationship is vast and surprisingly unexplored; and that there is a lot of fear about—even outright hostility to—talking about population and environment in the same breath, let alone in a publication for a mass audience.

In approaching this uncharted and rather dangerous territory, we were fortunate to have many excellent guides. We are extremely grateful to the longtime experts on population and environment who helped us identify key issues at a meeting at the UN Foundation in 2007, the youth activists who forced us to totally rethink those issues at another convening in 2008, and the authors who addressed these issues with honesty and creativity in this book.

What we have learned is far too complex to summarize adequately in these final pages, but here are the essentials:

- Demographic change—particularly population growth—has a significant impact on the natural environment, including the global atmosphere. The impact is neither linear nor uniform, and it is shaped by a wide range of mediating factors, including technology, consumption patterns, economic policies, and political choices. All of these must be addressed if we are to confront the environment and development challenges of the future.

- The most effective way to slow population growth is to expand access to family planning and other reproductive health services, and—more broadly—to promote gender equity and advance human development and rights, especially for the young men and women who are now coming of age, the largest generation in history.

- The means to slow population growth are all important ends in and of themselves. Each will advance human rights and social justice; together they will help ensure a sustainable, equitable future.

- Implementing these steps will require policy change and action at every level of society, from the grassroots to the international policy arena, in both the global north and the global south, in the long term and also right now.

THE FIERCE URGENCY OF NOW: A CALL TO ACTION

We call on policymakers to set and implement the following priorities:

Universal Access to Family Planning and Other Reproductive Health Services

Reproductive health services are key to slowing population growth and improving maternal and child health. Governments should ensure universal access to high-quality reproductive health services, including but not limited to family planning; prevention and treatment of sexually transmitted infections, including HIV/AIDS; comprehensive sexuality education; safe and legal abortion; and maternal and child health. These services should be provided in a manner that is respectful of human dignity and choice, and they should be held accountable to the highest standards of human rights.

Equal Rights for Girls and Women

Discrimination against girls and women is a colossal human tragedy and waste; it also contributes to rapid population growth. Therefore, population policy must emphasize legal and social measures to protect and enhance women's rights. These include primary and secondary education for girls; enforcing laws and human rights standards that prohibit child marriage and gender-based violence; and improving women's access to credit, land, employment, and training—in short, the full legal and social rights of citizenship.

A Just and Sustainable Global Economic System

Poverty and inequality drive rapid population growth in many parts of the world. Yet poverty and inequality are not part of the natural order of the universe; they can be mitigated or exacerbated by policies and practices. Unfortunately, much of current economic policy does the latter. Developing countries stagger under heavy burdens of debt, while the financial and trade policies imposed by international financial institutions often enrich elites and drive others deeper into poverty. Those who are concerned about population growth must make common cause with groups seeking to implement the Millennium Development Goals, as well as those working for debt relief, reform of international financial institutions (including the World Bank, International Monetary Fund, and World Trade Organization), fair trade, food sovereignty, and other forms of sustainable and equitable development.

Reduced Consumption in the Affluent Countries

Population and resource consumption are the yin and yang of environmental harm. Given that substantial future population growth is virtually inevitable, it is imperative that we use resources more efficiently and find ways to live more lightly on the planet—first and foremost in the affluent countries that consume the lion's share of the planet's resources. Reducing consumption in the United States and other developed countries will require sweeping technological and social changes. For example, it requires a wholesale shift from fossil fuels to renewable energy; greater efficiency in buildings, vehicles, and industrial processes; cities and towns built around reliable public transportation; a new materials economy that emphasizes recycling and reuse; and food and agriculture systems that encourage us to eat lower on the food chain. It requires, fundamentally, a reorientation of our economy—from the production and consumption of goods of dubious value to a new emphasis on meeting human needs.

At this pivotal moment, there is remarkable alignment between what we must do and what we should do, between the cause of social justice and the imperative to preserve a habitable planet for current and future generations. If you want sustainability, work for justice.

Acknowledgments

This book—more than most, I believe—is the product of a "public conversation" involving dozens of people. That conversation was often challenging, and the conversants may not always have seen eye to eye, but all contributed their ideas, their unique experiences, and their deeply felt convictions. The result is a rich array of insights—and the basis for cooperation and collaboration toward shared goals.

Three people have made such outsize contributions to this effort that I must thank them up front. This book would not have come to pass without the help of Shira Saperstein of the Moriah Fund—my friend, colleague, and mentor. Shira provided and leveraged support for the project; broadened the vision by bringing new voices to the table; helped organize the meetings that informed the book's development; and spent uncountable hours reading drafts, offering advice, and generally serving as the project's intellectual and moral compass. Nor would the book have seen the light of day without Edith Eddy of the Compton Foundation, another longtime friend and mentor. Edie helped hatch the original idea for the book and provided vital seed funding. She also shaped the project by asking all the right questions and by skillfully connecting people and ideas. And Jacqueline Nolley Echegaray, also of the Moriah Fund, altered the course of the project when she stood up at our conference in September 2007 and challenged us to consider the economic and trade policies that contribute to rising inequality. With that comment, and her subsequent work as a

397

writer, thinker, organizer, and translator, Jackie helped build the justice framework that is at the heart of this book.

I am enormously grateful to the foundations that made this work
possible: the Compton Foundation, the Jake Family Fund, the Moriah
Fund, the David and Lucile Packard Foundation, The Summit
Foundation, the United Nations Foundation/Better World Fund, the
Wallace Global Fund, the WestWind Foundation, and the Mary
Wohlford Foundation.

And I owe a special debt of gratitude to the project's advisors: George
Abar, Todd Baldwin, Carmen Barroso, Roger-Mark De Souza, Edith Eddy,
Robert Engelman, Sarah Fairchild, Susan Gibbs, Kathy Hall, Tamara
Kreinin, Jacqueline Nolley Echegaray, Suzanne Petroni, Martha Farnsworth
Riche, Shira Saperstein, Steven W. Sinding, and James Wagoner.

The many activists and scholars who attended the New Population
Challenge Conference in 2007 and the New Leaders' Convening on
Population, Justice and the Environment in 2008—though too numerous to list here—profoundly influenced my thinking and contributed
to the diversity of ideas offered in this volume. Timothy E. Wirth and
the wonderful staff of the UN Foundation generously hosted both of
those meetings. Laura Chasin and Maggie Herzig of the Public
Conversations Project designed and moderated the 2007 conference in
a way that enabled a large group of people with divergent views to genuinely listen to one another. Monica Palacio of the Management
Assistance Group skillfully facilitated the 2008 New Leaders' meeting.

Many thanks to the authors of this volume, who addressed the
issues at hand with intelligence and compassion. Together, they have
pushed the envelope of thinking about population, justice, and the
environment. Barbara Becker and Loren Siegel provided invaluable
help communicating the book's messages to a larger audience.

I want to thank Chuck Savitt, Todd Baldwin and others at Island Press
for seeing the book through from conception to publication. Thanks, also,
to Dave Peattie at BookMatters for shepherding the book through the
production process. Lou Doucette did a fine—and speedy—job copyedit-
ing the manuscript. And I am grateful to Josie Ramos of the International
Health Program for her invaluable help with Spanish translation.

Finally, on a personal note, I want to thank my husband, Bristow
Hardin, for providing the perfect balance of constructive criticism and
unflagging support; and my children, Ben Hardin and Sam Hardin,
for being patient with my absences (and absentmindedness) and for
motivating me to care ever more deeply about the future.

SUZANNE PETRONI is a senior program officer at The Summit Foundation in Washington, DC, where she manages the Global Population and Youth Leadership program.

SANDRA POSTEL is director of the Global Water Policy Project.

MARTHA FARNSWORTH RICHE served as Director of the U.S. Census Bureau between 1994 and 1998 and is now a fellow at the Center for the Study of Economy and Society, Cornell University.

DR. FRED T. SAI, a Ghanaian family health physician and Presidential Advisor on Reproductive Health and HIV/AIDS, served as Chairman of the United Nations International Conference on Population and Development in 1994 and as President of the International Planned Parenthood Federation from 1989 to 1995.

SHIRA SAPERSTEIN is the Moriah Fund's Deputy Director and Program Director for Women's Rights and Reproductive Health.

STEVEN W. SINDING, a senior fellow at the Guttmacher Institute, served as Director-General of the International Planned Parenthood Federation from 2002 to 2006.

JAMES GUSTAVE SPETH is a Professor at the Yale School of Forestry & Environmental Studies and author, most recently, of *The Bridge at the Edge of the World: Capitalism, the Environment, and Crossing from Crisis to Sustainability* (Yale University Press, 2008).

ALEX STEFFEN is the Executive Editor of worldchanging.com.

ELEANOR STERLING is Director of the Center for Biodiversity and Conservation (CBC) at the American Museum of Natural History.

ADRIANA VARILLAS is a journalist based in Cancún, Mexico, where she specializes in writing about the environment, gender, human rights, politics, and other issues.

JULIA VARSHAVSKY is the Reproductive Health Program Coordinator for the Collaborative on Health and the Environment (CHE).

ERIN VINTINNER is Biodiversity Specialist at the Center for Biodiversity and Conservation (CBC).

TIMOTHY E. WIRTH is the President of the United Nations Foundation and Better World Fund.

MALEA HOEPF YOUNG was a research associate at Population Action International (PAI) from 2007 to 2009 before joining the Peace Corps, where she is a volunteer in Rwanda.

Index

Abortion rights, 346–47
  driving force of nativism in opposition to, 358–59
  immigration and, 354, 358–61
  United States and, 281, 282, 285
  see also Reproductive rights
Abstinence, promotion of, 286
Acid rain, 137–39
Adolescent Health and Information
  Project (AHIP), 302–3, 305–7
Africa, 69, 218–19, 323–24
  a growing political commitment in, 223–25
  reproductive health, 221–23
  sub-Saharan, 47, 188, 221, 265–67, 313
Africa Health Strategy, 224
African agriculture, 220
  from compliance to defiance, 187–90
African population growth, 46–48, 219–21
  and progress toward economic and human development, 219–21
  reasons for rapid, 47, 221–22
Age distribution of population, 1, 115f, 262

and climate-change resilience, 115, 116–17f, 118
coming-of-age scenarios, 41
in more *vs.* less-developed countries, 31, 35, 41
population policy and, 267–68
in United States, 208
"youth dearth" *vs.* "youth bulge," 54
see also Youth
Agriculture, 126, 127, 212
  difficulties finding new cropland, 170
  the great transformation, 184–87
  water and, 167–68, 196–99
  see also African agriculture; Green Revolution
"Anthropocene era," dawn of the, 3–4
Anti-immigrant movement, 61, 62, 353–54
  driving force of nativism in, 358–59
  environmental degradation and, 356–58
  immigrant women, population politics, and, 354
  pronatalist movement and, 359–60
Aquatic resources, 212

HIV/AIDS *(continued)*
  sexual and reproductive health and
    prevention of, 250–52, 254–57,
    273, 304, 308
*How Many People Can the Earth Sup-
  port?* (Cohen), 34
Human rights, xvi, 132, 298, 342–44,
  384
  as fundamental, 390–91
  rights-based approach to popula-
    tion control, 14
Hunger:
  increasing, 171–72
  see also Food shortages
Hurricane Jeanne, 111–12
Hurricane Katrina, 59, 73
Hyde Amendment, 347

ICPD, *see* International Conference
  on Population and Development
Immigrant rights movement, 62–63
Immigrant women, 61
  population politics and, 354
Immigrants, discrimination and
  crime against, 61
Immigration, 16
  abortion rights and, 354, 358–61
  see also Anti-immigrant movement
India, 56–57, 185
  women in, 313–14
Inequality:
  climate change and social, 59 (*see
    also under* Climate change)
  economic, 56, 61–62, 150
  environmental impact, 5–8
  and global migration, 56, 61–62
  population growth and, 5, 8, 10–11
Infertility, environmental causes of,
  234–35
Integrated Population and Coastal
  Resource Management (IPOP-
  CORM), 324
International Conference on Population and Development (ICPD),
  xi–xii, 97, 258, 261, 309
  focus on national level, 257
  funding shortfall, 251–52

new ethical paradigm launched at,
  386–89
progress after, 248–49, 387
and sexual and reproductive health
  and rights, xiv–xv, 14, 118, 238,
  246–57, 300–305
significance, 245–47
successful opposition to the opposition, 309
twenty-year goals, 249
unfinished agenda, 249–51
United States and, xv, 252–53, 284
see also Sexual and reproductive
  health; Sexual and reproductive
  health and rights
International Covenant on Civil and
  Political Rights (ICCPR), 343
International Covenant on Economic, Social and Cultural Rights
  (ICESCR), 343
International Monetary Fund (IMF),
  6, 180, 182–84, 187, 189
International Planned Parenthood
  Federation (IPPF), 277, 282–83
IPAT equation, 82–83, 137, 138, 142
Irrigation, 168

Japan, population growth and age
  distribution in, 42–43
Johnson, Lyndon B., 279

Kemp-Kasten amendment, 283
Kenya, 315
Kerala, India, 56–57
Kissinger, Henry, 280

Land use, 209–10
  see also Agriculture
Latin America, 48–49
Liberation theology, 368
Low elevation coastal zone (LECZ),
  74
"Low-hanging fruit" phenomenon,
  141, 268

Maathai, Wangari, 59, 315
Malawi, 188, 189

Sexual and reproductive health
(continued)
integrating poverty and, 253–54
mobilizing support for, 301–4
see also under International
Conference on Population and
Development
Sexual and reproductive health and
rights (SRHR), 300, 388–90
Bush administration and, 252
current challenges, 304–5, 308–9
HIV/AIDS and, 250–52, 254–57,
273, 304, 308
information and services for young
people, 308–9
mobilizing support for, 301–4
policy support for comprehensive
services, 304
see also International Conference
on Population and
Development; Reproductive
rights
Sexual and reproductive rights
(SRR), 342, 343
see also Reproductive rights
Sexual coercion, responding to,
273–74
Sexual initiation, 265
Sexually transmitted infections
(STIs), 222–23
Sierra Club, 358
Singh, Manmohan, 96
Sixth Great Extinction, 19
South Africa, women in, 313
"Sprawl," 209–210
Steffen, Alex, 48–49
Sterilization policy, forced:
legacy of, 296–98
lessons learned from, 298–99
Peru in 1990s and the origins of,
293–95
Sterilizations, forced:
impact on Peruvian women, 296
in practice, 295–96
Su Wei, 95, 96
Suburban areas, growth in, 207–8
Suburban sprawl, 210

Sustainability, 158
from crisis to, 329–37
equity and, 8
justice and, 393
Sustainable Coastal Communities
and Ecosystems (SUCCESS),
328n8

Tamayo, Giulia, 295
Taylor, Adegoke, 69
Technology, 12–13, 87
Teleodynamics, 379
Thatcher, Margaret, 186
Thresholds, 137–39
see also "Tipping points"
Tiban, Lourdes, 103
"Tipping points," 8, 131, 139
see also Thresholds
Trade liberalization, 56, 180–81,
183–87
Trade Related Aspects of Intellectual
Property Rights (TRIPs), 186
Trash, 212–13

United Nations (UN), 98, 303, 343
Convention Relating to the Status
of Refugees, 57–58, 60
see also International Conference on
Population and Development;
Millennium Development Goals
United Nations Fund for Population
Activities (UNFPA), 280, 282–84
United Nations Population Fund
(UNPF), xv
United States:
age distribution of population, 208
environmental impacts of population growth and resource consumption, 13, 209–14
food consumption, 172
"footprint," 5, 61, 205
ICPD and, xv, 252–53, 284
international family planning
assistance, 278–84, 279f (*see also*
Religious Right; United States
population policy)
population growth and fertility, 44